T0321148

From Matter to Life

Recent advances suggest that the concept of information might hold the key to unraveling the mystery of life's nature and origin. Fresh insights from a broad and authoritative range of articulate and respected experts focus on the transition from matter to life, and hence reconcile the deep conceptual schism between the way we describe physical and biological systems. A unique cross-disciplinary perspective, drawing on expertise from philosophy, biology, chemistry, physics, and cognitive and social sciences, provides a new way to look at the deepest questions of our existence. This book addresses the role of information in life and how it can make a difference in what we know about the world. Students, researchers, and all those interested in what life is and how it began will gain insights into the nature of life and its origins that touch on nearly every domain of science.

From Matter to Life

Information and Causality

Edited by

SARA IMARI WALKER
Arizona State University

PAUL C. W. DAVIES
Arizona State University

GEORGE F. R. ELLIS
University of Cape Town

Shaftesbury Road, Cambridge CB2 8EA, United Kingdom

One Liberty Plaza, 20th Floor, New York, NY 10006, USA

477 Williamstown Road, Port Melbourne, VIC 3207, Australia

314–321, 3rd Floor, Plot 3, Splendor Forum, Jasola District Centre, New Delhi – 110025, India

103 Penang Road, #05–06/07, Visioncrest Commercial, Singapore 238467

Cambridge University Press is part of Cambridge University Press & Assessment, a department of the University of Cambridge.

We share the University's mission to contribute to society through the pursuit of education, learning and research at the highest international levels of excellence.

www.cambridge.org
Information on this title: www.cambridge.org/9781107150539

10.1017/9781316584200

First published 2017

A catalogue record for this publication is available from the British Library

Library of Congress Cataloging-in-Publication data
Names: Walker, Sara Imari. | Davies, P. C. W. | Ellis, George F. R. (George Francis Rayner)
Title: From matter to life : information and causality / Sara Imari Walker, Arizona State University, Paul C. W. Davies, Arizona State University, George F. R. Ellis, University of Cape Town.
Description: New York : Cambridge University Press, 2016. | Includes bibliographical references and index.
Identifiers: LCCN 2016040374 | ISBN 9781107150539 (hardback : alk. paper)
Subjects: LCSH: Life–Origin. | Consciousness–Physiological aspects. | Biochemistry.
Classification: LCC QH325 .W35 2016 | DDC 577–dc23 LC record available at https://lccn.loc.gov/2016040374

ISBN   978-1-107-15053-9   Hardback

Contents

About the authors

CHRISTOPH ADAMI is Professor of Microbiology and Molecular Genetics, as well as Professor of Physics and Astronomy, at Michigan State University. He obtained his Ph.D. and M.A. in theoretical physics from Stony Brook University and a Diploma in Physics from Bonn University. His main research focus is Darwinian evolution, which he studies at different levels of organization (from simple molecules to brains). He has pioneered the application of methods from information theory to the study of evolution, and designed the Avida system, which launched the use of digital life as a tool for investigating basic questions in evolutionary biology. When not overly distracted by living things, he studies the fate of classical and quantum information in black holes. He wrote *Introduction to Artificial Life* (1998) and is the recipient of NASA's Exceptional Achievement Medal as well as a fellow of the American Association for the Advancement of Science (AAAS).

LARISSA ALBANTAKIS is a postdoctoral researcher with Giulio Tononi at the Department of Psychiatry at University of Wisconsin–Madison. She received her degree in physics with distinction at the Ludwig-Maximilians Universitt, Munich, followed by a Ph.D. in computational neuroscience at the Universitat Pompeu Fabra, Barcelona, under the supervision of Gustavo Deco. Her research focuses on the theoretical formulation of the integrated information theory of consciousness and its implications for evolutionary adaptation, emergence, and meaning.

ANDREW D. BRIGGS is the inaugural holder of the Chair of Nano-materials at the University of Oxford. His research interests focus on materials and techniques for quantum technologies and their incorporation into practical devices. Current hot topics include vibrational states of nanotubes and charge transport through single molecules in graphene nanogaps. He has nearly 600 publications, with more than 16,000 citations. In 2016 his book *The Penultimate Curiosity: How Science Swims in the Slipstream of Ultimate Questions*, cowritten with Roger Wagner, was published.

PAUL C. W. DAVIES is Director for the Beyond Center for Fundamental Concepts in Science and a regents' professor at Arizona State University. He is a theoretical physicist, cosmologist, astrobiologist, and best-selling author. His research ranges from the origin of the universe to the origin of life and includes the properties of black holes, the nature of time, and quantum field theory. He is the recipient of numerous awards, including the 1995 Templeton Prize, the 2002 Michael Faraday Prize from the Royal Society, and the 2011 Robinson Prize in Cosmology.

SIMON DEDEO is an assistant professor in Complex Systems, and faculty in Cognitive Science, at Indiana University, and is an external professor of the Santa Fe Institute. At Indiana, he runs the Laboratory for Social Minds, which conducts research in cognitive science, social behavior, economics, and linguistics; recent collaborative work includes studies of institution formation in online social worlds, the emergence of hierarchy in animal conflict, competitive pricing of retail gasoline, and parliamentary speech during the French revolution. In 2017 he will join the social and decision sciences department at Carnegie Mellon University.

GEORGE F. R. ELLIS, FRS, is Professor Emeritus of Applied Mathematics at the University of Cape Town. He is a relativist and cosmologist who has worked on exact and perturbed cosmological models in general relativity theory and carefully studied observational limits in cosmology. He cowrote *The Large Scale Structure of Space Time* with Stephen Hawking. His more recent research relates to the philosophy of cosmology and the emergence of complexity. He has various honorary degrees and was awarded the Star of South Africa Medal by President Nelson Mandela in 1999 and the Templeton Prize in 2004.

KEITH D. FARNSWORTH graduated in 1984 from Queen Elizabeth College, London University, with a B.Sc. in physics with astrophysics, and after an M.Sc. in acoustics from Southampton, worked in the Institute of Cancer Research. In 1994 he received a Ph.D. in mathematical biology from the University of Edinburgh and later an M.Sc. in public health and epidemiology from Aberdeen University. He has been a lecturer in theoretical and applied ecology at the Queen's University Belfast since 2003, applying complex system theory to ecological problems such as fisheries management and biodiversity. He is active in developing information-theoretic concepts to integrate our understanding of all scales of life's organisation and coordinates an international academic network on that topic.

JESSICA FLACK is a professor at the Santa Fe Institute and the director of SFI's Collective Computation Group, which draws on evolutionary theory, cognitive science and behavior, statistical mechanics, information theory, and theoretical computer science to study information processing and collective computation in nature and their roles in emergence of robust structure and function in biological and social systems. Flack was previously the founding director of University of Wisconsin–Madison's Center for Complexity and Collective Computation in the Wisconsin Institutes for Discovery. Flack's work has been covered by scientists and science journalists in many publications and media outlets, including the *BBC, NPR, Nature, Science, The Economist, New Scientist*, and *Current Biology*.

JAY T. GOODWIN received his Ph.D. in bioorganic chemistry with David Lynn in the department of chemistry at the University of Chicago, and after a sojourn into industrial and entrepreneurial drug discovery, he is currently a senior research fellow in the department of chemistry at Emory, and was a 2013–2015 AAAS Science and Technology Policy Fellow at the National Science Foundation. His scientific focus is on dynamic chemical networks and chemical evolution as they inform our understanding of the potential origins of living matter, the possibility of alternative chemistries of life on Earth and elsewhere in the universe, and the design and implementation of intelligent materials and novel therapeutic systems. He is also a Senior Program Officer in the Fellows Program at the John D. and Catherine T. MacArthur Foundation. The views expressed herein are solely in Dr. Goodwin's personal capacity and do not reflect the views of the MacArthur Foundation.

PAUL GRIFFITHS is a philosopher of science with a focus on biology and psychology. He was educated at Cambridge and the Australian National University, receiving his Ph.D. in 1989. He taught at Otago University in New Zealand and was later director of the Unit for History and Philosophy of Science at the University of Sydney before taking up a Chair in History and Philosophy of Science at the University of Pittsburgh. He returned to Australia in 2004, first as an Australian Research Council Federation Fellow and from 2007 as University Professorial Research Fellow at the University of Sydney. At present he is also Academic Associate Director for Humanities and Social Sciences at the Charles Perkins Centre, a major new initiative at Sydney focused on interdisciplinary research into obesity, diabetes, and cardiovascular disease. He is a Fellow of the AAAS and the Australian Academy of the Humanities and was president of the International Society for History, Philosophy and Social Studies of Biology from 2011 to 2013. From 2006 to 2012 he was a member of the Australian Health Ethics Committee of the

National Health and Medical Research Council. He spends part of each year at the University of Exeter at the Egenis, the Centre for the Study of the Life Sciences.

ANNE-MARIE GRISOGONO is a physicist by training and has worked in experimental and theoretical atomic and molecular physics, and in lasers and nonlinear optics for 14 years in various universities, followed by 20 years with the Defence Science and Technology Organisation (DSTO) working on systems design, modeling, and simulation, developing DSTO's Synthetic Environment Research Facility for defense capability development, and initiating an enabling research program into applications of complex systems science to address defense problems and future warfare concepts, for which she won a three-year Strategic Long Range Research internal fellowship. Dr. Grisogono is a member of the Australian Research Council's College of Experts, director of research for Ionic Industries, and holds an adjunct professor appointment in the engineering faculty of Flinders University. Her current research interests include fundamental questions of complexity science and improving the methodologies and tools that can be applied to dealing with complex problems.

JOSHUA A. GROCHOW is an Omidyar Fellow at the Santa Fe Institute (SFI). His overarching research dream – a distant, perhaps unreachable beacon – is a general, mathematical theory of complex systems. His primary research area is theoretical computer science, where he applies algebraic geometry and representation theory to understanding the computational complexity of algorithms. Prior to SFI he was a postdoctoral student at the University of Toronto, and he received his Ph.D. from the University of Chicago.

LUC JAEGER is a professor of chemistry and biochemistry at the University of California, Santa Barbara (UCSB). He works on informational biopolymers to further the development of new

methodologies and materials with potential biomedical applications in the emerging fields of RNA nanobiotechnology and RNA
synthetic biology. His research involves an effort to decipher
the logic of RNA modularity and self-assembly and to unravel
how complex informational molecules evolved at the origins of
life. He is involved in the dialogue between science and religion
and teaches a class on "What Is Life?" from both scientific and
philosophical perspectives. A graduate of the University Louis
Pasteur (ULP) in Strasbourg, Dr. Jaeger went on to earn a master's degree in chemistry and biology there and then a Ph.D. in
structural biochemistry and biophysics at ULP in 1993 under
the supervision of Eric Westhof and François Michel. He was
awarded a postdoctoral research fellowship from NASA to work
with Gerald Joyce at the Scripps Research Institute in La Jolla,
and in 1995 returned to France as a research scientist at the
Institut de Biologie Moléculaire et Cellulaire in Strasbourg. He
joined the faculty of the UCSB in 2002 and was promoted to his
present position in 2012. Dr. Jaeger has been a visiting professor
at the National Cancer Institute and has held a grant from the
ULP–National Institute of Bioscience and Human Technology
and Information Services for work in Japan, as well as being the
recipient of a UCSB Junior Faculty Research Incentive Award. He
served as a member of the John Templeton Foundation board of
advisors from 2011 to 2013. He is the author or coauthor of more
than 70 papers published in scientific journals.

DAVID KRAKAUER is President and William H. Miller Professor of
Complex Systems at the Santa Fe Institute (SFI). His research
explores the evolution of intelligence on Earth. This includes
studying the evolution of genetic, neural, linguistic, social,
and cultural mechanisms supporting memory and information
processing, and exploring their generalities. At each level
he asks how information is acquired, stored, transmitted,
robustly encoded, and processed. This work is undertaken
through the use of empirically supported computational and

mathematical models. He served as the founding director of the Wisconsin Institute for Discovery, the codirector of the Center for Complexity and Collective Computation, and was professor of mathematical genetics at the University of Wisconsin–Madison. He has previously served as chair of the faculty and a resident professor and external professor at SFI, a visiting fellow at the Genomics Frontiers Institute at the University of Pennsylvania, a Sage Fellow at the Sage Center for the Study of the Mind at the University of Santa Barbara, a long-term fellow of the Institute for Advanced Study in Princeton, and visiting professor of evolution at Princeton University.

THOMAS LABAR is a graduate student in the laboratory of Christoph Adami at Michigan State University and a member of the dual Ph.D. program between the department of microbiology and molecular genetics and the program in ecology, evolutionary biology, and behavior. He studies topics concerning the evolution of complexity and evolutionary innovation using digital experimental evolution. Before joining the Adami laboratory, he completed his B.Sc. degree in mathematics at Pennsylvania State University. While an undergraduate, he worked in the laboratory of Katriona Shea and researched how extinctions alter plant–pollinator communities, using computational models.

SHA LI received her B.S. degree in chemistry from Wuhan University in 2010. She then began her Ph.D. studies at Emory University working in the field of supramolecular chemistry in David Lynn's group. Her research interests include designing and engineering asymmetric peptide materials and the use of molecular self-assembly codes to develop multicomponent systems with unique electrical and optical properties. She obtained her Ph.D. degree in August 2015.

ERIC LIBBY is an Omidyar Fellow at the Santa Fe Institute where he researches the evolutionary origins of biological complexity. He is particularly interested in origins of multicellularity and the structure and shape of organismal life cycles. He completed a postdoctoral fellowship in an evolutionary biology group at the New Zealand Institute for Advanced Studies and has a Ph.D. from McGill University in quantitative physiology.

JOSEPH LIZIER is an ARC DECRA Fellow and Senior Lecturer in complex systems at the University of Sydney (since 2015). Previously he was a research scientist and postdoctoral fellow at the Commonwealth Scientific and Industrial Research Organisation (CSIRO) Information and Communication Technology Centre (Sydney, 2012–2014), a postdoctoral researcher at the Max Planck Institute for Mathematics in the Sciences (Leipzig, 2010–2012), and has worked as a research engineer in the telecommunications industry for 10 years. He obtained a Ph.D. in computer science (2010) and bachelor's degrees in engineering (2001) and science (1999) from the University of Sydney. His research interests include information-theoretic approaches to studying dynamics in complex systems, complex networks, and neural systems, and he is a developer of the JIDT toolbox for measuring information dynamics.

DAVID G. LYNN has contributed in the general areas of molecular recognition, synthetic biology, and chemical evolution and has developed chemical and physical methods for the analysis of supramolecular self-assemblies, of signal transduction in cellular development and pathogenesis, of molecular skeletons for storing and reading information, and of the evolution of biological order. After a fellowship at Columbia University and teaching briefly at the University of Virginia and Cornell University, he served as professor of chemistry at the University of Chicago until 2000. In that year he moved to accept the Asa Griggs Candler Professorship in Chemistry and Biology at Emory University. In 2002, Lynn was awarded one of 20 inaugural Howard Hughes Medical Institute Professorships and the 2011 Emory Scholar Teacher Award for pioneering several science/arts collaborations for communicating science. He was appointed chair of the department of chemistry in 2006, as a fellow of the AAAS in 2011, and as the American Chemical Society Herty Awardee in 2013.

CHIARA MARLETTO is a quantum physicist at the University of Oxford. After having studied Italian literature, engineering science, and condensed matter physics, she specialized in quantum computing at the University of Oxford, with a D.Phil. under the supervision of Artur Ekert. She is now collaborating with David Deutsch on the project to develop constructor theory, a new fundamental theory of physics that Deutsch proposed, which has promise for expressing in exact terms within fundamental physics emergent notions such as information, life, knowledge, and the universal constructor.

ANIL K. MEHTA received his Ph.D. in physical chemistry at Yale University and during this and his postdoctoral training at Washington University developed novel solid-state nuclear magnetic resonance methods for atomic-level structural characterization. He is now a faculty fellow at Emory University where he has focused on understanding the rules and forces directing molecular and supramolecular assembly and how these assemblies can be harnessed to store information and gain novel function.

DENIS NOBLE is a physiologist and systems biologist who discovered electrical mechanisms in the proteins and cells that generate the rhythm of the heart. He constructed the first mathematical model of this process, published in *Nature* in 1960. Today, this work has grown into an international project, the Physiome Project, which constructs models of all organs of the body. He is president of the International Union of Physiological Sciences and is Professor Emeritus at Oxford University. His book, *The Music of Life*, is the first popular science book on the principles of systems biology and has been translated into many languages.

THEODORE P. PAVLIC is an assistant professor in the School of Computing, Informatics, and Decision Systems Engineering and the School of Sustainability at Arizona State University. He received his Ph.D. in electrical and computer engineering from the Ohio State University and has had advanced postdoctoral training in computer science, behavioral ecology, and evolutionary biology. Using a mixture of mathematical, computational, and empirical methods, he studies decentralized behaviors found in both natural and artificial complex systems. His broad range of interests include bio-inspired algorithms, intelligent transportation systems (e.g., self-driving cars), and the individual-level mechanisms that allow eusocial insects (e.g., ants and honeybees) to solve complex colony-level problems. With colleagues across the disciplines of physics, biology, engineering, and mathematics, Pavlic uses his insights into engineering and biology to help better understand the role of information in major evolutionary transitions, including the transition from nonlife to life itself.

DAWID POTGIETER is a program and communications officer of Templeton World Charity Foundation. He is involved with developing new grant proposals in a wide range of areas including research in the natural sciences, philosophy, and public outreach activities. Before joining Templeton World Charity Foundation, he studied biochemistry at the University of Oxford and stayed on to complete a D.Phil. in neuroscience at the department of physiology, anatomy, and genetics. His research has been published in various journals.

VIOLA PRIESEMANN studied physics in Darmstadt and Lisbon and completed her Ph.D. at the Max Planck Institute for Brain Research and the Frankfurt Institute for Advanced Studies in 2013, including research projects at the École Normale Supérieure in Paris and at Caltech. Currently she is a Bernstein Research Fellow and has established her own group at the Max Planck Institute for Dynamics and Self-Organization in Göttingen, focusing on the question of how the collective dynamics of spiking networks in silico and in vivo coordinates information processing.

ANGELIKA SCHMIDT studied biology at the Technical University of Darmstadt and was a visiting scholar at the Rockefeller University, New York. She obtained her Ph.D. in immunology from the University of Heidelberg on a project about regulatory T cells in Peter Krammer's group at the German Cancer Research Center. She is currently a postdoctoral and Marie Curie Fellow in the Unit of Computational Medicine, Science for Life Laboratory, and the Centre for Molecular Medicine at Karolinska Institute in Stockholm, focusing on regulatory T cell induction and suppressive function.

JILLIAN E. SMITH-CARPENTER received a B.S. in chemistry and environmental studies from East Stroudsburg University in 2007. She then obtained her Ph.D. from Washington University in St. Louis under the mentorship of John-Stephen Taylor, focusing on the structure–activity relationships of G-quadruplex interloop photocrosslinking. After two years as a postdoctoral research fellow at Emory University, where her research focus expanded to include self-assembling peptides, she joined Fairfield University as an assistant professor in 2015.

KAROLA STOTZ is a senior lecturer at Macquarie University, Sydney, and a Templeton World Charity Foundation Research Fellow, working on the project "The Causal Foundations of Biological Information." She has published on philosophical issues in developmental, molecular, and evolutionary biology, psychobiology, and cognitive science. Stotz received her master's in physical and cultural anthropology from the University of Mainz and her Ph.D. in philosophy from the University of Ghent in Belgium. She has worked at the Konrad Lorenz Institute for Evolution and Cognition Research in Austria, the Unit for History and Philosophy of Science at the University of Sydney, the Department of History and Philosophy of Science at the University of Pittsburgh, and the Cognitive Science Program at Indiana University. From 2008 to 2012 she worked as an Australian Research Fellow at the

University of Sydney on the project "Postgenomic Perspectives on Human Nature."

JESPER TEGNÉR is Chaired Strategic Professor in Computational Medicine at the Centre for Molecular Medicine and Sciences for Life Laboratory at Karolinska Institutet and Karolinska University Hospital in Stockholm. He heads a research group of 35 people, one-third working in the molecular biology laboratory and two-thirds working as computational experts. He was a visiting scientist and postdoctoral fellow (Harvard) on a Wennergren three-years research position, 1998–2001, with an Alfred P. Sloan Fellowship for research. He has authored more than 100 papers, including technical computational papers as well as medical publications. He holds three undergraduate degrees (MedSchool, M.Sc.; Mathematics, M.Sc.; Philosophy), and a Ph.D./M.D. (1997) in medicine.

GIULIO TONONI is professor of psychiatry at the University of Wisconsin–Madison, where he holds the David P. White Chair in Sleep Medicine and is the Distinguished Professor in Consciousness Science. His research focuses on the understanding of consciousness and the function of sleep. His integrated information theory is a comprehensive theory of consciousness, how it can be measured, and how it is realized in the brain, which is being tested both experimentally and computationally. He is the author of more than 200 scientific publications and 5 books (2 coauthored). He has received numerous honors, including the Pioneer Award from the Director of the National Institutes of Health, and holds several patents.

SARA IMARI WALKER is an assistant professor in the School of Earth and Space Exploration and the Beyond Center for Fundamental Concepts in Science at Arizona State University and a fellow of the Arizona State University/Santa Fe Institute Center for Biosocial Complex Systems. She also serves on the board of directors of the education and research nonprofit Blue Marble Space. Her previous appointments include a NASA postdoctoral program fellowship with the NASA Astrobiology Institute and a postdoctoral fellowship in the Center for Chemical Evolution at the Georgia Institute of Technology. Her research lies at the intersection of theoretical physics, complex systems, and astrobiology and focuses on understanding the origin and nature of life.

STEVEN WEINSTEIN is a professor of philosophy at U Waterloo, cross-appointed in physics. He is also an affiliate of Perimeter Institute. Previous appointments include visiting positions at Princeton University, Dartmouth College, and the University of British Columbia. He has done pioneering work on the topic of multiple time dimensions. He also has an active career writing and recording music.

MICHAEL WIBRAL studied physics at the universities of Cologne and Konstanz, and medical physics at the University of Kaiserslautern. After working in the semiconductor industry for some years, he returned to science and obtained his Ph.D. in neurobiology at the Max Planck Institute for Brain Research in Frankfurt and the Technical University of Darmstadt. Since 2012, he is a professor for magnetoencephalography at the Goethe University, Frankfurt, and teaches information theory at the department of physics. His research interests are in neural information dynamics and predictive coding, and he is a developer of the TRENTOOL toolbox for the analysis of information transfer and active information storage in neural systems.

DAVID WOLPERT is the author of three books and more than 200 papers, has three patents, is an associate editor at more than half a dozen journals, and has received numerous awards. He has more than 10,000 citations, in fields ranging from the foundations of physics to machine learning to game theory to information theory to distributed optimization. In particular, his machine learning technique of stacking was instrumental in both winning entries for the Netflix competiton, and his papers on the no-free-lunch theorems jointly have more than 4,000 citations. He is a world expert on extending game theory to model humans operating in complex engineered systems, on exploiting machine learning to improve optimization, and on Monte Carlo methods. He is currently a member of the resident faculty of the Santa Fe Institute. Previously he was the Ulam Scholar at the Center for Nonlinear Studies, and prior to that he was at NASA Ames Research Center and was a consulting professor at Stanford University, where he formed the Collective Intelligence group. He has worked at IBM and a data-mining startup and is external faculty at numerous international institutions. His degrees in physics are from Princeton and the University of California.

HECTOR ZENIL has held positions at the Behavioural and Evolutionary Lab, Department of Computer Science, University of Sheffield; at the Structural Biology Group at the Department of Computer Science, University of Oxford; and at the Unit of Computational Medicine, Science for Life Laboratory, Centre of Molecular Medicine, Karolinska Institute in Stockholm. He is also the head of the Algorithmic Nature Group and a member of the board of directors of LABoRES in Paris. He has been visiting graduate student at the Massachusetts Institute for Technology, invited visiting scholar at Carnegie Mellon University, and invited visiting professor at the National University of Singapore. He has also been a senior research associate and external consultant for Wolfram Research, an invited member of the Foundational Questions Institute, and a member of the National Researchers System of Mexico.

I Introduction

Sara Imari Walker, Paul C. W. Davies, and George F. R. Ellis

The concept of information has penetrated almost all areas of human inquiry, from physics, chemistry, and engineering through biology to the social sciences. And yet its status as a physical entity remains obscure. Traditionally, information has been treated as a derived or secondary concept. In physics especially, the fundamental bedrock of reality is normally vested in the material building blocks of the universe, be they particles, strings, or fields. Because bits of information are always instantiated in material degrees of freedom, the properties of information could, it seems, always be reduced to those of the material substrate. Nevertheless, over several decades there have been attempts to invert this interdependence and root reality in information rather than matter. This contrarian perspective is most famously associated with the name of John Archibald Wheeler, who encapsulated his proposal in the pithy dictum 'it from bit?' (Wheeler, 1999).

In a practical, everyday sense, information is often treated as a primary entity, as a 'thing in its own right' with a measure of autonomy; indeed, it is bought and sold as a commodity alongside gas and steel. In the life sciences, informational narratives are indispensable: biologists talk about the genetic code, about translation and transcription, about chemical signals and sensory data processing, all of which treat information as the currency of activity, the 'oil' that makes the 'biological wheels go round'. The burgeoning fields of genomic and metagenomic sequencing and bioinformatics are based on the notion that informational bits are literally vital. But beneath this familiar practicality lies a stark paradox. If information makes a difference in the physical world, which it surely does, then should we not attribute to it causal powers? However, in physics causation is invariably understood at the level of particle and field interactions, not in the realm of abstract bits (or qubits, their quantum counterparts). Can we have both? Can two causal chains coexist compatibly? Are the twin narratives of material causation and informational causation comfortable bedfellows? If so, what are the laws and principles governing informational dynamics to place alongside the laws of material dynamics? If not, and information is merely an epiphenomenon surfing on the underlying physical degrees of freedom, can we determine under what circumstances it will mimic autonomous agency? This volume of essays addresses just such foundational questions. It emerged from a 2014 workshop on 'Information, Causality and the Origin of Life' hosted by the Beyond Center for Fundamental Concepts in Science at Arizona State University as part of a Physics of Living Matter series and funded by the Templeton World Charity Foundation under their 'Power of Information' research programme. Contributors included physicists, biologists, neuroscientists, and engineers. The participants were tasked with addressing the question: Is information merely a useful explanatory concept or does it actually have causal power? Among the questions considered were:

- What is information? Is it a mere epiphenomenon or does it have causal purchase?
- How does it relate to top-down causation?
- Where does it come from?
- Is it conserved, and if so, when?
- How and when does it take on 'a life of its own' and become 'a thing' in its own right?
- Are there laws of 'information dynamics' analogous to the laws of material dynamics? Are they the same laws? Or can information transcend the laws of mechanics?
- How does information on the microscale relate to information on the mesoscale and macroscale (if at all)?
- Is information loss always the same as entropy?

Although participants agreed that the concept of information is central to a meaningful description of biological processes, opinions differed over whether the ultimate explanation for life itself necessarily rests on informational foundations. Related to this dichotomy is whether life in general, and its informational characteristics in particular, can be accounted for in terms of known principles or if an entirely new conceptual framework is required. One area of science in which this issue is thrown into stark relief is the origin of life. If life is defined in terms of informational processes, somehow its distinctive informational properties had to emerge from the relatively incoherent realm of chemical networks. It is in the transition from matter to life that our traditional approaches to physics, which accurately describe the predictability of the physical and chemical world, must yield to the novelty and historical-dependency characteristic of life. Many scientists have argued that there may be little hope of resolving the issue of life's origin unless we can better understand what information is and how it operates in the physical world (Deacon and Sherman, 2008; Küppers, 1990; Walker and Davies, 2013; Yockey, 2005).

The primary challenge in addressing the majority of the foregoing questions is that many of the concepts of 'information' as currently understood are not readily accommodated in our current approaches to physics. To see why this is the case, consider the fact that we typically formulate mathematical treatments of physical systems almost exclusively in terms of *initial states and fixed deterministic laws*. Information, on the other hand, by its very definition, requires a physical system to take on one of several possible states, e.g., to say something carries one bit of information means that it could have been in one of two possible states (and even more challenging that these two different states can carry different *meaning*). Fixed deterministic laws permit one outcome only at a given time t for a given initial state at $t = 0$. Several contributors in this volume recognise the need to reformulate physical theory in a manner that better accommodates the different nature of information, and the need to break free of the 'tyranny of initial conditions' (e.g., Chapters 2, 3, and 8).

The concept of causation is not well defined in physical science. In the realm of particle physics and quantum field theory it merely means the absence of superluminal interactions: there is no directionality attached. In everyday affairs we have little difficulty in understanding the nature and asymmetry of cause and effect, and the concept of a causal chain. Complex systems theory has a number of definitions of causation, among which is the notion of 'causal intervention', which involves studying the downstream effects of a small perturbation on the system that directs it into alternative states. This view of causation, like information, requires the existence of counterfactuals – roughly speaking, alternative possibilities that do not happen but could. To appreciate the distinction, contrast the causal-neutral perspective of an external observer who might reason 'if I had initialised the system in state X_1 rather than Y_1, it would now be in state X_2 rather than Y_2', with the stronger claim that Y_2 rather than Y_1 could have been *caused* to occur. Removing the notion of an external observer elevates counterfactual statements to the status that they must be intrinsic to the operation of reality for such a statement to hold. In this stronger view, information and causation are fundamentally at odds with an explanatory framework based solely on initial states and fixed dynamical laws. The tension constitutes what Deutsch calls the 'nightmare of counterfactuals' in theoretical physics (Deutsch, 2013). The above somewhat abstract considerations lead naturally to the 'hard problem of life', as articulated by Walker and Davies (Chapter 2), thus termed by analogy with the 'hard problem of consciousness' identified by Chalmers (1995). He made the case that even if we eventually uncover a complete mechanistic understanding of the wiring and firing of every neuron in the brain, we will still be unable to explain the phenomenon of qualia (roughly speaking, the subjective experience of the observer) and the impression that neural information (loosely speaking, thoughts and urges) seemingly possess causal power. Scientists sympathetic to the dualism implied by Chalmers might seek to invoke new physical laws and principles, on the basis that the physics of the brain tells us nothing about thoughts, feelings, and what it is like to experience something. Walker and Davies similarly argue that a complete mechanistic account of living matter will still fail to capture what it means to be living. By implication, some new physical laws or principles will be needed. Although these laws and principles are not known at this time, Walker and Davies argue that a promising area to seek them is precisely in the realm of information theory and macrostates, specifically that macrostates have causal power. To elaborate on this quest, their chapter discusses the issue of what exists and why the world is complex in a manner that a causal theory of information could potentially explain. Tackling the hard problems of both life and consciousness within the framework of information theory and causal counterfactuals holds the promise of an eventual unification of biology and neuroscience at a fundamental conceptual level.

Another contribution along these general lines is Chapter 3, by Marletto, in which she rejects the standard formulation of information being a facet of probability distributions or inferential knowledge. Instead, she appeals to constructor theory to argue that 'information' implicitly refers to certain interactions being possible in nature (e.g., copying) and that the properties we associate with information are constraints on the laws of physics themselves. Constructor theory is an entirely new mode of explanation, which attempts to reformulate all of science by introducing information as a foundational concept. Thus, one can discuss what is necessary of the laws of physics in order for processes such as self-replication and adaptation – central to life – to exist. Marletto considers in particular how the principles of constructor theory can account for the very existence of life (defined as accurate self-reproduction) under 'no-design laws' (Wigner, 1961) – that is, laws of physics that do not explicitly include the design of an organism at the outset. Marletto concludes that accurate constructors are indeed permitted under no-design laws, provided that the laws of physics allow the physical instantiation of modular, discrete replicators, which can encode algorithms or 'recipes' for the construction of the replicator. A novelty of the foregoing approaches is that they invert the usual assumption that physics informs biology. Since life is obviously permitted by the laws of physics, we can ask how the existence of life can inform our understanding of physical law.

While it is crucial to understand how the existence of life is consistent with (possibly new) laws of physics, establishing that fact would still leave open the question of *how life emerges from nonlife*. (To say that B is consistent with A does not imply that A explains B.) It is an open question of whether *bio emerged from bit*, that is, whether a better understanding of the concepts of information and causation have anything to say about the transition from the nonliving to living state. In Chapter 4, Grisogono asks whether there was a time before information itself emerged. While there is an elementary sense in which information existed prior to life, in terms of the Shannon information describing the configurations of matter (which Grisogono refers to as 'latent' information), the nature of biological information goes beyond the mere syntactics of Shannon. The 'added value' implicit in biology is semantics and is eloquently captured by Bateson's famous dictum (1972):

> What we mean by information – the elementary unit of
> information – is a difference that makes a difference.

The physical aspects of information do not touch the key issue that all living systems have purpose or function (Hartwell et al., 1999), and that purpose is realised by use of information deployed so as to enable the organism to attain its goals, enabled by architectures that encode, decode, and utilise information to sense the environment and respond in a suitable way. Grisogono addresses this distinction by pointing out three significant features of information: (1) it differs qualitatively from matter or energy, (2) it can have causal effects, and (3) not all differences make a difference. It is in the emergence of 'differences that make a difference', she argues, that the key to understanding the origin of life lies. A scenario is envisaged where the steps towards life are initiated with the emergence of an autocatalytic set of molecules that can collectively reproduce and create as a by-product membrane molecules that enclose the set (e.g. form a boundary), a set of ideas proposed in Deacon's autocell model (Deacon, 2006). While these structures would certainly contain information, the key transition that makes a difference is the origin of meaning – defined as when an autonomous agent emerges that can make an informed rather than random choice between at least two actions (counterfactuals), i.e., when it can take action on differences that make a difference. This chapter therefore connects the emergence of meaning (semantic information) and *coded* information to the emergence of causally effective information (Walker and Davies, 2013), a critical step in the origin of life.

Although a major thrust in this volume is to understand information as a distinct category, it is agreed that it cannot be 'free-floating'. As proclaimed by Landauer, 'information is physical!' (Landauer, 1996), by which he meant that every bit of information must be instantiated in a physical degree of freedom. Physical and chemical constraints on how information is processed therefore do matter. In particular, life employs both digital and analogue information, both of which may have emerged early in the transition from matter to life. In Chapter 5, Smith-Carpenter et al. consider chemistries that could permit emerging codes through the processes of chemical evolution, using self-assembling peptide β-sheets as an explicit example of a more general theoretical framework. They identify three necessary dimensions: (1) structural order, (2) dynamic networks, and (3) function. It is at the intersection of these dimensions that pathways to increasing complexity are possible, they argue, suggesting new modes for thinking about chemical evolution that are neither strictly digital nor analogue. An interesting feature of the peptide assemblies discussed is that there exists both a digital and an analogue aspect to their information content: digital in the sequence of amino acids composing the peptide, but analogue in the conformations that macroassemblies can assume.

The digital nature of life's information processing systems is familiar – DNA, for example, is best described as a digital (discrete) sequence of nucleobases, and gene regulatory networks are often described using the operations of binary logic (Davidson and Levin, 2005), such that they may be likened to circuit boards (Nurse, 2008). Less appreciated are analogue aspects of information as it operates in organisms. These are explored in Chapter 6 by Noble, who asks: 'Are organisms encoded as purely digital molecular descriptions in their gene sequences?' The answer he provides is a resounding No! One argument in favour of analogue information as a major contributor to biological function is the sheer magnitude of the information encoded in structural versus genomic degrees of freedom within a cell: a back-of-the-envelope calculation reveals that it is easy to represent the three-dimensional image structure of a cell as containing much more information than is possible in the genome. This perspective is parsimonious if one considers the evolutionary advantage of *not* encoding everything the cell does – why code for what physics and chemistry can do for you? Noble invokes a computer analogy, suggesting that not only do we need the 'program' of life; we also need the 'computer' of life (the interpreter of the genome), i.e. the highly complex egg cell, to have a full explanatory account of information in living systems.

The analogy between biology and computation brings to light the question of how much of life can be understood in terms of informational software or programs (be they genetic or in other forms) that transcend the chemical substrates of life. In Chapter 7, Adami and LaBar take one extreme, considering a purely informational definition for life as 'information that copies itself' and explore the consequences for our understanding of the emergence of life, utilising the digital life platform Avida as a case study. Based on information-theoretic considerations, they demonstrate that it is rare, but not impossible, to find a self-replicating computer program purely by chance. However, the probability is significantly improved if the resource distribution is biased in favour of the resource compositions of self-replicators, that is, if one uses a 'biased typewriter'. The conclusion is that the composition of the prebiotic milieu really does matter in determining how likely it is to stumble across a functional replicator by chance.

Flipping this narrative on its head, it suggests new information-theoretic approaches to determining the optimal environments from which life is *most probable* to emerge.

New approaches to information are necessary for understanding not only the origin of life but also all levels of description within the biological (information) hierarchy – from cells to societies. It is likely the case that insights from other (higher) scales of organisation in living systems will ultimately inform our understanding of life's emergence, and in particular that new principles will be necessary to unify our understanding of life as it exists across different scales of space and time. One such 'hidden principle' proposed by Krakauer in Chapter 8 suggests that life can be thought of as a collection of evolved inferential mechanisms dependent on both information (memory storage) and information dynamics – e.g., 'computation'. Computation provides an advantage for adaptive search in the quest for more efficient means to utilise available free energy gradients. Many examples of what he calls cryptosystems are provided, ranging from parasites such as *Trypanosoma brucei*, which uses 'noise' to evade its host's immune system by combinatorially rearranging its surface proteins, to combinatorial ciliates, which hide information by encrypting their own genome. It seems that once one starts looking for encrypted informational systems in biology, they are everywhere.

It is a surprising twist that noise could find a constructive role in biology by encrypting valuable information; noise is usually regarded as the antithesis of information in our standard physical interpretations. Another area where noise takes on a surprising role in the informational narrative of living systems is with respect to function. In nonlinear systems introducing noise can preferentially amplify functional aspects of a signal above a threshold for detection, as occurs in the phenomenon of *stochastic resonance*, a concept derived from engineering due to its utility in increasing the signal-to-noise ratio of a signal. In Chapter 9, Weinstein and Pavlic explore how biological systems might equivalently utilise noise to execute biological function. Examples are provided, such as nest-site selection by fire ants. The examples of Krakauer and Weinstein and Pavlic emphasise the contrast between 'information' as we view it in physics and 'information' as it is useful and functional in biology. Noise by definition contains no (syntactic) information as quantified by Shannon (1949) but can take on meaning in biology if it serves a function (its semantic aspect).

Sensing and response require determining what information possesses a fitness value to an organism (Donaldson-Matasci et al., 2010) – that is, *what information is necessary to attain knowledge for adaptive success.* Organisms are able to build high-level descriptions (models) of their environments, e.g., to identify relevant coarse-grainings or 'macrostates' (Flack et al., 2013). This process in itself can often seem subjective. In physics there is continuing debate about whether thermodynamic macrostates are objective or subjective (Shalizi and Moore, 2003). Often the answer boils down to the purpose of building a high-level description, i.e., the function or goal of the organism. In Chapter 10, Wolpert et al. tackle the challenge of building a quantitative framework for addressing one purpose organisms have for coarse-graining their world. Specifically, they focus on a goal that may be intrinsic to many living systems – that of predicting observables of interest concerning a high-dimensional system, which must be done with as high an accuracy as possible while minimising the computational cost of doing so. The framework that emerges – which they call state space compression – provides a guide to objectively determine macrostates of interest, with the aforementioned purpose. An intriguing question is how the framework might adapt to identifying macrovariables that fulfil other notions of purpose.

Once one moves to a description of physical reality that is cast in terms of macrostates, rather than microstates, concepts such as 'information' and 'computation' become much more natural descriptors of reality. In physical terms, for example, computation may most simply be thought of as a transformation performed on a macrostate. It is an interesting question in the foundations of logic as to whether the computational nature of the world is in fact unknowable. Be that as it may, the theme of computation and the programmability of nature can provide important insights into the unity of matter and life, dubbed the 'algorithmiticity of nature', as discussed in Chapter 11 by Zenil et al. From this picture emerges a perspective where natural selection itself may be thought of as 'reprogramming' biology, and where solutions to notoriously difficult diseases to treat, such as cancer, might come from 'hacking' the function of cells. Cancer, for example, could be a reversion to a more ancestral software or operating system for the cell (Davies and Lineweaver, 2011).

A computational view of nature could also explain biological information hierarchies as they exist across multiple functional space and time scales, a view advocated by Flack in Chapter 12. Again, the concepts of coarse-graining and computation take centre stage. Flack goes one step further, proposing a mechanism for new levels of biological organisation to emerge. She argues that this occurs when coarse grained (slow) variables become better predictors than microscopic behaviour (which is subject to fluctuations). Convergence of these variables to predict macroscopic behaviour is argued to permit consolidation of new levels of organisation. These macrostates are conjectured to be intrinsically subjective and should be regarded as local optima. Such processes can give the appearance of top-down causation being deployed by the relevant macrovariables, but it is an open question whether it is real or illusory.

A different perspective on life's information hierarchy is adopted in Chapter 13 by Farnsworth et al., who explicitly consider top-down causation to explain the hierarchical organisation of the biosphere (see also Walker et al., 2012). It is plausible that most information processing in biology is at the 'top level'. New levels of organisation can emerge when new levels of information processing undergo a transition and become causally effective over lower levels. The origin of this higher-level biological information is adaptive selection, a key process whereby meaningful information can be created from a disordered jumble of data by discarding that which is meaningless in a specific context. This process, leading to the storing of genetic information in the DNA of cells, underlies the emergence of complex life. One can make the case that each of the major inventions of evolutionary history was the discovery in this way of new means of deploying information to control material outcomes in such a way as to enhance survival prospects. Farnsworth and colleagues argue that top-down causal control and the resulting appearance of autonomy are hallmarks of life.

Adopting a framework where one explicitly focuses on causal structure – rather than dynamics through state space – may help to elucidate some of the debate. This is the perspective provided in Chapter 14 by Albantakis and Tononi, who consider the distinction between 'being' and 'happening', utilising cellular automata (CA) as a case study. Most prior work on dynamical systems, including CA, focuses on what is 'happening' – the dynamical trajectory of the system through its state space – that is, they take an extrinsic perspective on what is observed. Often, complexity is characterised using statistical methods and information theory. In a shift of focus to that of causal architecture, Albantakis and Tononi consider what the system 'is' from its own *intrinsic* perspective, utilising the machinery of integrated information theory (IIT), and demonstrate that intrinsic (causal) complexity (as quantified by integrated information Φ in IIT) correlates well with dynamical (statistical) complexity in the examples discussed. These and similar approaches could provide a path forward for a deeper understanding of the connection between causation and information as hinted at in the beginning of this chapter.

Further insights into unifications of the concept of information and causation are provided in Chapter 15 by Stotz and Griffiths, who focus on the concept of biological specificity to illuminate the relationship between biological information and causation. They propose that causal relationships in biology are 'informational' relationships simply when they are highly specific, and introduce the idea of 'causal specificity', adopted from the literature on the philosophy of causation, as a way to quantify it. An example is 'Crick information', defined such that if a 'cause' makes a specific difference to the linear sequence of a biomolecule, it contains Crick information for that molecule (Griffiths and Stotz, 2013). Nucleic acids are one example of causal specificity. However, the general theoretical framework is expected to apply to a wide array of biological systems where causal specificity plays an important role.

Another example concerns animal nervous systems, which are hard-wired to collect information about the world through multiple sensory modalities. Brains are exquisitely structured to search for patterns in that information in the light of the current context and expected future events, so as to extract what is meaningful and to ignore the rest. Language has been developed to store, analyse, and share that information with others, thus permitting the transfer of cultural information. The ability of humans to transmit information qualitatively distinguishes them from the rest of the animal kingdom. Hence information acquisition, analysis, and sharing through the use of language are core aspects of what it means to be human. The information is *specifically* encoded via symbolic systems such as writing. Thus, many aspects of the role of information and causation in life and its origins in the preceding discussion are apparent in social and technological systems. Despite more than 3.5 billion years obscuring reconstruction of the events surrounding life's origin(s), these connections suggest that perhaps common principles might underlie the transition from matter to life and the current transitions we are undergoing in human social and technological systems.

In Chapter 16, DeDeo details major transitions in the structure of social and political systems, drawing insights from major evolutionary transitions more broadly. He identifies three critical stages in the emergence of societies, each of which relies on the relationship of human minds to coarse-grained information about their world. The first is awareness and use of group-level social facts (e.g., a social hierarchy as provided by the example of the monk parakeet). In a second transition, these facts eventually become norms, forming a notion of the way things 'should be' (i.e., such as thanking a shopkeeper after visiting their shop). In the third transition, the norms aggregate into normative bundles, which establish group-level relationships among norms. It is intriguing to speculate that the role of top-down causation in the origin of society as outlined by DeDeo could parallel that conjectured to occur in the emergence and evolution of life (Ellis, 2011; Noble, 2012; Walker and Davies, 2013), particularly through major evolutionary transitions (Maynard Smith and Szathmary, 1997; Walker et al., 2012).

Among all life's information processing, none evokes more appreciation for the causal power of information than that of the human mind. Explaining behaviour in terms of information processing has been a fundamental commitment of cognitive science for more than 50 years, leading to the huge strides made in psychology, linguistics, and cognitive neuroscience. Although the success of these sciences puts the reality of neural information processing beyond serious doubt, the nature of neural information remains an unanswered foundational question. This is a topic addressed in Chapter 17 by Wibral et al., who discuss how implementing techniques from information theory can aid in identification of the algorithms run by neural systems, and the information they represent. A computational principle adopted by many neural information-processing systems is to continuously predict the likely informational input and to carry out information processing primarily on error signals that record the difference between prediction and reality. This predictive coding actually shapes the way we experience the world (Frith, 2013). Wibral et al. provide insights for identifying the algorithms that are applicable to both natural and artificial systems – perhaps inspiring the design of artificial systems.

The implications of these kinds of approaches are potentially profound. A foundational, deep understanding of living systems, of the kind we currently enjoy in other domains of science such as quantum theory or general relativity, would undoubtedly dramatically change our perceptions of the world and our place in it, just as was the case for previous scientific revolutions. A hint at the implications of a fundamental understanding of the role of information in life and mind is the topic of discussion in Chapter 18 by Briggs and Potgeiter (Chapter 18), who consider the specific example of the scientific and ethical implications of machine learning. They ask: To what extent can the mechanisms that underlie machine learning mimic the mechanisms involved in human learning? An important distinction between the human mind and any machine we have yet created is that while the human brain can learn algorithms, its natural *modus operandi* is holistic pattern recognition based in neural networks (the information is integrated [Kandel, 2012]). The laws of physics themselves also do not appear to be algorithmic in nature (but instead appear to be descriptions of interactions between fields/forces and particles with fixed dynamical laws). It is therefore unclear at present how far our current approaches can take us in realising 'machines that think' (or feel) without addressing both the hard problems of consciousness and life.

REFERENCES

Bateson, G. (1972). *Steps to an ecology of mind: collected essays in anthropology, psychiatry, evolution, and epistemology.* University of Chicago Press.

Chalmers, D. J. (1995). Facing up to the problem of consciousness. *Journal of Consciousness Studies,* 2(3):200–219.

Davidson, E., and Levin, M. (2005). Gene regulatory networks. *Proceedings of the National Academy of Sciences of the United States of America,* 102(14):4935.

Davies, P. C. W., and Lineweaver, C. H. (2011). Cancer tumors as metazoa 1.0: tapping genes of ancient ancestors. *Physical Biology,* 8(1):015001.

Deacon, T. W., (2006). Reciprocal linkage between self-organizing processes is sufficient for self-reproduction and evolvability. *Biological Theory,* 1(2):136–149.

Deacon, T., and Sherman, J. (2008). The pattern which connects pleroma to creatura: the autocell bridge from physics to life. In Jesper Hoffmeyer (ed), *A legacy for living systems: Gregory Bateson as precursor to biosemiotics,* pages 59–76. Springer.

Deutsch, D. (2013). Constructor theory. *Synthese,* 190(18):4331–4359.

Donaldson-Matasci, M. C., Bergstrom, C. T., and Lachmann, M. (2010). The fitness value of information. *Oikos,* 119(2):219–230.

Ellis, G. F. R. (2011). Top-down causation and emergence: some comments on mechanisms. *Interface Focus*, 2:126–140.

Flack, J., Erwin, D., Elliot, T., and Krakauer, D. (2013). Timescales, symmetry, and uncertainty reduction in the origins of hierarchy in biological systems. In Kim Sterelny, Richard Joyce, Brett Calcott, and Ben Fraser (eds), *Cooperation and its evolution*, pages 45–74, MIT Press.

Frith, C. (2013). *Making up the mind: how the brain creates our mental world*. John Wiley & Sons.

Griffiths, P., and Stotz, K. (2013). *Genetics and philosophy: an introduction*. Cambridge University Press.

Hartwell, L. H., Hopfield, J. J., Leibler, S., and Murray, A. W. (1999). From molecular to modular cell biology. *Nature*, **402**:C47–C52.

Kandel, E. (2012). *The age of insight: the quest to understand the unconscious in art, mind, and brain, from Vienna 1900 to the present*. Random House.

Küppers, B. (1990). *Information and the origin of life*. MIT Press.

Landauer, R. (1996). The physical nature of information. *Physics Letters A*, **217**(4):188–193.

Maynard Smith, J., and Szathmary, E. (1997). *The major transitions in evolution*. Oxford University Press.

Noble, D. (2012). A theory of biological relativity: no privileged level of causation. *Interface Focus*, **2**(1):55–64.

Nurse, P. (2008). Life, logic and information. *Nature*, **454**(7203):424–426.

Shalizi, C., and Moore, C. (2003). What is a macrostate? Subjective observations and objective dynamics. arXiv preprint cond-mat/0303625.

Shannon, C. (1949). Communication theory of secrecy systems. *Bell System Technical Journal*, **28**(4):656–715.

Walker, S., and Davies, P. (2013). The algorithmic origins of life. *J. R. Soc. Interface*, **10**(79):20120869.

Walker, S. I., Cisneros, L., and Davies, P. C. W. (2012). Evolutionary transitions and top-down causation. arXiv preprint arXiv:1207.4808.

Wheeler, J. A. (1999). Information, physics, quantum: the search for links. In A. J. G. Hey (ed), *Feynman and computation*. Perseus Books.

Wigner, E. P. (1961). The probability of the existence of a self-reproducing unit. In *The logic of personal knowledge: essays presented to M. Polanyi on his seventieth birthday, 11th March, 1961*, pages 231–238.

Yockey, H. P. (2005). *Information theory, evolution, and the origin of life*. Cambridge University Press.

Part I Physics and Life

2   The "Hard Problem" of Life

Sara Imari Walker and Paul C. W. Davies

There are few open problems in science as perplexing as the nature of life and consciousness. At present, we do not have many scientific windows into either. In the case of consciousness, it seems evident that certain aspects will ultimately defy reductionist explanation, the most important being the phenomenon of qualia – roughly speaking, our subjective experience as observers. It is a priori far from obvious why we should have experiences such as the sensation of the smell of coffee or the blueness of the sky. Subjective experience isn't necessary for the evolution of intelligence (we could, for example, be zombies in the philosophical sense and appear to function just as well from the *outside* with nothing going on *inside*). Even if we do succeed in eventually uncovering a complete mechanistic understanding of the wiring and firing of every neuron in the brain, it might tell us nothing about thoughts, feelings, and what it is like to experience something. Our phenomenal experiences are the only aspect of consciousness that appears as though they cannot, *even in principle*, be reduced to known physical principles. This led Chalmers to identify pinpointing an explanation for our subjective experience as the "hard problem of consciousness." The corresponding "easy problems" (in practice not so easy) are associated with mapping the neural correlates of various experiences. By focusing attention on the problem of subjective experience, Chalmers highlighted the truly inexplicable aspect of consciousness, based on our current understanding. The issue, however, is by no means confined to philosophy. Chalmers' proposed resolution is to regard subjective consciousness as an irreducible, fundamental property of mind, with its own laws and principles. Progress can be expected to be made by focusing on what would be required for a theory of consciousness to stand alongside our theories for matter, even if it turns out that something fundamentally new is not necessary.

The same may be true for life. With the case of life, it seems as though we have a better chance of understanding it as a physical phenomenon than we do with consciousness. It may be the case that new principles and laws will turn out to be unnecessary to explain life, but meanwhile their pursuit may bring new insights to the problem (Cronin and Walker, 2016). Some basic aspects of terrestrial biology – for example, replication, metabolism, and compartmentalization – can almost certainly be adequately explained in terms of known principles of physics and chemistry, and so we deem explanations for these features to belong to the "easy problem" of life. Research on life's origin for the past century, since the time of Oparin and Haldane and the "prebiotic soup" hypothesis, has focused on the easy problem, albeit with limited progress. The more pressing question, of course, is whether all properties of life can in principle be brought under the "easy" category, and accounted for in terms of known physics and chemistry, or whether certain aspects of living matter will require something fundamentally new. This is especially critical in astrobiology; without an understanding of what is meant by "life" we can have little hope of solving the problem of its origin or to provide a general-purpose set of criteria for identifying it on other worlds. As a first step in addressing this issue, we need to clarify what is meant by the "hard problem" of life, that is, to identify which aspects of biology are likely to prove refractory in attempts to reduce them to known physics and chemistry, in the same way that Chalmers identified qualia as central to the hard problem of consciousness. To that end we propose that *the hard problem of life is the problem of how "information" can affect the world*. In this chapter we explore both why the problem of information is central to explaining life and why it is hard, that is, why we suspect that a full resolution of the hard problem will not ultimately be reducible to known physical principles (Walker, 2015).

WHY THIS "HARD PROBLEM"?

There is an important distinction between the hard problem of life and that of consciousness. With consciousness it is obvious to each of us that we experience the world – to read this page of text you are *experiencing* a series of mental states, perhaps a voice reading aloud in your head or a series of visual images. The universal aspects of experience are therefore automatically understood to each of us: if intelligent aliens exist and are also conscious observers like us, we might expect the objective fact that they experience the world to be similar (even if the experience itself is subjectively different), despite the fact that we can't yet explain what consciousness is or why it arises. By contrast, there is no general agreement on what features of life are universal. Indeed, they could be so abstract that we have yet to identify them (Davies and Walker, 2016).

As Monod (1974) emphasized, biological features are a combination of chance and necessity, combining both frozen accidents and law-like evolutionary convergence. As a result of our anthropocentric vantage point (Carter and McCrea, 1983) (thus far observing life only here on Earth), both astrobiology and our assumptions about nonhuman consciousness tend to be biased by our understanding of terrestrial life. With only one sample of life at our disposal, it is hard to separate which features are merely accidental, or incidental, from the "law-like" features that we expect would be common to *all life* in the universe.

Discussions about universal features of life typically focus on chemistry. In order to generalize "life as we know it" to potential universal signatures of life, we propose to go beyond this emphasis on a single level (chemistry) and recognize that *life might not be a level-specific phenomenon* (Walker et al., 2012). Life on Earth is characterized by hierarchical organization, ranging from the level of cells to multicellular organisms to eusocial and linguistic societies (Szathmary and Maynard Smith, 1994). A broader concept of life, and perhaps one that is therefore more likely to be universal, could be applied to multiple levels of organization in the biosphere – from cells to societies – and might in turn also be able to describe alien life with radically different chemistries. The challenge is to find universal principles that might equally well describe any level of organization in the biosphere (and ones yet to emerge, such as speculated transitions in social and technological systems that humanity is currently witnessing, or may one day soon witness). Much work has been done attempting to unify different levels of organization in biological hierarchies (see, e.g., Campbell, 1974; Szathmary and Maynard Smith, 1994), and although we do not yet have a unified theory, many authors have pointed to the concept of information as one that holds promise for uncovering currently hidden universal principles of biology at any scale of complexity (e.g., Davies and Walker, 2016; Farnsworth et al., 2013; Flack et al., 2013; Jablonka and Szathmáry, 1995; Smith, 2008; Szathmary and Maynard Smith, 1994; Walker et al., 2012; to name but a few) – ones that in principle could be extrapolated to life on other worlds.

Although we do not attempt in this chapter to define "biological information," which is a subject of intense debate in its own right (Godfrey-Smith and Sterelny, 2008), we wish to stress that it is not a passive attribute of biological systems, but plays an active role in the execution of biological function (see, e.g., Chapter 15). An example from genomics is an experiment performed at the Craig Venter Institute, where the genome from one species was transplanted to another and "booted up" to convert the host species to the foreign DNA's phenotype – quite literally reprogramming one species into another (Lartigue et al., 2007). Here it seems clear that it is the *information* content of the genome – the sequence of bits – and not the chemical nature of DNA as such that is (at least in part) "calling the shots." Of course, a hard-nosed reductionist might argue that, in principle, there must exist a purely material narrative of this transformation, cast entirely in terms of microstates (e.g., events at the molecular level). However, one might describe this position as "promissory reductionism," because there is no realistic prospect of ever attaining such a complete material narrative or of its being any use in achieving an understanding of the process even if it were attained. On practical grounds alone, we need to remain open to the possibility that the causal efficacy of information may amount to more than a mere methodological convenience and might represent a new causal category not captured in a microstate description of the system. What we term the "hard problem of life" is the identification of the actual physical mechanism that permits information to gain causal purchase over matter. This view is not accommodated in our current approaches to physics.

WHAT IS POSSIBLE UNDER THE KNOWN LAWS OF PHYSICS?

Living and conscious systems attract our attention because they are highly remarkable and very special states of matter. In the words of the Harvard chemist George Whitesides,

> How remarkable is life? The answer is: very. Those of us who deal in networks of chemical reactions know of nothing like it. How could a chemical sludge become a rose, even with billions of years to try?
>
> *(Whitesides, 2012)*

The emergence of life and mind from nonliving chemical systems remains one of the great outstanding problems of science. Whatever the specific (and no doubt convoluted) details of this pathway, we can agree that it represents a transition from the mundane to the extraordinary.

In our current approaches to physics, where the physical laws are fixed, any explanation we have for why the world is such as it is ultimately boils down to specifying the initial state of the universe. Since the time of Newton, our most fundamental theories in physics have been cast in a mathematical framework based on specifying an initial state and a deterministic dynamical law. Under this framework, while physical states are generally time dependent and contingent, the laws of physics are regarded as timeless, immutable, and universal. Although the immutability of the laws of physics is occasionally challenged (Davies, 2008; Peirce, 1982; Smolin, 2013; Wheeler, 1983), it remains the default assumption among the vast majority of scientists. To explain a world as complex as ours – which includes things like bacteria and at least one technological civilization with *knowledge* of things like the laws of gravity (Walker, 2016) – requires that the universe have a very special initial state indeed. The degree of "fine-tuning" necessary to specify this initial state is unsatisfactory and becomes ever more so the more complex the world becomes.[1] One should hope for a better explanation than simply resorting to special initial conditions (or for that matter stochastic fluctuations, which in many regards are even less explanatory). Indeed, there are serious attempts, such as constructor theory, which aim to remove the dependency on initial conditions from our explanations of the world (Deutsch, 2013; see also Chapter 3). Resolving this problem of fine-tuning is essential to explaining life, since the most complex structures we know of – those least explicable from an arbitrary initial state – are precisely those of interest in our discussion here, that is, living, technological, and conscious physical systems (e.g., things like us).

A new framework for explanation may come from reconsidering the kinds of dynamical trajectories permitted by our current theories. Herein we follow an argument outlined by one of us (in Walker, 2016), where the focus was on why our use of mathematics, as an example of particularly powerful information encoding utilized by living systems, can so effectively describe much of reality. In physics, particularly in statistical mechanics, we base many of our calculations on the assumption of metric transitivity, which asserts that a system's trajectory will eventually explore the entirety of its state space – thus everything that is physically possible will eventually happen. It should then be trivially true that one could choose an arbitrary "final state" (e.g., a living organism) and "explain" it by evolving the system

---

[1] The alternative explanation that everything interesting has arisen as a result of quantum fluctuations is equally unsatisfactory.

backwards in time, choosing an appropriate state at some "start" time t_0 (fine-tuning the initial state). In the case of a chaotic system the initial state must be specified to arbitrarily high precision. But this account amounts to no more than saying that the world is as it is because it was as it was, and our current narrative therefore scarcely constitutes an explanation in the true scientific sense.

We are left with a bit of a conundrum with respect to the problem of specifying the initial conditions necessary to explain our world. A key point is that if we require specialness in our initial state (such that we observe the current state of the world and not any other state), metric transitivity cannot hold true, as it blurs any dependency on initial conditions – that is, it makes little sense for us to single out any particular state as special by calling it the "initial" state. If we instead relax the assumption of metric transitivity (which seems more realistic for many real-world physical systems, including life), then our phase space will consist of isolated pocket regions, and it is not necessarily possible to get to any other physically possible state (see, e.g., Figure 2.1 for a cellular automata example). Thus, the initial state must be tuned to be in the region of phase space in which we find ourselves, and there are regions of the configuration space our physical universe would be excluded from accessing, even if those states may be equally consistent and permissible under the microscopic laws of physics (starting from a different initial state). Thus, according to the

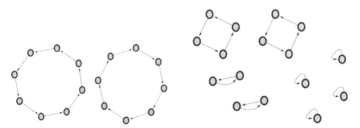

FIG. 2.1 Example of a finite, deterministic, time-reversible model. These reversible causal graphs demonstrate the concept that "you can't always get there from here," since the state space is composed of many disconnected regions.

standard picture, we require special initial conditions to explain the complexity of the world, but also have a sense that we should not be on a particularly special trajectory to get here (or anywhere else), as it would be a sign of fine-tuning of the initial conditions. Stated most simply, a potential problem with the way we currently formulate physics is that you can't necessarily get everywhere from anywhere (see Walker, 2016 for discussion).

A real living system is neither deterministic nor closed, so an attempt to attribute life and mind to special initial conditions would necessarily involve fixing the entire cosmological initial state to arbitrarily high precision, even supposing it were classical. If instead one were to adopt a quantum cosmological view, then the said pathway from the mundane to the extraordinary could, of course, be accommodated within the infinite number of branches of the wave function, but again this is scarcely a scientific explanation, for it merely says that anything that can happen, however extraordinary, will happen somewhere within the limitless array of histories enfolded in the wave function.

Leaving aside appeal to special initial conditions, or exceedingly unusual branches of cosmological wave functions, one may still ask whether pathways from the mundane to the extraordinary are problematic within the framework of known physics. Here we wish to point out a less appreciated fact with respect to the problem of fine-tuning and explaining the complexity of our world. Just because every intermediate state on a pathway to a novel state is physically possible does not mean that an arbitrary succession of states is *also possible*. If we envisage the route from mundane chemistry to life, and onward to mind, as a trajectory in some enormous state space, then not every trajectory is consistent with the known laws of physics. In fact, it may well be that almost all trajectories are inconsistent with the known laws of physics (this could be true even if individual steps taken along the way are compatible with known laws).

To justify this claim we explore a toy model inspired by cellular automata (CA), which are often used as computational models for exploring aspects of living and complex systems. However, we note that our arguments, as presented here, are by no means exclusive to CA and could apply to any discrete dynamical system with fixed rules for its time evolution. We note that it is not necessarily the case that the physical laws governing our universe are completely deterministic (for example, under collapse interpretations of quantum theory) and that reality is not necessarily discrete. However, by demonstrating a proof-of-principle for the more conservative case of discrete deterministic systems we expect that at least some aspects will be sufficiently general to apply to physical laws, as they might describe the real universe under assumptions more relaxed than those presented herein.

CA are examples of discrete dynamical systems that consist of a regular grid of cells, each of which can be in a finite number of states – in particular, we focus on systems with cells that can be in one of two possible states: "0" or "1." For simplicity, let's also assume our universe is one-dimensional with a spatial size of w cells. The configuration space of the system then contains 2^w possible states. If we restrict ourselves to deterministic systems, saying nothing yet about the laws that operate on them, each state may appear exactly once on any given trajectory, prior to settling into an attractor (otherwise, the system would not be deterministic). Under this constraint, the total number of deterministic trajectories of length $r \leq 2^w$, $n_t(r)$, is just the number of possible permutations on a subset r chosen from 2^w elements:

$$n_t(r) = \frac{2^w!}{(2^w - r)!} \tag{2.1}$$

which quantifies the number of ways to obtain an *ordered* subset of r elements from a set of 2^w elements. The total number of unique possible trajectories is just the sum over all possible trajectory lengths r:

$$N = \sum_{r=1}^{2^w} \frac{2^w!}{(2^w - r)!} = e\Gamma(1 + 2^w, 1) - 1 \tag{2.2}$$

where $\Gamma(x, a)$ is the incomplete gamma function. The above includes enumeration over all nonrepeating trajectories of length 2^w and trajectories of shorter length that settle to an attractor at a time $r < 2^w$. Here, N *should be interpreted as the number of total possible deterministic trajectories through a configuration space, where states in the space each contain w bits of information.* So far, our considerations are independent of any assumptions about the laws that determine the dynamical trajectories that are physically realized. We can now consider the number of possible trajectories for a given class of *fixed deterministic laws.*

A natural way to define a "class" of laws in CA is by their locality. For example, elementary cellular automata (ECA) are some of the simplest one-dimensional CA studied and are defined by a nearest-neighbor interaction neighborhood for each cell, where cell states are defined on the two-bit alphabet $\{0, 1\}$. Nearest-neighbor interactions define a neighborhood size $L = 3$ (such that a cell is updated by the fixed rule, according to its own state and two nearest neighbors), which we define as the locality of an ECA rule. For ECA there are $R = 2^{2^L} = 256$ possible fixed "laws" (ECA rules) (see Wolfram, 2002). We can therefore set an upper bound for the number of trajectories contained within *any* given rule set, defined by its neighborhood L as:

$$f_L \leq 2^{2^L} \times 2^w \qquad (2.3)$$

where f_L is the total number of possible realized trajectories that *any* class of *fixed*, deterministic laws operating with a locality L could realize, starting from a set of 2^w possible initial states (i.e., any initial state with w bits of information).[2] Eqs. 2.2 and 2.3 are not particularly illuminating taken alone; instead we can consider the ratio of the upper bound on the number of trajectories possible

---

[2] This is an upper bound, as it assumes each trajectory in the set is unique, but as it happens it is possible, for example, that the application of two different ECA rules in the set of all ECA could yield the same trajectory, so this constitutes an absolute upper bound.

under a given set of laws to the total number of deterministic trajectories:

$$\frac{f_L}{N} = \frac{2^{2^L} \times 2^w}{e\Gamma(1 + 2^w, 1) - 1} \tag{2.4}$$

Taking the limit as the system size tends toward infinity, that is, as $w \to \infty$, yields:

$$\lim_{w \to \infty} \left[\frac{2^{2^L} \times 2^w}{e\Gamma(1 + 2^w, 1) - 1} \right] = 0 \tag{2.5}$$

Note that this result is *independent* of L. *For any class of fixed dynamical laws that one chooses (any degree of locality), the fraction of possible physically realized trajectories rapidly approaches zero as the number of bits in states of the world increases.*[3] Thus, the set of all physical realizations of universes evolved according to local, fixed deterministic laws is very impoverished compared with what could potentially be permissible over all possible deterministic trajectories (and worse so if one considers adding stochastic or nondeterministic trajectories in the summation in Eq. 2.2). Only an infinitesimal fraction of paths are even realizable under *any* set of laws, let alone under a particular law drawn from any set.

If we impose time-reversal symmetry on the CA update rules, by analogy with the laws of physics, there is an additional restriction: a small subset of the 256 ECArules are time-reversal invariant. For these laws, there is no single trajectory that includes all possible states (see Figure 2.1). Thus, we encounter the problem that "you can't get there from here," and even if you are in the right regime of configuration space, there is only one path (ordering of states) to follow.

Of course, explanations referring to real biological systems differ from CA in several respects, not least of which is that one deals with macrostates rather than microstates. However, the conclusion is unchanged if one considers the dynamics of macrostates rather than microstates: the trajectories among all possible macrostates

---

[3] This gets worse if states contain more information, that is, if the alphabet size $m > 2$.

will also be diminished relative to the total number of trajectories (this is because the information in macrostates is less than in the microstates, e.g., will be $< 2^W$ macrostates for our toy example).[4] This toy model cautions us that in seeking to explain a complex world that is ordered in a particular way (e.g., contains living organisms and conscious observers), based on fixed laws that govern microstate evolution, we may well need to fine-tune not only the initial state but also the laws themselves in order to specify the particular ordering of states observed (constraining the universe to a unique past and future). Expressed more succinctly, if one insists on attributing the pathway from mundane chemistry to life as the outcome of fixed dynamical laws, then (our analysis suggests) those laws must be selected with extraordinary care and precision, which is tantamount to intelligent design: it states that "life" is "written into" the laws of physics ab initio. There is no evidence at all that the actual known laws of physics possess this almost miraculous property.

The way to escape from this conundrum – that "you can't get anywhere from here" – is clear: we must abandon the notion of fixed laws when it comes to living and conscious systems.

LIFE ... FROM NEW PHYSICS?

Allowing both the states *and the laws* to evolve in time is one possibility for alleviating the problems associated with fine-tuning of initial states and dynamical laws, as discussed in the previous section. But this cannot be done in an ad hoc way and still be expected to be consistent with our known laws of physics. A trivial solution would be to assume that the laws are time dependent and governed by meta-laws, but then one must explain both where the laws and meta-laws come from, and whether there are meta-meta-laws that govern the meta-laws, and meta-meta-meta-laws, ad infinitum. This is therefore no better an explanation than our current framework,

---

[4] This holds even if one considers that the number of possible partitions of our state space for 2^W possible states is given by the Bell number B_n, where $n = 2^W$, which approaches ∞ more slowly than the denominator in Eq. (2.4).

as it just pushes the problem off to be one of meta-laws rather than laws. As it stands right now, there exists no compelling evidence that any of our fundamental laws depend on time (although claims are occasionally made to the contrary; Webb et al., 2011).

A better idea is to assume that there exist conditions under which the dynamical rules are a function of the states (and therefore are in some sense an emergent property) (Adams et al., 2016; Pavlic et al., 2014). Indeed, this seems consistent with what we know of life, where the manner in which biological systems evolve through time is clearly a function of their current state (Goldenfeld and Woese, 2011). Biological systems appear to be incredibly path dependent in their dynamical trajectories, as exemplified by the process of biological evolution. Starting from the same initial state (roughly speaking a genome), biological systems trace out an enormous array of alternative trajectories through evolutionary adaptation and selection (see Figure 2.2). It is difficult, if not impossible, to write an equation of motion with a fixed rule for such historical processes due to their state-dependent nature. State-dependent laws are a hallmark of self-referential systems such as life and mind (Goldenfeld and Woese, 2011; Hofstadter, 1979; Walker and Davies, 2013). It is a contrast with our current views of immutable laws in physics keenly appreciated by

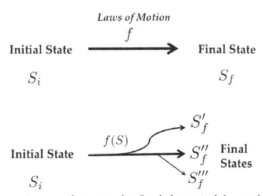

FIG. 2.2 Contrast between the fixed dynamical laws of physics (top) and the apparent historical, state-dependent trajectories characteristic of living systems (bottom).

Darwin, as eloquently expressed in the closing passage of his book *On the Origin of Species* (Darwin, 1859):

> whilst this planet has gone cycling on according to the fixed
> law of gravity, from so simple a beginning endless forms most
> beautiful and most wonderful have been, and are being,
> evolved.

It seems that one resolution to the above conundrum, consistent with what we know of biology, is therefore to introduce *state-dependent dynamical laws*. However, there is a trivial sense in which state-dependent dynamical laws might alleviate the fine-turning problem, that is, by assuming that each state uniquely specifies the next state. One could then construct a state-dependent algorithm unique to each possible trajectory such that there is exactly one algorithm that would "explain" each path taken to reach any arbitrary final state. However, this is inconsistent with what we know of physics: such laws would not be algorithmically compressible, and therefore it would be impossible to write such a succinct equation as $F = ma$ to describe *anything* about a universe governed by these kinds of laws.

How then can we reconcile the biological narrative with our current understanding of physics? When we externally describe a system, we articulate the counterfactual possibilities and assign a quantity of "information" to them. However, in order for these counterfactual possibilities to be physically realized, the information specifying them must be contained within the system and contribute to specifying which dynamical path through state space is taken (consistent with the underlying physical laws). This should be a local property, yielding an effective description that is state dependent. The challenge is that we do not have a physical theory for information that might explain how information could "call the shots." There are some indications for a potentially deep connection between information theory (which is not cast as a physical theory and instead quantifies the efficacy of communication through noisy channels) and thermodynamics, which is a branch of physics[5] due to the mathematical relationship between Shannon and Boltzmann entropies. Substantial work over the last decade has attempted to make this connection explicit; we point the reader to Lutz and Ciliberto (2015) and Parrondo et al. (2015) for recent reviews. Schrödinger was aware of this link in his deliberations on biology, and famously coined the term "negentropy" to describe life's ability to seemingly violate the second law of thermodynamics.[6] Yet he felt that something was missing, and that thermodynamic considerations alone are insufficient to explain life (Schrödinger and Schroedinger, 2004):

> living matter, while not eluding the "laws of physics" as
> established up to date, is likely to involve "other laws of physics"
> hitherto unknown …

We suggest one approach to get at these "other laws" is to focus on the connection between the concept of "information" and the equally ill-defined concept of "causation" (Davies and Walker, 2016; Kim et al., 2015; Walker et al., 2016). Both concepts are implicated in the failure of our current physical theories to account for complex states of the world without resorting to very special initial conditions. In particular, we posit that the manner in which biological systems implement state-dependent dynamics is by utilizing information encoded *locally* in the current state of the system, that is, by attributing causal efficacy to information. It is widely recognized that coarse-graining (which would define the relevant "informational" degrees of freedom) plays a foundational role in how biological systems are structured (Flack et al., 2013), by defining the biologically relevant macrovariables (see, e.g., Chapters 10, 12, and 16). However, it is not clear how those macrostates arise, if they are objective or subjective

---

[5] Stating that information theory is not a physical theory is not the same as saying that information is not physical – a key insight of information theory is that information is a measurable physical quantity. "Information is physical!" in the words of Rolf Landauer (1996).

[6] "Schrödinger's paradox" with regard to life's ability to generate "negative entropy" is quickly resolved if one considers that living systems are open to an environment.

(Shalizi and Moore, 2003), or whether they are in fact a fundamental aspect of biological organization – *intrinsic to the dynamics* (i.e., such that macrostates are causal) rather than merely a useful phenomenological descriptor. A framework in which coarse-grained information-encoding macrostates are causal holds promise for resolving many of the problems discussed herein. This is the key aspect of the hard problem of life.

CONCLUSIONS

There are many difficult open problems in understanding the origin of life – such as the "tar paradox" (Benner, 2014) and the fact that prebiotic chemistry is just plain hard to do. These problems differ qualitatively from the "hard problem of life" as identified here. Most open problems associated with life's origin such as these, while challenging right now, will likely ultimately reduce to known principles of physics and chemistry, and therefore constitute, by our definition, "easy problems." Here we have attempted to identify a core feature of life that won't similarly be solved based on current paradigms – namely, that life seems distinct from other physical systems in how information affects the world (i.e., that macrostates are causal). We have focused on the problem of explaining the pathway from nonliving chemical systems to life and mind to explicate this problem and have attempted to motivate why new principles and potentially even physical laws are necessary. One might regard this as too a radical step; however, it holds potential for resolving deep issues associated with what life is and why it exists. Previous revolutions in our understanding of physical reality, such as general relativity and quantum mechanics, dramatically reshaped our understanding of the world and our place in it. To quote Einstein, "One can best feel in dealing with living things how primitive physics still is" (letter tio L. Szilard, quoted in Prigogine and Stengers, 1997). Given how much more intractable life seems, we should not immediately jump to expecting anything less of a physical theory that might encompass it. If we are so lucky as to stumble on new fundamental understanding of life, it could be such a radical departure from what we know now that it might be left to the next generation of physicists to reconcile the unification of life with other domains of physics, as we are now struggling to accomplish with unifying general relativity and quantum theory a century after those theories were first developed.

ACKNOWLEDGMENTS

This work was made possible through support of a grant from Templeton World Charity Foundation. The opinions expressed in this publication are those of the author(s) and do not necessarily reflect the views of Templeton World Charity Foundation.

REFERENCES

Adams, A., Zenil, H., Davies, P. C. W., and Walker, S. I. (2016). Formal definitions of unbounded evolution and innovation reveal universal mechanisms for open-ended evolution in dynamical systems. *arXiv:1607.01750.*

Benner, S. A. (2014). Paradoxes in the origin of life. *Origins of Life and Evolution of Biospheres*, **44**(4):339.

Campbell, D. T. (1974). Downward causation in hierarchically organised biological systems. In *Studies in the philosophy of biology*, pages 179–186. Springer.

Carter, B., and McCrea, W. H. (1983). The anthropic principle and its implications for biological evolution [and discussion]. *Philosophical Transactions of the Royal Society of London A: Mathematical, Physical and Engineering Sciences*, **310**(1512):347–363.

Chalmers, D. J. (1995). Facing up to the problem of consciousness. *Journal of Consciousness Studies*, **2**(3):200–219.

Cronin, L., and Walker, S. (2016). Beyond prebiotic chemistry. *Science*, **352**:1174–1175.

Darwin, C. (1859). *On the origin of species by means of natural selection.* Murray.

Davies, P. (2008). *The Goldilocks enigma: why is the universe just right for life?* Houghton Mifflin Harcourt.

Davies, P. C. W., and Walker, S. I. (2016). The hidden simplicity of biology: a key issues review. *Rep. Prog. Phys.* **79**(10):102601.

Deutsch, D. (2013). Constructor theory. *Synthese*, **190**(18):4331–4359.

Farnsworth, K. D., Nelson, J., and Gershenson, C. (2013). Living is information processing: from molecules to global systems. *Acta Biotheoretica*, **61**(2):203–222.

Flack, J., Erwin, D., Elliot, T., and Krakauer, D. (2013). Timescales, symmetry, and uncertainty reduction in the origins of hierarchy in biological systems. In Kim Sterelny, Richard Joyce, Brett Calcott, and Ben Fraser (eds), *Cooperation and its evolution*, pages 45–74. MIT Press.

Godfrey-Smith, P., and Sterelny, K. (2008). Biological information. In *The Stanford Encyclopedia of Philosophy* (summer 2016 edition), Edward N. Zalta (ed.), http://plato.stanford.edu/archives/sum2016/entries/information-biological/.

Goldenfeld, N., and Woese, C. (2011). Life is physics: evolution as a collective phenomenon far from equilibrium. *Annu. Rev. Condens. Matter Phys.*: 375–399.

Hofstadter, D. R. (1979). *Godel Escher Bach*. Basic Books.

Jablonka, E., and Szathmáry, E. (1995). The evolution of information storage and heredity. *Trends in Ecology & Evolution*, **10**(5):206–211.

Kim, H., Davies, P., and Walker, S. I. (2015). New scaling relation for information transfer in biological networks. *Journal of the Royal Society Interface*, **12**(113):20150944.

Landauer, R. (1996). The physical nature of information. *Physics Letters A*, **217**(4):188–193.

Lartigue, C., Glass, J. I., Alperovich, N., Pieper, R., Parmar, P. P., Hutchison, C. A., Smith, H., and C., (2007). Genome transplantation in bacteria: changing one species to another. *Science*, **317**(5838):632–638.

Lutz, E., and Ciliberto, S. (2015). Information: From Maxwell's demon to Landauer's eraser. *Physics Today*, **68**(9):30–35.

Monod, J. (1974). *On chance and necessity*. Springer.

Parrondo, J., Horowitz, J., and Sagawa, T. (2015). Thermodynamics of information. *Nature Physics*, **11**:131–139.

Pavlic, T., Adams, A., Davies, P., and Walker, S. (2014). Self-referencing cellular automata: a model of the evolution of information control in biological systems. *arXiv:1405.4070*.

Peirce, C. S. (1982). *Writings of Charles S. Peirce: a chronological edition, volume 1: 1857–1866*, volume 4. Indiana University Press.

Prigogine, I., and Stengers, I. (1997). *The end of certainty*. Simon and Schuster.

Schrödinger, E., and Schroedinger, E. (2004). With mind and matter and autobiographical sketches. In *What is life*, Cambridge University Press, Cambridge UK. Reprinted 2012.

Shalizi, C., and Moore, C. (2003). What is a macrostate? Subjective observations and objective dynamics. *arXiv preprint cond-mat/0303625*.

Smith, E. (2008). Thermodynamics of natural selection. I: Energy flow and the limits on organization. *Journal of Theoretical Biology*, **252**(2):185–197.

Smolin, L. (2013). *Time reborn: from the crisis in physics to the future of the universe*. Houghton Mifflin Harcourt.

Szathmary, E., and Maynard Smith, J. (1994). The major evolutionary transitions. *Nature*, **374**:227–232.

Walker, S., and Davies, P. (2013). The algorithmic origins of life. *J. R. Soc. Interface*, **10**(79):20120869.

Walker, S. I. (2015). Is life fundamental? In Aguirre, A., Foster, B., and Merali, Z., editors, *Questioning the foundations of physics: which of our fundamental assumptions are wrong?* Springer: 259–268.

Walker, S. I. (2016). The descent of math. In Aguirre, A., Foster, B., and Merali, Z., editors, *Trick of truth: the mysterious connection between physics and mathematics?* Springer: 183–192.

Walker, S. I., Cisneros, L., and Davies, P. C. W. (2012). Evolutionary transitions and top-down causation. *arXiv preprint arXiv:1207.4808.*

Walker, S. I., Kim, H., and Davies, P. C. W. (2016). The informational architecture of the cell. *Phil. Trans. A*, page 20150057.

Webb, J. K., King, J. A., Murphy, M. T., Flambaum, V. V., Carswell, R. F., and Bainbridge, M. B. (2011). Indications of a spatial variation of the fine structure constant. *Physical Review Letters*, **107**(19):191101.

Wheeler, J. A. (1983). On recognizing without law, oersted medal response at the Joint APS-AAPT meeting, New York, 25 January 1983. *American Journal of Physics*, **51**(5): 398–404.

Whitesides, G. (2012). The improbability of life. In Barrow, J., Morris, S., Freeland, S., and Harper, C. Jr., editors, *Fitness of the cosmos for life*, volume 1: xi–xii. Cambridge: Cambridge University Press.

Wolfram, S. (2002). *A new kind of science*, volume 5. Wolfram Media.

3   Beyond Initial Conditions and Laws of Motion

Constructor Theory of Information and Life

Chiara Marletto

Constructor theory (Deutsch, 2013) is a new mode of explanation in fundamental physics, intended to improve on the currently most fundamental theories of physics and explain more of physical reality. Its central idea is to formulate *all laws of physics as statements about what transformations are possible, what are impossible, and why*. This is a sharp departure from what I call the *prevailing conception of fundamental physics*, which formulates its statements exclusively in terms of predictions given the initial conditions and the laws of motion. For instance, the prevailing conception of fundamental physics aims at predicting where a comet goes, given its initial conditions and the laws of motion of the universe; instead, in constructor theory the fundamental statements are about where the comet *could be made to go*, with given resources, under the dynamical laws.

This constructor-theoretic, task-based formulation of science makes new conceptual tools available in fundamental physics, which resort to *counterfactual statements* (i.e., about possible and impossible tasks). This has the potential to incorporate *exactly* into fundamental physics notions that have so far been considered as highly approximate and emergent – such as information and life.

The *constructor theory of information* (Deutsch and Marletto, 2015) has accommodated the notion of (classical) information within fundamental physics and has unified it exactly with what currently goes under the name of 'quantum information' – the kind of information that is deemed to be instantiated in quantum systems.

This theory provides the conceptual basis for the *constructor theory of life* (Marletto, 2015a). Here constructor theory is applied to produce the explanation of how certain physical transformations that are fundamental to the theory of evolution by natural selection – such as *accurate* replication and self-reproduction – are compatible with laws of physics that do not contain the design of biological adaptation. It also shows what features such laws must have for this to be possible: in short, they must allow the existence of *information media*, as rigorously defined in the constructor theory of information.

Before going into the details of these theories, one has to introduce the foundations of constructor theory. Constructor theory consists of 'laws about laws': its laws are *principles* – conjectured laws of nature, such as the principle of conservation of energy. It underlies our currently most fundamental theories of physics, such as general relativity and quantum theory, which in this context we call *subsidiary theories*. These are not to be derived from constructor theory; in contrast, constructor theory's principles constrain subsidiary theories, ruling out some of them. For example, as we shall see, nonlocal variants of quantum theory are not compatible with the principles.

The fundamental principle of constructor theory is that *every physical theory is expressible via statements about what physical transformations, or tasks, are possible, what are impossible, and why*. Therefore, constructor theory demands a description of the world in terms of transformations, defined as physical processes involving two kinds of physical systems, with different roles. One is the *constructor*, whose defining characteristic is that it causes the transformation to occur and remains *unchanged in its ability to cause it again*. The other consists of the subsystems – called the *substrates* – which are transformed from having some physical attribute to having another. Schematically:

Input attribute of substrates $\xrightarrow{\text{Constructor}}$ Output attribute of substrates,

where the constructor and the substrates jointly constitute an isolated system. For example, the catalyst in a chemical reaction approximates a constructor, and the chemicals being transformed are its substrates. By *attribute* here one means a set of states of a system in which the system has a certain property according to the subsidiary theory describing it – such as being red or blue. The basic entities of constructor theory are *tasks*, which consist of the specifications of only the input–output pairs of a transformation, with the constructor abstracted away:

Input attributes of substrates → Output attributes of substrates.

Therefore, a task A on a substrate S is a set:

$$A = \{x_1 \to y_1, \ x_2 \to y_2, \ldots\},$$

where the $x_1, x_2, \ldots$ and the $y_1, y_2, \ldots$ are attributes of S. The set $\{x_i\} = In(A)$ are the legitimate input attributes of A and the set $\{y_i\} = Out(A)$ its legitimate output attributes. Tasks may be composed into networks, by serial and parallel composition, to form other tasks.

Quite remarkably, this is an explicitly local framework, requiring that individual physical systems have states (and attributes) in the sense described. Indeed, another cardinal principle of constructor theory is Einstein's principle of locality (Einstein, 1949): *There exists a mode of description such that the state of the combined system* $S_1 \oplus S_2$ *of any two substrates* S_1 *and* S_2 *is the pair* (x, y) *of the states* x *of* S_1 *and* y *of* S_2, *and any construction undergone by* S_1 *and not* S_2 *can change only* x *and not* y. In quantum theory, the Heisenberg picture is such a mode of description (see Deutsch, 2000).

A constructor is *capable of performing a task* A if, whenever presented with substrates having an attribute in $In(A)$, it delivers them with one of the corresponding attributes in $Out(A)$. A task A is *impossible* if it is forbidden by the laws of physics. Otherwise it is *possible* – which means that the laws of nature impose no limit, short of perfection, on how accurately A could be performed, nor on how well things that are capable of approximately performing it could retain their ability to do so again. However, it is crucial to bear in mind that *no perfect constructors exist in nature*, given our laws of physics. Approximations to them, such as catalysts, living cells, or robots, do make errors and also deteriorate with use. However, when a task is possible, the laws of nature permit the existence of an approximation to a constructor for that task to any given finite accuracy. The notion of a constructor is shorthand for the infinite sequence of these approximations.

Therefore, in this framework a task either is categorically impossible or is possible. Both are *deterministic* statements: in the worldview of constructor theory, probabilistic theories can only be approximate descriptions of reality. Probabilities are indeed emergent in constructor theory – see Marletto (2015b) – just like in unitary quantum theory (Deutsch, 2000; Wallace, 2003). For how such a theory can be testable, see Deutsch (2015).

Although a task refers to an isolated system of constructor and substrates, one is sometimes interested in what is possible or impossible irrespective of the kind of resources required. So if it is possible to perform the task A in parallel with some task T on some generic substrate that is preparable – see Deutsch and Marletto (2015) – one says that A is *possible with side effects*, which we write as $A^{\angle}$. The task T represents the side effect.

So in constructor theory everything important about the world is expressed via statements about the possibility and impossibility of tasks, *without mentioning constructors*. One might wonder what difference switching to this formulation can possibly make. After all, it is perfectly possible to express the possibility or impossibility of a task in the prevailing conception, as a conditional statement about the composite system of the constructor and the substrates, given certain initial conditions and the laws of motion. However, as we are about to see, the constructor-theoretic approach (where one can abstract away the constructor) makes all the difference in the case of information.

Whether or not information is physical (Landauer, 1961) and what this can possibly mean has been at the centre of a long-standing debate. Information is widely used in physics, but appears to be very different from all the entities appearing in the physical descriptions of the world. It is not, for instance, an observable – such as the position or the velocity of a particle. Indeed, it has properties like no other variable or observable in fundamental physics: it behaves like an abstraction. For there are laws about information that refer directly to it, without ever mentioning the details of the physical substrates that instantiate it (this is the *substrate-independence* of information), and moreover it is *interoperable* – it can be copied from one medium to another without having its properties qua information changed. Yet information can exist only when physically instantiated; also, for example, the information-processing abilities of a computer depend on the underlying physical laws of motion, as we know from the quantum theory of computation. So, there are reasons to expect that the laws governing information, like those governing computation, are laws of physics. How can these apparently contradictory aspects of information be reconciled?

The key to the answer is that the informally conceived notion of information implicitly refers to *certain interactions being possible* in nature; it refers to the existence of certain regularities in the laws of physics. A fundamental physical theory of information is one that expresses such regularities. As an example of what these regularities are, consider interoperability – as we said above, this is the property of information being *copiable* from one physical instantiation (e.g., transistors in a computer) to a different one (e.g., DNA). This is a regularity displayed by the laws of physics of our universe, which is taken for granted by the current theories of information and computation. However, one could imagine laws that did not have it – under which 'information' (as we informally conceive it) would not exist. For example, consider a universe where there exist two sectors A and B, each one allowing copying-like interactions between media inside it, but such that no copying interactions were allowed between A and B. There would be no 'information' (as informally conceived) in the composite system of the two sectors. This is an example of how whether or not information can exist depends on the existence of certain regularities in the laws of physics. These regularities have remained unexpressed in fundamental physics; the constructor theory of information precisely expresses them in the form of new, conjectured laws of physics. This is how information can be brought into fundamental physics: one does so by expressing in an exact, scale-independent way what constraints the laws of physics must obey in order for them to instantiate what we have learnt informally to call 'information'.

It is not surprising that constructor theory proves to be particularly effective to this end. As Shannon and Weaver (1949) put it, information has a counterfactual nature:

> this word 'information' in communication theory relates not so much to what you do say, as to what you could say.

The constructor theory of information differs from previous approaches to incorporating information into fundamental physics, e.g. Wheeler's 'it from bit' (Wheeler, 1990), in that it does not consider information itself as an a priori mathematical or logical concept. Instead, it requires that the nature and properties of information follow entirely from the laws of physics.

The logic of how the theory is constructed is elegant and simple. The first key step is that in constructor theory information is understood in terms of computations, not vice versa as is usually done. So, first one defines a *reversible computation* C_Π (S) as a task – that of performing, with or without side effects, a permutation Π over some set S of at least two possible attributes of some substrate:

$$C_\Pi (S) = \bigcup_{x \in S} \{x \to \Pi(x)\} .$$

By a 'reversible computation' C_Π is meant a logically reversible, i.e., one-to-one, task, but the process that implements it may be physically irreversible, because of the possibility of side effects. A *computation variable* is a set S of two or more possible attributes for which C_Π for all permutations Π over S, and a *computation medium* is a substrate with at least one computation variable. A quantum bit in any two

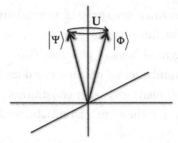

FIG. 3.1 Pictorial representation of the swap U of two nonorthogonal states $|\Psi\rangle$ and $|\Phi\rangle$ of a quantum bit (e.g., a $\frac{1}{2}$ spin). U is unitary as prescribed by quantum theory, i.e., it preserves the inner product; applying it twice is equivalent to the identity.

nonorthogonal states is an example of a computation medium: these two states can be swapped by a unitary transformation, despite their not being distinguishable by a single-shot measurement (Deutsch and Marletto, 2014). See Figure 3.1.

The next step is to introduce the notion of an *information medium*, which requires one to consider computations involving two instances of the same substrate **S**. The *cloning task* for a set S of possible attributes of **S** is

$$R_S(x_0) = \bigcup_{x \in S} \{(x, x_0) \rightarrow (x, x)\} \tag{3.1}$$

on **S** $\oplus$ **S**, where x_0 is some fixed (independent of x) attribute with which it is possible to prepare **S**. A set S is *clonable* if R_S is possible (with or without side effects) for some such x_0.

An *information variable* is then defined as a clonable computation variable. An *information attribute* is one that is a member of an information variable, and an *information medium* is a substrate that has at least one information variable.

We are now in the position to express exactly what it means that a system contains information. In particular, a substrate **S** *instantiates information* if it is in a state belonging to some attribute in some information variable S of **S** and if it could have been given any of the other attributes in S. The constructor-theoretic mode of explanation has allowed this to be expressed as an *exact, intrinsic* property of the substrate by switching to statements about possible/impossible tasks. In the prevailing conception, instead, it would have to be expressed as a property of the composite system of substrate and constructor. This is an example of the constructor-theoretic tools at work: bringing counterfactual into fundamental physics does make all the difference in this case.

On this ground, one can then conjecture *principles about information media* that have elegant expressions in terms of possible and impossible tasks. As an example of such principles, consider the *interoperability principle*: *The combination of two substrates with information variables S_1 and S_2 is a substrate with information variable $S_1 \times S_2$*, where $\times$ denotes the Cartesian product of sets. This principle states that information can be copied from any kind of information medium to any other kind – a property that cannot even be stated in the prevailing conception of fundamental physics but has this elegant expression in constructor theory. Similar elegant expressions are found for the other principles conjectured in Marletto (2015b), which represent the physical laws that underlie the existence of (classical) information. This provides a purely constructor-theoretic notion of classical information that is free from its dependence on classical physics and is expressed in a totally general framework.

Since quantum computation is so much more powerful than classical computation, one might then think that it is necessary to conjecture some additional law, declaring some counterintuitive tasks to be possible, in order to incorporate in this framework what we currently call quantum information – i.e., the kind of information that can be instantiated in quantum media. But it turns out that one can define a new kind of medium, called *super-information medium*, that obeys a simple, elegant constraint: that it has two information variables, the union of which is not an information variable. Remarkably, this constraint requires some tasks to be *impossible*. Yet from this one can derive all the most distinctive, qualitative properties of quantum information (Deutsch and Marletto, 2015):

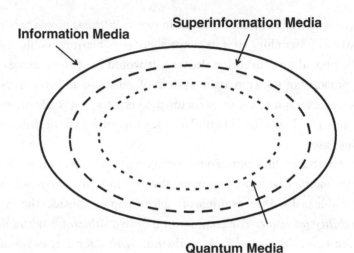

FIG. 3.2 Information media in the constructor theory of information.

- Existence of pairs of attributes that cannot be cloned
- Irreducible perturbation caused by measurement
- Unpredictability of the outcome of certain experiments
- Locally inaccessible information in composite systems, as in entangled systems (on the additional requirement of reversibility).

These properties find an elegant expression in constructor-theoretic terms because they are all expressed naturally in terms of possible/impossible tasks. This unifies *exactly* our understanding of quantum information and classical information, in a way that was not previously hinted at. The picture that emerges is that quantum information is an instance of superinformation, as depicted in Figure 3.2.

In summary, the constructor theory of information explains the regularities in physical systems that are associated with information, including quantum information. In this respect, it is a challenge to the worldview that maintains that quantum theory is 'random' and 'nonlocal': constructor theory is a local and deterministic framework. It also provides the foundations for theories such as Shannon's information theory (Shannon, 1948), giving a base for the notion of distinguishability of attributes that implicitly underpins the communication scheme investigated by Shannon (as discussed in Deutsch and Marletto, 2015). This theory also provides a framework, independent of quantum theory, where one can investigate information under a broad range of theories (including 'postquantum' theories) if they obey the principles. Finally, it provides an exact, scale-independent physical characterisation of what it means for laws of physics to allow for 'information'.

CONSTRUCTOR THEORY OF LIFE

The constructor theory of life (Marletto, 2015a) builds on this to tackle a problem relevant to the foundations of both physics and biology. To understand what the problem is, let me first clarify the connection between evolutionary theory and fundamental physics.

The problem the theory of evolution was proposed to address is the existence of living entities. That living entities are particularly remarkable physical objects has been known since antiquity. They struck the curiosity of early humans, as the superb example of cave paintings testifies (see Wagner and Briggs, 2016). Early attempts to classify the properties of living things, to distinguish them from inert matter, date back to Socrates and Aristotle. However, only in modern times was it possible to express the objective property that makes their existence a *scientific problem*. This property is rooted in physics.

In modern biology, living entities are characterised by the *appearance of design* displayed in their biological adaptations. As William Paley put it, they have *several, different subparts all coordinating to the same purpose* (Paley, 2006). For instance, trunks in elephants appear as highly designed objects serving a specific function.

Darwin's *theory of evolution* was proposed precisely to explain how the appearance of design in living things can have come about without an intentional design process, via the undesigned process of variation and natural selection. It is notable that Darwin's theory is based on an (informal) constructor-theoretic reasoning. Despite its predictive power – see, e.g., the famous case of Darwin's moth (Kritsky, 1991) – the core statement of the theory is not in the form of a prediction. It does not state that, say, elephants' trunks must exist; it states that it is *possible* that living things have come about via natural selection and variation, without an intentional design process, and explains how. Recasting this statement in terms of predictions would not serve the purpose. Having a prediction (probabilistic or not) that maintains that, say, elephants will occur (or will probably occur) at some point in our universe does not rule out the possibility that the laws or the initial conditions contain design and thus would not serve the purpose of understanding how elephants can have come about without an intentional design process.

In constructor theory, one can express more precisely how the problematicity of living things is rooted in physics: living things are problematic because, in sharp contrast with inert matter, they *approximate accurately the behaviour of a constructor*. They perform tasks to a *high accuracy*, *reliably*, and they maintain this property in time, displaying *resiliency*. This is problematic because of how we conjecture the laws of physics of our universe are: under such laws, the *generic resources* – defined as the physical objects that occur in effectively unlimited numbers (such as atoms, photons, and simple chemicals) – are elementary. In particular, they do not contain accurate constructors: if they ever perform tasks, they do so only to a low, finite accuracy. Moreover, under such laws it is impossible that an accurate constructor arises from generic resources only, acted on by generic resources only. I shall call laws of this kind *no-design laws* (Marletto, 2015a).

Thus the problem about the existence of living entities – the problem that the theory of evolution aims at solving – is better expressed as that of how accurate constructors such as living entities *can* emerge from generic resources, without an intentional design process, given laws of physics that are no-design. This reveals the connection between evolutionary biology and fundamental physics.

In the modern *neo-Darwinian synthesis* (Dawkins, 1976, 1999), the explanation of Darwin's theory was merged with molecular biology, where the connection with physics becomes more explicit. The centerpiece of the explanation is a physical object, the *replicator*, that is copied in the following pattern:

$$(R, N) \xrightarrow{C} (R, R, W)$$

where R is the replicator and C is a constructor for the copying (a copier), acting on some generic resources N (possibly producing waste products W).

In nature this process occurs to different accuracies. Short RNA strands and simple molecules involved in the origin of life (Szathmáry and Maynard Smith, 1997) are poor, inaccurate replicators. In those cases, the copier C is implicit in the dynamical laws of physics. In cells, at the other extreme, the replicator R is the DNA strand, which is copied very accurately by various enzymes.

The replication of replicators in cells relies crucially on the ability of a cell to undergo *accurate self-reproduction* – the construction where an object S (the *self-reproducer*) brings about another instance of itself, in the schematic pattern:

$$(S, N) \rightarrow (S, S, W)$$

Here W represents products; the raw materials N do not contain the means to create another S; and the whole system could be isolated. Thus a self-reproducer S cannot rely on any mechanism other than itself to cause the construction of another S, unlike a replicator R that is allowed to use an external copying mechanism, such as C.

That evolutionary theory relies on both these processes being possible constitutes the problem addressed by the constructor theory of life. Indeed both replication and self-reproduction, which is essential to replication, occur in living things with remarkable accuracy. Thus it is necessary, for the theory of evolution fully to explain the appearance of design in the presence of no-design laws, to provide an additional argument of how and why these processes are compatible with underlying dynamical laws that are no-design, i.e., that do not contain the design of biological adaptations. The constructor theory of life provides precisely this explanation.

The compatibility of accurate self-reproduction with the laws of physics has indeed been contested, along the lines advocated by Wigner, who proposed the claim that accurate self-reproduction, as it occurs in living cells, requires laws of physics that are 'tailored for self-reproduction to occur' (Bohm, 1969; Wigner, 1961).

Wigner uses technical quantum theory to make his case. But his claim is actually simpler and broader than that and expresses the problem: how can self-reproduction be so accurate in the presence of no-design laws – laws of physics that are simple and do not contain any reference to accurate self-reproducers or replicators? His statement would have wide implications were it true. Not only would it require our laws of physics to be complemented with ad hoc ones, containing the design of biological adaptations, but also the theory of evolution would, after all, rely on laws of physics containing the design of biological adaptations.

The constructor theory of life shows that accurate self-reproduction and accurate replication are possible under no-design laws, thus rebutting that claim and vindicating the compatibility of Darwin's theory of evolution with no-design laws. It also shows what other features no-design laws must have to allow those processes; in particular, they must allow the existence of information media, as defined in constructor theory. In addition, it shows what logic accurate self-reproducers must follow, under such laws; it turns out that an accurate replicator must rely on a self-reproducer, and vice versa.

The logic of the argument is as follows. First, one notes that replicators are already expressed naturally in the constructor theory of information; see Equation (3.1): substrates allowing a set of attributes that can be permuted in all possible ways and replicated (i.e., cloned) are information media. Moreover, self-reproducers are characterised as constructors for another instance of themselves, as we can see by rewriting the self-reproduction of a self-reproducer S, in the presence of generic resources N only (with possible waste products W) in a constructor-theoretic notation:

$$N \xrightarrow{S} (S, W)$$

Thus, the problem can be reformulated naturally and exactly, in constructor theory, as: Are accurate self-reproducers and replicators possible under no-design laws?

Furthermore, the appearance of design and the notion of no-design laws can both be expressed, exactly, within constructor theory. Here it is crucial that constructor theory allows one to avoid using the notion of probability to model those concepts. In particular, we are interested here in defining no-design intending the design being *that of biological adaptations.* Laws of physics might be fine-tuned in other senses (see Davies, 2000), but here we are interested only in design of living things. In the prevailing conception, resorting to probabilities, it is not possible to model this concept. For example, one could say that some dynamical law that is nontypical under some natural measure is a designed one, as Wigner conjectured. But clearly this is a non sequitur: the unitary of our universe is indeed 'nontypical' because, e.g., it is local. But this gives no indication as to whether it contains the design of biological adaptations. In constructor theory, instead, one can characterise precisely, within physics and without resorting to probability, what 'no-design laws of physics' are. They are, as expressed above, laws whose generic resources do not contain accurate constructors, nor do they allow the sudden arising of accurate constructors out of generic resources only.

A similar approach allows one to express the appearance of design. The latter also cannot be modelled by being 'improbable'. For probabilities are multiplicative, but the appearance of design of the composite system of two objects with the appearance of design need not have more of an appearance of design than either of the two separately. In constructor theory one can give, instead, an elegant constructor-theoretic expression that has the required property, in terms of *programmable constructors* – constructors that have, among their input substrates, an information medium holding one of its information attributes, which in turn can also act as a constructor (Marletto, 2015a).

Once this conceptual ground has been set, it is then possible to develop the argument for why *accurate self-reproducers and replicators are possible under no-design laws*, by demonstrating what follows.

1. One first establishes the *necessary features of an accurate constructor* for a generic task, under no-design laws. In short, every accurate constructor must contain a modular replicator, containing a recipe with all the information about how to perform the task and how to error-correct it. The logic of the argument goes as follows. Since no-design laws are not allowed to contain any accurate constructor, the constructor must implement a *recipe*: this is a sequence of elementary, nonspecific steps, each of which is simple enough to be compatible with no-design laws. The recipe contains all the information to perform the task, whilst the rest of the constructor – that we may call the *vehicle* – is nonspecific to the task, as its function is to implement the elementary steps of the recipe. Moreover, since errors do occur during the process under no-design laws, the recipe must also implement step-wise *error-correction* to achieve a high enough accuracy. For instance, consider a car factory constructing a car out of elementary components (iron, silicon, etc.). Its action must be decomposable in a sequence of elementary steps, *nonspecific* to the construction of the car, and implementing quality control. Also, the constructor must be resilient to last long enough to achieve the given accuracy. But under no-design laws, every constructor wears out after some time. Thus, to meet the requirement, the recipe must be copied, instruction by instruction, from one instance of the constructor to the next one, when the former is about to wear out. The necessity of high-quality error-correction on this replication requires the information in the replicator to be instantiated in *discrete units*, because otherwise error-correction would have an upper bound on the achievable accuracy. Therefore, the recipe must be instantiated in a *modular replicator*.

2. It follows as a special case that an accurate self-reproducer, under no-design laws, must obey von Neumann's *replicator-vehicle logic* (von Neumann, 1951). In particular, under such laws a self-reproducer S must contain a modular replicator R, containing the recipe to construct the rest of S, that we call the vehicle V. The latter, in turn, consists of two subconstructors: B and C (see Figure 3.3, top). Self-reproduction occurs in *two steps* (see Figure 3.3, centre). First, the replicator R is copied blindly ('bit by bit') by the copier C in the former instance of S, and error-correction is implemented too, blindly. This is the DNA copy phase in living cells. Then, the new vehicle is constructed by the constructor B, following the recipe, subunit by subunit. This corresponds in actual cells to the ribosome mechanism. The newly constructed vehicle and the copy of the recipe finally constitute the new instance of the self-reproducer (see Figure 3.3, bottom).

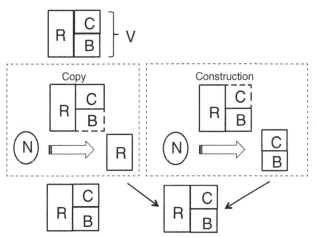

FIG. 3.3 The replicator-vehicle logic for self-reproduction. The self-reproducer S contains a vehicle, made of two subconstructors, C and B, and the replicator R instantiating the recipe (top). The self-reproduction occurs in two phases: in the copy phase the copier C copies R bit by bit (C and R [solid outline] act as constructors); in the construction phase, B executes the recipe in R to construct the new vehicle (B and R [solid outline] act as constructors). The newly constructed vehicle and the copy of the replicator form the new self-reproducer (bottom).

Since each step is elementary, this process is compatible with no-design laws of physics. Thus, von Neumann's original discovery about the logic of self-reproduction in the purely computational context of cellular automata is extended here to the actual laws of physics. It is shown, in particular, that this logic (formerly proven to be sufficient in that context) is *necessary* for accurate self-reproducers to be possible, under no-design laws of physics, e.g., the ones conjectured to rule our universe. As a result, this also implies that an *accurate replication*, as it occurs in living entities, *requires* a vehicle that can perform error-correction – and thus a self-reproducer. This is an interesting spin-off, subverting the assumption that most neo-Darwinian theorists would take, that 'The only thing that is really fundamental to Darwinian life is self-replicating, coded information – genes, in the terminology of life on this planet' (Dawkins, 1976).

3. The last step of the argument is to explain how it is *possible* that accurate self-reproducers (and accurate replicators) have arisen from naturally occurring resources, under no-design laws. The theory of evolution by natural selection and variation provides the explanation for how this occurs: constructor theory shows that this explanation is indeed compatible with no-design laws, establishing two points. The first one is that the logic of evolution by natural selection and variation operates by nonspecific, elementary steps that are *not systematically directed to improvement*. Indeed, the variations caused by the environment in the populations of replicators on which the selection operates are *nonspecific* to the end product, and natural selection is *blind and undirected*. Indeed, the whole process is a highly inaccurate construction for the emergence of accurate self-reproducers from inaccurate ones, given enough time and resources. This construction is so inaccurate and unreliable that it requires no further explanation, as it is compatible with no-design laws. The second point is that natural selection, *to get started*, does not require accurate self-reproducers to be in the initial generic resources. It is sufficient that the latter contain *only inaccurate ones*, such as short RNA strands capable of achieving highly approximate

replication without a vehicle. This concludes the proof that accurate self-reproducers and replicators are possible under no-design laws.

Note that here the problem was not that of predicting with what probability accurate self-reproducers would arise, given certain initial conditions and laws of motion – a problem that has been tackled in Walker and Davies (2013). The problem was a constructor-theoretic one: whether, and how, accurate self-reproducers are *possible, under no-design laws*, and how accurate they are. This problem can be addressed without explicitly formulating any predictions. The final conclusion is that those accurate constructors are permitted under such laws, provided that these laws allow the possibility of modular, discrete replicators to be physically instantiated. In constructor-theoretic terms, *it is necessary that the laws allow information media*. This is also what Darwin's theory of evolution requires of the laws of physics. Rather crucially, this is a requirement that is nonspecific to life. Note also that this is not the usual claim that a vague notion of information is needed for life. The statement of the constructor theory of life is an exact, scale-independent one, based on the rigorous notion of information media provided by the constructor-theory of information.

The recipe in the self-reproducer can be characterised in constructor theory as a *special kind of information*, which *causes itself to remain instantiated in physical systems and can act as a constructor*. We call it *knowledge*, in the sense of Popper's objective knowledge (Popper, 1992). Notably, not all information acts as a constructor, and not all information that can act as a constructor is knowledge. To explicate the distinction, one can consider the difference between a generic sequence of letters (which simply instantiates information, as it is copiable); a syntactically meaningful, but faulty, computer program (which can act as a constructor when executed, but lacks the ability to cause itself to remain instantiated in physical systems, as it is 'fruitless'); and a computer program implementing an effective algorithm (which does, indeed, instantiate knowledge). This distinction between information that can act as a constructor and information that does not is once more an exact one – and it is constructor theory that allows one to express it in fundamental physics.

In the constructor theory of life one also proves that any accurate constructor must contain knowledge in the form of a modular replicator containing the information about the task it performs. More generally, knowledge is implicitly at the foundations of constructor theory (Deutsch, 2013): if a task is not forbidden by the laws of physics, it is possible, which means, under no-design laws, that it is possible, given the requisite knowledge is brought about. Thus knowledge-creators, such as human minds, are the main cause for the realisation of the overwhelming majority of the possible tasks. This is how constructor theory brings knowledge-creators centre-stage in fundamental physics, as shapers of the landscape of physical reality.

Acknowledgements

The author thanks Paul Davies and Sara Walker, for fruitful discussions on early life and replicators, and especially David Deutsch, for providing criticism and helpful comments on earlier drafts of this chapter. This publication was made possible through the support of a grant from the Templeton World Charity Foundation. The opinions expressed in this publication are those of the author and do not necessarily reflect the views of Templeton World Charity Foundation.

REFERENCES

Bohm, D. 1969. Some remarks on the notion of order. Further remarks on order. In C. H. Whaddington (ed.), *Towards a theoretical biology*, vol. 2, pp. 18–60. Aldine.

Davies, P. 2000. *The fifth miracle: the search for the origin and meaning of life*. Simon and Schuster.

Dawkins, R. 1976. *The selfish gene*. Oxford University Press.

Dawkins, R. 1999. *The extended phenotype: the long reach of the gene*. Oxford Paperbacks.

Deutsch, D. 2013. Constructor theory. *Synthese*, **190**(18), 4331–4359.

Deutsch, D. 2015. The logic of experimental tests, particularly of Everettian quantum theory. *arXiv preprint arXiv:1508.02048*.

Deutsch, D., and Hayden, P. 2000. Information flow in entangled quantum systems. Pages 1759–1774 of *Proceedings of the Royal Society of London A: Mathematical, Physical and Engineering Sciences*, vol. 456. The Royal Society.

Deutsch, D., and Marletto, C. 2014. Reconstructing reality. *New Scientist*, **222**(2970), 30–31.

Deutsch, D., and Marletto, C. 2015. Constructor theory of information. In *Proceedings of the Royal Society of London A: Mathematical, Physical and Engineering Sciences*, vol. 471. The Royal Society.

Kritsky, G. 1991. Darwin's Madagascan hawk moth prediction. *American Entomologist*, **37**(4), 206–210.

Landauer, R. 1961. Irreversibility and heat generation in the computing process. *IBM Journal of Research and Development*, **5**(3), 183–191.

Marletto, C. 2015a. Constructor theory of life. *Journal of the Royal Society Interface*, **12**(104), 20141226.

Marletto, C. 2015b. Constructor theory of probability. *arXiv preprint arXiv:1507.03287*.

Paley, W. 2006. *Natural theology* [1802]. New York: The American Tract Society, p. 19.

Popper, K. 1992. *Conjectures and refutations: the growth of scientific knowledge*. New York: Routledge Classics.

Shannon, C. E. 1948. BA mathematical theory of communication. *Bell System Tech J.*, 623656.

Shannon, C. E., and Weaver, W. 1949. *The mathematical theory of communication*. University of Illinois Press.

Szathmáry, E., and Maynard Smith, J. 1997. From replicators to reproducers: the first major transitions leading to life. *Journal of Theoretical Biology*, **187**(4), 555–571.

Von Neumann, J. 1951. The general and logical theory of automata. *Cerebral Mechanisms in Behavior*, **1**(41), 1–2.

Wagner, R., and Briggs, A. 2016. *The penultimate curiosity*. Oxford University Press.

Walker, S. I., and Davies, P. C. W. 2013. The algorithmic origins of life. *Journal of the Royal Society Interface*, **10**(79), 20120869.

Wallace, D. 2003. Everettian rationality: defending Deutsch's approach to probability in the Everett interpretation. *Studies in History and Philosophy of Science Part B: Studies in History and Philosophy of Modern Physics*, **34**(3), 415–439.

Wheeler, J. A. 1989. Information, physics, quantum: the search for links. In *Proceedings III International Symposium on Foundations of Quantum Mechanics*, pp. 354–368. Tokyo.

Wigner, E. P. 1961. The probability of the existence of a self-reproducing unit. In *The logic of personal knowledge: essays presented to M. Polanyi on his seventieth birthday, 11th March, 1961*, 231–238. Routledge and Kegan Paul.

Part II Bio from Bit

4 (How) Did Information Emerge?

Anne-Marie Grisogono

WAS THERE A TIME BEFORE INFORMATION EXISTED? Immersed as we are in an ocean of information constantly battering our senses from every possible angle, there can be little doubt of its existence now, but was it always so?

Information comes in many forms. The most obvious is all the explicit information – text, images, video, sound – that appears in the media, in libraries, on the net, in e-mails, on signs, in archives, exchanged in conversations or overheard, searched for, or unsolicited.

But there is also implicit information in body language, in actions taken or not taken, in tone of voice, in the design of everyday objects that suggest purpose, invite or constrain ways of interacting, in smells and tastes and sounds, in geological formations, in fossils, and most of all in all living things – the networks of their interactions, their forms and behaviour patterns, and their internal structure and functions at every scale down to the subcellular, where research in the last few decades has unearthed not only an immense store of implicit biological information in the exquisite structures and processes of life, but also explicitly encoded data in complex information-bearing molecules such as the nucleic acids.

Information can be concrete or intangible, fleeting or durable, meaningful or content free – a passing thought, a privately held idea, a barely perceptible shrug, a conceptual abstraction, a string of random digits, or a literary masterpiece, a film, a beautiful engineering design. Information can also exist at multiple levels; for example, while a message's explicit content may be trivial or misleading, that fact might itself be meaningful. It depends on the receiver's prior information and ability to interpret the message.

In recognising these examples as information, we inevitably consider them from the subjective perspective of observers or receivers of the information and look for the essential aspect of information, that it could convey something, at least in principle. Most of these examples also highlight the dual roles of sources and receivers – generating and sending, on the one hand, and receiving and interpreting the information, on the other. Even in the privacy of one's own mind, it is possible to distinguish the conceiving of an idea and later the recall and recapitulation, revision, or application.

Generating, processing, and using information is central to conscious (and unconscious) human activity, but also at the heart of all life processes, and the thesis to be explored in this chapter is that there is a continuum from the most basic and primitive to the most sophisticated forms of information, not just in degrees, but also historically, in evolutionary terms.

On the other hand, from a physicist's perspective, information does not have to be communicated or processed. It can simply *be*, apparent in patterns and inherent in the statistics of the microstates of a physical system. The essence of information from this objective perspective is simply descriptive of a system's configuration, but as it turns out, precisely quantifiable, and bearing deep relationships to other precisely defined physical properties: work, free energy, and entropy. These links will be important in framing a tentative story about the origins of information, complexity, and life.

THERE IS NO DOUBT THAT INFORMATION MATTERS
Actions and events at every scale of life, from the subcellular to transnational and global, are shaped by the presence or absence of information and by how it is interpreted, and generate real-world consequences that spawn further information and trigger more actions and events.

Of course all these actions, events, and consequences are governed by the same physical laws that apply to everything else that happens, where there is no apparent role for information, so how is it that information is able to exert such a powerful (dare we say causal?) influence on the unrolling trajectories of living systems and their environments?

What exactly is this extraordinary and pervasive phenomenon that we call information?

These are hard but very important questions, and they are to some extent interdependent, so to get traction we first consider what we really mean by information. The examples encountered so far are so diverse that identifying elements that they share and finer distinctions that differentiate them both are necessary, in order to organise and reason about what we know.

Unsurprisingly, the literature on information is dauntingly enormous, but useful definitions are somewhat sparser with many being unhelpfully circular, too anthropocentric, or not general enough to include all the forms of information that need to be considered in examining the essential nature and origin of information in the universe. For example, one widely quoted definition – 'information is that which causes one to revise one's estimates of likelihoods' (Tribus and McIrvine, 1971) – is clearly not only human-centred but very subjective, since the same message might cause one individual to revise their estimates of likelihoods but another individual to yawn – so that would make the property of being or not being information dependent on who was paying attention to it. On the other hand, information theory (Shannon, 1948) takes up the opposite pole with an abstract definition of information as a string of symbols with no reference at all to what it may signify.

As it turns out, both of these concepts are very fruitful in their respective fields, leading to advances in decision and estimation theory, on the one hand, and mathematical models of communication over noisy channels, on the other. But for the present purpose, a better place to start is Bateson's celebrated dictum, 'What we mean by information – the elementary unit of information – is a difference that makes a difference' (Bateson, 1972), which he originally enunciated in the course of delivering the 19th Annual Korzybski Memorial Lecture.

Korzybski, founder of the field of general semantics, is now most remembered for his assertion that *the map is not the territory* (Korzbyski, 1933). Following Korzybski, Bateson posited that the mind creates maps of reality, and in seeking to identify the 'unit of mind', asked the question, 'What is it in the territory that gets onto the map?':

> [I]f the territory were uniform, nothing would get on the map except its boundaries, which are the points at which it ceases to be uniform against some larger matrix. What gets on the map, in fact, is difference, be it a difference in altitude, a difference in vegetation, a difference in population structure, difference in surface, or whatever. Differences are the things that get on to a map.

In subsequent paragraphs Bateson made three important observations:

1. Differences are abstract; they are not of the same stuff as the things being compared, which differ.
2. Differences that have entered a mind can have a causal effect.
3. The selection of *which* differences to notice, out of the virtually infinite set of possible differences available for noticing, depends on the characteristics of the organism – the neural pathways that 'are ready to be triggered', as he puts it – and this constitutes the mechanism for making a difference.

Bateson was specifically addressing the role of information in the human mind and in human actions, but his observations can be paraphrased in much more general terms, prompting further questions:

1. Information is qualitatively different from matter or energy.
   So then what is it exactly? Noting that information is intimately connected with matter and energy, since both are required for its creation, instantiation, transmission, and effect and, moreover, that information has a physical basis through thermodynamics, how should these notions be reconciled?
2. Information can reside in, interact with, and influence physical systems, and so can have causal effects.

Can we construct rigorous physics-based models of the mechanisms of such processes that fully account for the observed phenomena? Can such models be extended to fully account for the further open-ended processes that arise when the causal effects of the information include generation of further information that generates further consequences?

3. Not all differences make a difference. A distinction needs to be made between the vast amount of information that is implicit in the detailed configuration of matter and the tiny fraction that makes it into a mechanism for making a difference. For clarity in the present discussion we will refer to the former as *latent* information. This term includes all the aspects needed to describe how configurations of matter differ from the lowest energy arrangement possible, in which there are no correlations or significant inhomogeneities.

This distinction focuses attention on the characteristics of the mechanism that acquires the information, in particular its role in determining *what* latent information is acquired, which in turn must depend on the design of that mechanism, in other words on the information that already exists and is instantiated in the mechanism. How should we account for the existence and nature of such mechanisms for transforming latent information into a form that has causal power? Are there further distinctions needed in the forms that information can take?

The inherent circularity implied in the use of preexisting instantiated information to transform further latent information into instantiated information was also recognised by Bateson's contemporary, von Weiszacker, in his somewhat recursive definitions: (a) information is only that which is understood, and (b) information is only that which produces information (von Weizsäcker, 1980). This line of thinking skirts the question of origins but stresses the dynamic and pragmatic aspect of information and also offers a welcome escape from the cacophony of the contemporary examples listed earlier, by pointing us to the past – the earlier information that generated the later and that permitted its interpretation. Following this thread successively back through time, the din of human communication fades away, and the crucial role of information in the evolution and life processes of all living things comes into focus.

And what if we wind the clock further back, to the very dawn of life, and the prebiotic night? Can we, by examining the earliest plausible forms of life and information, in fact address the skirted question and frame some testable conjectures about the interplay of their origins?

WHAT FORMS CAN INFORMATION TAKE AND WHAT ROLES CAN IT PLAY?

To address these questions, an organising framework for encompassing the various concepts associated with information and life is needed. By separating differences that exist from the possibility of their consequences, Bateson's deceptively simple definition suggests a place to start.

The foundation layer of such a framework has to be the differences that existed in the prebiotic universe – inhomogeneous distributions of matter and energy, arising from early fluctuations and amplified by natural processes with feedbacks, such as gravitational accretion of matter into stars, and the violent events in their lifecycles. By definition, there are no observers yet to make measurements and to transform that latent information into what Bateson or von Weiszacker would have acknowledged as information. But we can surely be confident that the laws of physics were operating. Therefore we can retrospectively talk about past structures and processes, using the language of physics to describe sequences of events from hypothesised earlier configurations to the resulting later configurations of the young universe. Entropy, the physical variable that quantifies both disorder (and its converse, structure) in thermodynamics and uncertainty (and its converse, information) in information theory is well defined and calculable for a given configuration. So there is no ambiguity in talking about the latent information in the prebiotic universe as a way of talking about its structure.

But in the absence of life, can those differences make a difference? The physicist will answer with a resounding Yes! The laws of physics act on configurations of matter and energy and drive their development in time, so the latent information in the details of the initial conditions, or arising through quantum fluctuations, does make a difference to the subsequent configurations.

Of course this is not what either Bateson or von Weiszacker had in mind, but it is an important observation because we are seeking the abiotic roots of the role of information in life when it arises.

The point is, inhomogeneities in the distribution of matter and energy amount to local departures from the lowest energy and least structured state, and therefore represent metastable accumulations of free energy, which can do work if released under certain conditions and redistributed within the wider system into configurations with the same total energy (since energy is conserved) but lower free energy (less inhomogeneity) overall. According to the laws of thermodynamics, free energy will ultimately be dissipated into thermal energy at the temperature of the environment, at which point it can no longer do any work. However, many interesting things can happen on the way down from its origin at unimaginably high temperatures in the Big Bang, and therefore, as it gradually thermalises, there is a continuous flux of free energy undergoing transformations from higher to lower concentrations.

In particular, on our own planet, not only is the Earth bathed in a constant flux of free energy radiating from the sun, but there is also a continual inner source of free energy in the Earth's core, where nuclear fission of heavy nuclei (which themselves were the long-ago product of higher-temperature nuclear reactions in supernovae) generates geothermal energy and drives geochemical reactions forming energetic molecules, some of which can escape through cracks in the Earth's crust to become available for further breakdown of their chemical free energy.

But if that free energy is trapped in some kind of potential well, as in chemical bonds, then dissipating it first requires some work to be done to provide the additional energy needed initially to overcome the potential barrier in order to release the trapped energy. As is well known, chemical energy can be released through providing the initial energy in the form of heat, and then the process can continue by virtue of the energy released. But another way to release chemical energy is by coupling to a complex scaffold of self-organising and self-maintaining chemical interactions that can lower the potential barrier – which is exactly what occurs at the core of all life! (See Hoelzer et al., 2006; Morowitz and Smith, 2007.)

Seen from this perspective, life appears as a thermodynamically necessary mechanism to relieve the continuous production of free geochemical energy on Earth, and later, when photosynthesis arose, to also degrade the incident flux of energy from the sun, more efficiently than abiotic processes could. If this astonishing hypothesis is correct, then it could be that wherever sufficiently steep free energy gradients exist and conditions permit, life will arise not as a rare, fragile, and contingent accident, but inexorably as a driven phase transition to a lower-energy nonequilibrium steady state.

Lightning discharges and convective storms offer simpler examples of such nonequilibrium steady states arising to dissipate excess free energy, which in these cases take the form of electrical and thermal energy gradients in the atmosphere. The storms peter out once the energy gradients are sufficiently reduced and reinitiate when they build up again. Life, on the other hand, is apparently not so easy to ignite, and may have petered out after an unknown number of false starts. But at least once, it did take off and keeps going continuously for over 3 billion years, sustained by the continual supply of free energy, and evolving ever better ways to dissipate it, with information in various forms playing increasingly significant roles along the way.

We do not yet know how easily and frequently life could arise in the universe, but science can investigate whether there are robust routes to it that can be initiated and facilitated by natural processes and the chemistry and physics of the available precursors.

There are many efforts under way exploring the plausibility of such routes based on different conjectures about the prebiotic environment, about the sequence of steps, and about the roles of contributing processes such as Darwinian selection, self-assembly, and self-organisation. But what we are interested in here are approaches that develop systematic conceptual frameworks that are deeply grounded in the relevant scientific disciplines and that draw from them universal, or at least very broadly applicable, generalisations that shed light on the nature and origin of different forms of information.

INFORMATION IS PHYSICAL

The fundamental discipline with which all such frameworks must reconcile is physics, which in particular addresses flows of energy, entropy, and information in various processes and how these may result in constructive work being performed. Thermodynamics tells us the relevant relationships. The first law is essentially a statement of the conservation of energy: we have to keep track of heat energy entering or leaving a system, work done on or by, the system, and its internal energy. The famous second law, that entropy can be created but not destroyed, is a constraint on which processes are actually possible, since conservation of energy is necessary but not sufficient. The second law allows entropy to be moved but forbids processes that would destroy it, for example, by *just* moving heat from a cooler to a hotter body, or equivalently, by *just* converting heat into work. It *is* possible to move heat to a hotter body or to use heat to do work, but such processes are necessarily accompanied by other effects that ensure overall increases in entropy.

Entropy is a measure of the number of possible detailed configurations of the elements of a system, in other words its microstates, given its energy and any other macroscopic parameter needed to specify it, such as constraints. It is temperature dependent, because the higher the temperature, the more energy there is to distribute between the elements and the more possible states they can access. When only the macroscopic parameters are observed there is therefore a corresponding uncertainty about the detailed microscopic configuration. The best estimate one can make is to assign equal probability to each accessible state, but which of them is actually occupied at a given time? This is the basis for the association of entropy with uncertainty.

Processes such as self-organisation and self-assembly that generate structure in the system reduce this uncertainty because now the microstates are no longer equally probable – some previously accessible microstates may no longer be available, and others may have a higher probability of being occupied, so the number of possibilities – the entropy – is reduced. Creating structure therefore requires entropy to be produced to at least[1] balance the reduction of uncertainty in the system, for example, by releasing heat. This is the basis for association of entropy with information and with structure. Any process that generates structure increases the latent information inherent in that structure, which corresponds to a decrease in entropy (reduced number of microstates). Self-assembly and self-organisation are so called because the structures they generate are not imposed by any external source of information, which is not to say that there is no external influence at all – merely that it is insufficient to account for the order that appears. What distinguishes them from each other is the thermodynamics of their mechanisms (Halley et al., 2008). Self-assembly proceeds spontaneously because the state in which the components are not bound is a higher energy state than the assembled state, for example, monomers assembling to form polymers, and crystal formation in supersaturated solutions. So self-assembly is an equilibrium-seeking process that releases energy (an exergonic process) and does not require anything to support it other than availability of the component molecules themselves. The energy released is dissipated as heat and, as required by the second law, thereby produces an

---

[1] The 'at least' condition holds only in the ideal case of reversible processes, which will not produce any waste heat. Real processes are generally irreversible and so additional entropy must be generated through some energy being wasted and dissipated as heat.

amount of entropy equivalent to the increase in latent information that the self-assembly creates.

Self-organisation, on the other hand, is a nonequilibrium process, characterised by a stable pattern of coordination in the interactions between the elements of the system and requiring a continual input of energy (an endergonic process) as well as the material resources needed to keep it going. In this case, the stability of the structure produced stems from the dynamics of the interactions, which require a continuous flux of energy to maintain them, rather than the statics of falling to the lowest energy state. If the energy flux decreases sufficiently, the self-organised order decays. So both self-assembly and self-organisation spontaneously produce structure, and therefore latent information – but self-organisation has the additional requirement of external energy needing to be supplied to do the work implied in the creation and maintenance of the coordination.

The concept of work is quite precise in physics and is of particular interest for both the origin of life and the emergence of information, since life clearly needs much constructive and coordination work to be done, and work is so closely tied to the creation of structure and therefore of latent information.

But what is really needed is a concept of how the required work can be embedded in processes that can operate continuously and reliably to create the persistence and dynamic stability that would permit further development and evolution to occur (Pascal et al., 2013). The physics of thermodynamic cycles offers one such concept. The basic idea is to identify a cyclic sequence of states of the system and its environment, and the processes that can take the system from one state to the next in the sequence, such that the system ends up in the same state as it started, but with some work having been done on the way, either by the system on the environment or by the environment on the system. Since the system is back in the same state as it started, the cycle can be repeated over and over again, so work can indefinitely continue to be done, obviously at the cost of importing free energy from the environment, and being able to export entropy back to it.

The best-known thermodynamic cycle is the ideal reversible Carnot cycle, named for its French inventor, which is demonstrably the most efficient work cycle possible for converting thermal energy from a higher-temperature source into work at a lower tempera-ture. It is therefore a benchmark against which other more realistic irreversible cycles can be compared. In terms of energy flow, a key feature of thermodynamic cycles is that the sequence of processes in the cycle alternates between importing energy into the system by work being done on it (applying constraints to the system) and work being done by the system using its free energy to relax the constraints. Some of the imported energy is converted into the work done by the system, and the rest is dissipated into the environment as waste heat, producing an amount of entropy to compensate for both the entropy reduction achieved by the work and the degree of irreversibility. A thermodynamic cycle therefore transduces free energy from an exergonic process into a form that can then supply the energy needed to drive an endergonic process, accompanied by the inevitable production of entropy as waste heat.

All these relationships make it clear that information is a physical quantity on an equal footing with, and closely related to, energy, not some ghostly abstraction that somehow overlays the material world.

Apart from the obvious fact that energy is required to do the physical work of modifying a material in order to encode information into it, for example, chiselling an inscription in stone, there is also an energy transaction associated with the reduction in entropy that information implies. There is a precisely defined energy equivalent of one binary bit of information (Landauer, 1961) and that energy must be released as heat if the information is destroyed, that is, if the bit is randomised. Moreover, the reverse has now also been demonstrated experimentally – the conversion of information to free energy (Toyabe et al., 2010). This is not using information to release energy from a higher temperature source, which would hardly be newsworthy; it is actually using information to extract useful energy from thermal equilibrium, which is a remarkable result.

So, equipped with the basic physics concepts, and having identified the latent information inherent in the patterns of distribution of matter and energy as the starting point for emergence of further information, the next questions are: How does it come about that networks of self-organising and self-maintaining chemical interactions arise and feed on free energy gradients, and what further forms of information (if any) arise in those processes?

SELF-ORGANISING CHEMICAL NETWORKS

The theory of autocatalysis suggests some plausible conjectures. Autocatalysis is a natural phenomenon in chemistry, whereby self-organising and self-maintaining chemical reaction cycles can arise in an environment containing freely available small precursor molecules, if the reactants and their products can all be formed from each other and the precursors, by mutually catalysed reactions – that is, the reaction producing each molecule is accelerated by another one of the produced molecules without consuming it. The way this happens is by the catalyst molecule being able to form a short-lived complex with the reactants, which brings them into closer proximity to each other than they could achieve independently, so facilitating their reaction. Once the new bonds are created, the product molecule is released, and the catalyst molecule is ready to repeat the process with fresh reactants.

These and similar concepts were developed over several decades (Eigen and Schuster, 1978; Kauffman, 1986) and have since been confirmed both theoretically (Hordijk, 2013; Hordijk and Steel, 2004; Mossel and Steel, 2005) and experimentally (Lincoln and Joyce, 2009; Sievers and Von Kiedrowski, 1994; Vaidya et al., 2012). In fact, there has been a growing realisation that catalytic properties in organic molecules are far more prevalent than once thought and that so-called RAF[2] sets can occur with reasonable frequency in a mixture of chemicals that are potential substrates for biological processes and/or catalysts.

When such a closed set of reactions arise, each molecule in the set is produced at an accelerated rate, and in turn accelerates the production of the other molecules that it catalyses. As a result, all the molecules in the set grow in concentration relative to molecules not in the set, because they outcompete them for the available precursors. Thus differences in concentrations are amplified, setting the stage for the next steps in the emergence of primitive biological processes to elaborate the reaction networks utilising the free energy harvested from the precursors, and now stored in the concentrated catalysed molecules.

An equivalent way of describing the outcome of autocatalysis is to say that it creates constraints on the phase space available to the system. What we mean by this is that as the available precursors get drawn into the cycle to produce more of the autocatalytic set molecules, the system is constrained to occupy only that region of its phase space that corresponds to high concentrations of those molecules, and is unable to access the more numerous configurations that correspond to other distributions of species and concentrations. It takes work to create constraints, and the result is a reduction in entropy (by definition since fewer microstates are available), and therefore an increase in free energy – which is accounted for by part of the energy input to do the work (or all of it in the limiting case of reversibility). Moreover, the effect of the constraint is to channel more of the available chemical energy in the precursors into further increases in the concentrations of the autocatalytic set members, in other words, to channel it into doing constructive work creating

---

[2] RAF stands for reflexively autocatalytic and food-generated, meaning each molecule in the set is catalytically produced by other molecules in the set (acting as reagents or catalysts) from 'food' – a freely available external supply of small precursor molecules such as CO_2, N_2, or H_2.

more free energy–containing structure, as opposed to being dissipated as heat.

As we will see shortly, it turns out that such roles of constraints are important more generally – not only because they amount to an imposition of structure (latent information) on the system, but because they are essential for channelling free energy into doing work. This theme of constraints as information, as intimately related to work and entropy, and as associated with the origin of life, will recur in each of the perspectives that we will examine below and will help in synthesising them into a coherent overview, albeit incomplete in many ways.

Experimental verification of the theoretical possibility of RAF sets was a significant advance in origin of life research because it established the plausibility of such sets being a springboard from which a route to life might emerge in an energy-rich abiotic environment. In fact, metabolism, the fundamental life processes that build the molecules needed by the organism and that provide the energy needed to do that, is a complex network of autocatalytic cycles, and much evidence now points to the emergence of metabolic networks as one of the earliest steps towards life (Copley et al., 2010; Trefil et al., 2009).

However, taking the next step towards something not yet living, but displaying some rudimentary life-like properties, requires a little more: first, containment to individuate the set and to preserve it from dilution through diffusion in an open environment, and, second, some way of replicating the individuated sets so that selection can start shaping them.

A TOY MODEL

Many models (e.g., Gánti, 2003) have been put forward for these next steps, but Deacon's autocell (Deacon, 2006; Deacon and Sherman, 2008) is a particularly instructive and satisfying one because it demonstrates in a systematic way the minimal essence of what is required to initiate evolvability, and because it naturally links to other perspectives that illuminate our questions, as we discuss below.

Deacon coined the term 'autocell' to describe a hypothetical system in which a RAF set is able to generate its own containment through one of its products being a molecule that can spontaneously aggregate and grow into large regular structures by self-assembly. Crystal growth is a familiar example of self-assembly, but, depending on the properties of the molecule, self-assembly can also result in linear structures (polymers) and in two-dimensional sheets, tiles, or tubular structures, which might then close to form a natural container, and thereby enclose some of the RAF set that generated it. In such a case, autocatalysis would continue until the enclosed precursor molecules were depleted, and then a stable 'dormant' phase would ensue until the container happens to rupture, releasing its concentrated catalysts into the environment where fresh precursors are available, which could then spawn further cycles of autocatalysis, self-assembly of containers, growth to depletion, and then dormancy until ruptures launch yet further cycles.

This is a very interesting scenario for several reasons. First, it describes a primitive form of replicating entity, which arises spontaneously from self-organised chemistry (the RAF set) under a very plausible condition (at least one of the RAF set molecules undergoes self-assembly to create a closed structure). Separately these processes are well understood and documented, so postulating their occasional conjunction with the right mix of molecules is not unreasonable.

Second, from the perspective of what kinds of information exist and what roles they play, we observe that the linking of the two processes implies a qualitative increase in latent information, which can justifiably be labelled as emergent since it is not possessed by systems displaying either process alone. RAF sets create more struc-ture in molecular composition; self-assembly creates more structure in spatial configuration; and these are both examples of quantitative increases – more of the same kind of information. However, the linking of the processes creates a new kind of regularity – cycles, and coordination of the processes in time.

Moreover, the fact that one self-organising process (autocatalysis) can in principle initiate another (self-assembly), by virtue of creating conducive conditions for it (accelerated rate of producing the components for assembly in proximity to each other), and that the second process in turn could create conducive conditions for the first (by containment preventing loss of the RAF set by diffusion), means that two such processes would be mutually reinforcing in a similar way to the reflexively catalytic dynamic within a RAF set, and this is precisely the kind of self-organising increase in complexity prevalent in living systems that one would hope to find in origin-of-life theories.

A third point of interest is that the linking of the processes in this way amounts to each providing a supportive function for the other, with the net functional result of enabling the persistence and proliferation of the new entity, the autocell. The concept of function implies a purpose that is served, and prior to the initiation of self-assembly, there was nothing in the system to which either purpose or function could be ascribed.[3] The simple fact of a RAF set arising in which one of the members is able to self-assemble in this way creates an intrinsic relationship between the two processes (Deacon calls it reciprocal linkage), and inherent in that relationship is the emergence of function, and therefore purpose, again features that are signatures of living systems.

Of course the autocell as described is clearly not a living system, but a fourth observation points further in the right direction. Once a primitive form of replication occurs, it opens the possibility of selection since some of the autocells generated might have slightly

---

[3] It would be stretching the concept of function too far to ascribe it to the mere growth of the RAF set in an open environment, since without containment this can be only a transient phenomenon until local resources are depleted and the set diffuses away. By contrast, in the context of an autocell, the autocatalytic and self-assembly processes are functional because they increase the probability of persistence of the autocell form (Deacon and Sherman, 2008).

different RAF sets through accidental incorporation of other molecules (in general there will be more than one possible RAF set in a given collection of molecular species), and now these variants can compete for precursors as individual entities – as autocells, rather than the molecules in a RAF competing for precursors with molecules not in the set. The dynamics of competition between autocells is quite different from the competition between freely moving molecules in an open environment because of the modulation introduced by the autocell's temporal structure.

For example, a variant of an original RAF set, wherein a modified molecule is able to catalyse the production of the molecules that self-assemble even faster than before, would build its new enclosures more quickly, or perhaps build larger enclosures if the modification of the molecule resulted in a suitable change to its geometry. Such variations would be inherited by the autocells it builds and passed on to the autocells they build (perhaps with further variations arising). If this line of autocells commandeers the precursors at a faster rate as a result, the logic of selection is that it will displace the original.

So now we recognise the beginnings of evolvability, not yet involving any information-bearing templates such as RNA, but operating purely on the basis of the functions of energy harvesting and construction, which we might construe as a primitive form of metabolism. Moreover, this selection dynamic implies the emergence of a further level of information – a population of individuated sets, with a distribution of frequencies of the variants, besides the individual elements that constitute the population.

A fifth feature, self-repair – the ability to reconstitute itself after partial disruption – is such a ubiquitous aspect of life that its first appearance here is worth remarking on, even though it is really an extension of the already noted self-assembly building of enclosures.

THE ORIGIN OF MEANING
The forms of information we have so far identified as arising in the autocell model are mostly due to amplification and proliferation of the

far-from-equilibrium patterns of concentration of chemical species, and of spatial patterns, and therefore still fall into the overall category of latent information as we have defined it here. But we have also noted some interesting and novel informational structures: temporal patterns; the embryonic emergence of individuation, purpose, and function; and population-level information about frequencies of variants. To the extent that these combined elements were able to proceed down an evolutionary path, we might anticipate the appearance of yet more interesting forms of information – for example, information *about* the environment or *about* how the autocells reproduce.

These constitute two very significant classes of information that represent seemingly abrupt departures from the passive and nonreferential nature of the latent information discussed so far, and arguably underlie the emergence of causal power in information. The fact that such information carries meaning, implications about something other than its own instantiation, places it in a unique category, which we might call *semantic* information, and its existence raises many deep and important questions and considerations, which we will return to armed with further insights from exploring some of the other perspectives.

But before we move on from the simple autocell toy model, let us take one last look at the implications of the classes of information that it has thrown up and see if we can glean any hints about the origin of meaning. Obviously the composition of individual autocells will depend on the composition of the material that was drawn from the context and became enclosed, but the autocell has no way of using that information – it is simply descriptive of what is – so such correlations are still what we would want to call latent information.

However, there is a glimmer of something more interesting in the information contained in the frequency distribution of variations in the population of autocells, which admittedly could carry some implicit information about the environment, since the more frequent variants are by definition better able to reproduce in that environment than the less frequent ones, by virtue of the details of their composition. The transmission and proliferation of those variations (i.e., that latent information) therefore results in accelerated conversion of precursor molecules into more of those variants.

With an eye on what we know to be the future consequences of evolution we might recognise this as an exemplar of how natural selection could begin to encode environmental information in a way that causes cumulative future impact on the environment. But with an eye on the antecedent abiotic self-organising processes we could equally argue that there is nothing significant about this compositional variation to distinguish it from similar variations in other nonequilibrium systems, such as weather patterns. There is still something missing in the story, to which we will return below. For now, we cannot answer the question of whether the emergence of semantic or causally effective information is a discontinuous event like a phase change, or whether it is a long and smooth transition.

Beguiling as this autocell model is, with so many suggestively life-like emergent features, it still falls well short of life itself. The exergonic self-assembly process and the endergonic autocatalysis are linked, but in a very limited way. The energy released by self-assembly is dissipated thermally rather than used to drive the endergonic process, as in a thermodynamic cycle. And since autocatalysis requires a continual influx of energetic precursor molecules (the food set) to proceed, once the local precursors trapped in the enclosure are depleted, it stops, whereas life has evolved all kinds of further elaborate and interconnected catalytic cycles, to use the energy and the products of the core metabolic cycle to build more and more effective replicators that can actively seek, compete for, and exploit resources. Could there be a pathway from such a simple model to the kind of open-ended evolution that is needed for complex life to emerge?

The critical issue here is evolvability, so we now take an evolutionary perspective to consider what that means and what is required. Noting that the core features of evolvability – replication with heredity, variation, and selection – mean that the copies are similar to the originals, but can also have some differences, and those *differences make a difference* to whether the copies will themselves be copied, we immediately recognise Bateson's hallmark of information. We will also find a clue to the missing link in the autocell story. But first let us reflect on two profoundly important angles that the evolutionary perspective offers on the roles of information.

We have already anticipated the notion of semantic information, and here we see it in action, in that the difference that made a difference represents more than just information about the system that gets copied; it also represents something about the factors in the environment that contributed to the system's success in getting copied. This is arguably the genesis of semantic information. The counterargument stems from the observation that in the simple autocell model, it is still no more than latent compositional information, admittedly correlated with some relevant aspects of the environment such as the autocell now including particular molecules that can exploit chemical species peculiar to that particular environment, but the pathway to making a difference is no more than the physical propertes of those molecules catalysing useful reactions, rather than the latent information they imply being processed and *used as information* to make a difference. Nevertheless, it is a first step.

The second angle is that the differences that made the difference to whether a copy got copied again are part of what gets copied, so the information that those differences constituted is getting transmitted from one generation to the next. Transmission of information *as* information, that is, coupled to a mechanism for making a difference, is action not only at a spatial distance, but also at a temporal distance once the ability to store the information for later use arises. We are so accustomed to information being transmitted that we rarely pause to reflect on how extraordinary it is that such a process should have spontaneously arisen. Arguably this is its wellspring. Why exactly does it appear here? Because of the combination of selection and heredity. Selection is the ratchet filter through which the noise of random variation must pass, and the signal that gets through does so *by* being transmitted to the next generation through heredity. Transmission through heredity *is* the mechanism of selection.

So once evolvability gets traction, in principle we have the seeds of information that carries meaning and of information transmission – two necessary and momentous steps towards life – yet so subtle at this stage that they could easily be dismissed. However, getting traction is not trivial, and while the core features of evolvability are necessary, they are not sufficient. There is another question we need to ask.

WHAT MORE DOES IT TAKE TO BOOT UP LIFE?

Evolution is an information generating and transmitting process, but the path to greater evolvability (faster accumulation of more useful information) is a precarious balancing act between two abysses. On one side, Eigen's error paradox (Eigen and Schuster, 1978) threatens to rapidly extinguish any nascent protolife if the rate of error (or introduction of differences from the original) is too high, for the simple reason that whatever latent information exists gets randomised by errors faster than it can propagate and accumulate. On the other hand, evolvability *requires* the introduction of differences (or copying errors) in order to discover better ways to persist and replicate. Maximising the error rate to accelerate exploration of new possibilities, while remaining just below the error threshold that terminates the process in a dead end, is a very fine balance. Moreover, since the number of errors per molecule replicated is the product of the number of pieces it is assembled from and the error rate per piece, then a given error rate places a hard limit on the size and complexity (amount of structure or latent information) of molecules that can be assembled and replicated indefinitely.

So how can evolution get to discover and exploit more complex molecules and structures? It has to first discover ways to reduce the error rate, but it has to do so using the limited palette of molecules it can currently produce and replicate. This is a very tough call (Czárán et al., 2015; Kun et al., 2015; Maynard Smith et al., 1995; Vasas et al., 2015), and in a sense it is the ultimate chicken-and-egg problem if we equate the chicken with the currently feasible set of molecules and the egg with the current information limit. More complex molecules are needed to evolve mechanisms that can reduce the error rate, but such complex molecules cannot be built until the error rate is reduced sufficiently to permit them.

There are other perils too on the path to open-ended evolution, such as the delicate balance of the autocatalytic set collapsing under the parasitic onslaught of mutant catalysts that benefit their own replication but do not play their part in the cycle, and a competitive race between catalysts that require the same precursors – ultimately there can be only one winner (Gause and Witt, 1935), which then leaves the system in an evolutionary dead end because there is not enough diversity to spawn innovation.

It is interesting that both of these processes are essentially informational problems – ways in which the evolving latent information gets trapped prematurely on a local entropy minimum (one dominant replicator) from which there is no longer any path to the more complex structures that could continue to evolve in an unlimited way.

There are solutions to all these difficulties, of course, else we would not be here to discuss them. The information bottleneck imposed by the replication error rate is bridged by building the more complex molecules in two or more steps, each of which combines components that are short enough to reproduce sustainably. This is an example of a well-known motif in the emergence of complex structures, so-called stable intermediate forms (Simon, 1991), which naturally lead to a multiscale hierarchical architecture in complex systems. Interestingly again, we see that the resolution to the informational impasse creates yet another new form of information – an additional level of description to specify the links between the component replication processes and to detail the hierarchical structure that results.

Robustness to parasitic mutants is achieved by confining the emerging protolife systems in enclosures or on surfaces of suitable substrates, so that any mutants that do arise are isolated and cannot spread to neighbouring systems. Ultimately they commit suicide by killing off the system they are confined to. Alternately they may mutate further to become 'domesticated' by the system, by evolving features that are useful to the system and reducing their harmful properties. Could this be the original 'cooperate or die!' storyline? Yet again the solution can be seen as an informational one, since confinement implies constraints, which reduces the number of available microstates, and therefore the entropy.

The diversity necessary for ongoing open-ended evolution is also maintained by a cooperative strategy, in this case overcoming the competitive exclusion problem when replicators depend on the same food source, by recruiting previously untapped food sources in the environment for the autocatalytic cycles, through evolution of new catalysts that can exploit them. The benefits of this strategy are twofold. It accesses new resources of energy and materials to accelerate growth, and it initiates functional specialisation in the autocatalytic networks, since such catalysts would essentially be primitive metabolic enzymes, a novel role that is differentiated from the more generalised roles of participants in RAF sets. What makes this a cooperative strategy is that, first, specialisation of functions means that replicators and metabolic enzymes are playing different but mutually supportive roles, and, second, specialisation of food sources means they are no longer in competition. We discern in this scenario the earliest shades of the decoupling of metabolism and reproduction, ultimately leading to coded replication, the first instance of genuinely semantic information, about which we will have more to say shortly.

To recap, a closer look at the dynamics of evolution has shed some light on what it takes to break out of the limitations of compositional inheritance in simple models such as the autocell, and start down the path to robust open-ended evolution. In particular we have seen that evolution is essentially an informational process; that it generates, accumulates, and transmits information; that it has to avoid a number of informational pitfalls; and that in doing so, it creates new forms of latent information – hierarchical structure, constraints, functional specialisation and cooperative relationships, and, ultimately, semantic information with all its future ramifications for causality.

ABOUT 'ABOUTNESS'

What else can be said about semantic information? For information to be *about* something there first of all must be some correlation between a pattern in the information and a pattern in the 'something' that it purports to be *about*. Correlations in the physical world are not unusual; they abound precisely because of interactions between some aspect of the physical world and some physical process that generates patterns. But correlation alone is not a sufficient condition for *aboutness*.

Imagine, for example, a scenario on a planet surface illuminated by its star. As a result of multiple reflections from the structure of the surface, there will be ambient light bouncing around in all directions. If one were to select a subset of that ambient light that was travelling from a particular feature of the surface towards an arbitrary point and apply suitable processing to it such as focusing it into an image, one would find correlations between some aspects of the patterns in the feature and the patterns in the selected light flux, simply because the light that was incident on that feature had been modified by interaction with the feature when it was reflected from it.

Would we construe this as information about the feature? Yes, but surely only in the case that such selection and processing actually occurs, which then relies on the existence of particular mechanisms to do the selecting and processing, which then also requires a prior explanation for how the mechanisms came about to exist.

In the absence of such mechanisms, we simply have a scenario bathed in a flux of radiation with some distribution of frequencies, intensities, and directions, from which it would be hypothetically possible to establish a vast number of different correlations with various aspects of the scenario, given suitable mechanisms for doing so. But clearly this is still no more than latent information.

Furthermore, assuming the necessary mechanisms do exist, there is arguably another condition we would want to see satisfied before we would attribute *aboutness* to the correlations that the mechanisms select and process, particularly if we are interested in the connection from *aboutness* to causality. As Bateson recognised, what is needed is a pathway to making a difference – the processed correlations must be potentially utilised so as to make a difference.

Such mechanisms and pathways are evident in all known forms of life and are broadly understood as products of evolution, so we would not expect to see them already in the very earliest forms of life and their nonliving precursors. Nevertheless, thinking about what is required for aboutness to emerge helps to clarify where to look next in seeking insights into the emergence of causally effective information.

What these speculations suggest is that for semantic information about the environment to exist and be causally effective, mechanisms to sense and respond need to have evolved first. The emergence of this kind of semantic information is therefore bound up with the emergence of complex systems that can sense and make use of it. The search for an origin of semantic information points us back therefore to the earliest forms of sensing and responding in the emergence of life, and indeed to the mechanism of evolution itself.

AUTONOMOUS AGENCY

So let us now ask, what is the fundamental essence of sensing and responding mechanisms? This brings us to another perspective, in which there is unambiguous utilisation of information by a simple hypothetical system acting autonomously on its own behalf. Acting autonomously on its own behalf is a hallmark of any living system, in how it seeks to survive, feed, reproduce, and avoid danger. But this capacity for autonomous agency – for sensing relevant information and using it to choose actions that enhance its own interest – is such an extraordinary phenomenon to have emerged from physics and chemistry that it bears deep examination. Is an autonomous agent necessarily living? What exactly is required for a system to display autonomous agency? What does 'choosing' mean in a physical system? And what are the underlying mechanisms?

Kauffman tackled these questions head-on in his attempt to distil the essence of an autonomous agent (Kauffman, 2002; Kauffman and Clayton, 2006). He proposed that the minimal conditions necessary for an entity to possess agency were that it should be an autocatalytic system, be able to carry out thermodynamic work cycles and thereby to detect and measure free energy (i.e., displacements of external systems from equilibrium) and to construct constraints to release that free energy so as to do work, in other words to release it into a few degrees of freedom as opposed to releasing it as heat into the maximum number of degrees of freedom.

The 'work' done powers all these activities – the self-production and reproduction, the detection of free energy, and the construction of constraints. Provided sufficient free energy is detected and harvested, these processes can be chained into ongoing propagating work cycles. All these themes have appeared already in our conceptual explorations, but Kauffman adds one more that is absolutely critical to our most fundamental questions. An autonomous agent must be able to make a choice between at least two actions, in the sense of exhibiting different and appropriate behaviours under different circumstances.

This does increasingly look like a living system that has the ability to sense opportunities and threats, and choose appropriate actions to preserve its existence and promote its reproduction. But what exactly is happening when an agent makes a 'choice'?

Obviously, a choice implies that more than one action is possible. From a physics perspective this might suggest a system that was in metastable equilibrium, like a pencil balanced vertically on its point. It could fall in any direction, so many 'actions' are possible. However, such a system would be at the mercy of noise, random fluctuations that would determine in which direction it 'fell' – hardly a candidate for autonomous choice in its own interests!

The key to autonomous choice is not in equilibrium physics but in that the agent must be a far-from-equilibrium system, poised in such a way that the internal states corresponding to different actions being initiated are reliably inhibited yet able to be triggered by essentially tiny signals that bear some correlation with a relevant aspect of the environment. Tiny because it is not the energy or magnitude of the signal that produces the effect, but the specificity of a pattern that can be interpreted as meaningful. Ambient light suddenly occluded provokes fleeing; sensing of a food molecule stimulates feeding; a particular sound pattern triggers freezing. It is not hard to imagine how such linkages between what can be sensed and what adaptive action is thereby triggered can evolve, and further to speculate how such linkages might in principle gradually evolve to become more complex and incorporate more conditionals – in other words, to evolve more information-processing capability – to provide more 'intelligent' nuanced choices.

A system that has these properties is necessarily an evolved complex system, characterised by a network of interactions through which inputs are processed and connected to the action mechanisms, and displaying both high selectivity and sensitivity to certain inputs and robust insensitivity to others. The inputs can also include internally generated signals, for example, from subsystems that monitor internal processes, from memory subsystems, and from complex feedback circuits.

Such systems are obviously well down the path of evolution and far beyond the origins scenarios on which we have been focusing. But the main points we take away from this perspective are, first, that it is the design, and particularly the control architecture of the agent, that enables the existence of semantic information about the world and its causal effectiveness, and, second, that the design must possess complex dynamics of selectivity, robustness, and sensitivity to enable information-based choices.

FINALLY, WE COME FULL CIRCLE

We need to refocus on the mechanism of evolution itself to understand how evolvability evolved to become potent enough to produce such systems.

We left the story of evolvability at the very earliest stage of circumventing the informational pitfalls of simple replicating systems. In fact there were many further breakthrough steps along that journey (Maynard Smith et al., 1995; Ridley, 2000), but the single most significant was undoubtedly the emergence of the genetic code enabling coded template replication rather than just straight compositional replication.

There is a lot known about the genetic code, about how the transcription and translation processes work, and about the evolutionary advances they enabled. In modern lifeforms these are incredibly complex processes, so clearly they must have evolved through many intermediate forms. But the actual origins of coding are obscure and controversial (Koonin and Novozhilov, 2009; Vetsigian et al., 2006), perhaps inevitably so, since there is scant evidence to be found of what may have occurred nearly four billion years ago. Nevertheless the result has been a complete game changer. In brief, the transition from compositional inheritance to coded inheritance was essentially a transition from analogue life to digital life.

One immense advantage of digital coding is that reliable transmission can occur in spite of noise, because the distinguishability between the possible symbols (for example, 0 and 1 in binary code, or the four nucleotide bases of DNA in the genetic code) is greater than the average noise, and also because error-correcting mechanisms become possible. The reduction in error rate permitted more complex molecules to be replicated and opened up unlimited new possibilities for evolution to work on by giving birth to the protein world, which all known life is based on and which enabled the evolution of vastly more efficient enzymes, highly specialised molecular machinery to perform those complex processes, and the emergence of regulatory functions, which in turn enabled the emergence of information processing in biological systems.

There is far more to this story than we can delve into here, but let us pull a couple of threads that particularly resonate with our reflections so far.

One thread has to with replication becoming decoupled from selection. When replication is just based on catalytic activity in a RAF set, the reaction rate depends on the extent to which the physical form (shape and affinities) of the catalyst correlates with those of the substrates. So there is a great deal of variability in rates for different reactants and catalysts, as well as some likelihood of variations in which of the possible reactions actually happen, resulting in a range of similar but not identical products. The specificity of such reactions is therefore not high, and efficiency is correspondingly lower. With template replication, however, because highly efficient specific enzymes have evolved for each of the twenty or so amino acids that living systems use, the replication rate, efficiency, and accuracy or specificity are high and essentially independent of the actual sequence on the template. What this means is that any possible change to the sequence will be replicated as well and fast as any other, and it is only the impact that change has on the performance of the system possessing it that will affect its selection. This both sharpens the power of selection and preserves more neutral (nonharmful but not currently beneficial) variations that can be the feedstock for further variations (since nothing critical depends on them), with the net result of more effective exploration of the possibility space.

Furthermore, coded replication is a two-step process because the sequence has first to be transcribed from DNA to the intermediary messenger RNA, and then translated from RNA to the protein-building machinery that assembles the protein coded for.

Here is what is interesting about that with regard to our questions. In the first stage, there is unbounded variety in the possible sequences on the DNA, and furthermore, it is easy for mutations to generate further variations there. Similarly the third stage consisting of the protein products also has unbounded possible variety. But the middle stage is well defined, highly conserved, and operates with high fidelity. This constrained middle stage is the neutral and efficient interface translating huge variety in possible sequences on the genetic material, into huge variety in the possible products.

This is an example of another well-known motif in the architecture of complex systems, the so-called *bow tie* (Csete and Doyle, 2004), which can take into account a great diversity of inputs (fanning into the knot of the bow tie), process them with much smaller diversity in the protocols (the central part of the knot) to elaborate these inputs, and produce wide diversity of outputs (fanning out of the bow tie). This design pattern is ubiquitous because it provides robustness and efficiency in transforming inputs into outputs, while also retaining, and indeed maximising, the capacity to continue to evolve.

Bow-tie structures have been shown (Friedlander et al., 2015) to spontaneously evolve when two conditions are met: first, the information in the evolutionary goal can be compressed, and, second, there is an active process to reduce the number of possible interactions. These are both informational conditions, arguably strongly met in the evolution of life, since the complexity of finding resources and evading threats can be reduced to the simple evolutionary goal of reproduction, and selection will weed out costly unnecessary interactions. Thus the bow-tie architecture is yet another informational solution to an informational problem on the route to life, with the now-familiar subthemes of constraints, efficiency, and evolvability.

But the thread with the strongest resonance with our information reflections is that because of the digital genetic code, a DNA sequence is semantic information. The triplets of nucleotides each 'stand for' a particular amino acid, so the sequence can literally be read out as the sequence of amino acids that needs to be stitched together to make the protein that the gene codes for.

What is this semantic information about, then? Primarily and directly, it is information about the proteins that are needed to build and operate the organism, so, indirectly, it is information about (what has been useful in) the past environments in which it evolved. But there is much more information here than meets the eye, since the genes (or proteins) per se are not sufficient to specify the system. The proteins are necessary components either as end products or as tools for building the organism, but the actual structure and functions that emerge imply a vast amount of additional information that needs to be generated in the process. Where does it come from?

The key is in the interactions between the genes, via the interactions between their protein products. Many genes code for proteins that can act as promoters or suppressors to switch other specific genes on or off, by which we mean initiating or preventing the transcription of particular other genes. As a result, a cascade of genes turning other genes on or off generates complex patterns of protein production in time, which can be influenced by internal and external factors as well as by the system's current state. Taken together, and embedded in the right environment, the genome, protein products, and the gene expression machinery are a self-organised interactive information processing and materials processing system that not only reproduces a very complex system, and operates it during its lifecycle, but also thereby generates far more complex latent information – structure – than the genes alone can account for.

And some of these structures will be mechanisms for sensing, storing, and using semantic information about the environment during the system's lifecycle – implying yet more complex forms of both latent and semantic information.

CODA

Did information emerge? Yes. We have seen that latent information abounds in the universe and that there are plausible pathways from abiotic latent information and free energy gradients to functional structures and patterns in prebiotic systems. We have learned that there are plausible informational solutions to the viability and evolvability challenges to those systems and that they result in the emergence of more complex forms of latent information.

We have argued that meaningful semantic information about any aspect of the world could emerge only when systems that could sense and respond to that aspect had evolved, and that therefore the question of the fundamental origin of semantic information had to point back to earlier questions about how such systems could have evolved, and, in particular, how evolvability could itself have evolved the necessary level of creative power to enable that. We found that the degree of complexity that can be reached is limited until coded replication arises to decouple heredity – the business of information transmission from one generation to the next – from metabolism and the business of living. And once that happened, the consequences of digital coding have reinforced and amplified the adaptive power of evolution, which in turn has produced the mechanisms that create yet more complex structural and functional information and elicit semantic information from the latent information in the environment.

But how and why did semantic information in the form of coded replication arise in the first place?

We cannot point to the benefits that flowed from coded replication, because evolution can operate only with hindsight, multiplying what has worked well and tolerating what has not killed it. The picture is far from clear, but we have seen some tantalising hints.

Increasing evolvability and specificity seem to play a part. As more efficient catalysts evolved, they also became more specific, since the closer a catalyst's fit to one set of reactants, the less it would fit alternate reactants with slightly different properties. A set of highly specific catalysts that vastly accelerate a particular process and have little effect on others lays a foundation for the transition from analogue to digital processing, and for the emergence of complex control structures and the kind of self-organised criticality behaviour (Adami, 1995; Halley and Winkler, 2008) that seems to be required for autonomous choice.

Each step along the way must have offered sufficient advantage for the system possessing it that selection could act on it effectively, and there must have been many steps, each plausibly generated by exploration from the previous, to arrive at the extraordinarily complex processes we see today, But evolution has had the advantages of extraordinary numbers to get there – billions of years and massively parallel searches in possibility space since every member of a population is a simultaneous experiment with its particular set of variations from the previous generation.

To close, let us reflect on what this fragmented sketch of a story suggests about the emergence of causally effective information and its relation to the origin of life. The boundary between life and nonlife seems clear when one looks at unambiguous instances of each, but the closer one gets to the boundary, the greyer it becomes. A crisply defined criterion for life has been elusive, but perhaps one of the best candidates that has been offered so far is one based on the presence of causally effective information (Walker and Davies, 2013). We have linked the emergence of causally effective information to the emergence of semantic information, in turn linked to the emergence of coded replication. And while we have been able to focus on clear examples and models that are far removed from earlier or later times, the closer one attempts to focus on the transition itself, the greyer it too becomes.

The origin of life, the origin of semantic information, the origin of information processing or computation, and the origin of the explosion of diversity and complexity that flowered as a result – our story suggests that these are intimately linked, but we still have much to learn.

REFERENCES

Adami, C. 1995. Self-organized criticality in living systems. *Physics Letters A*, **1**(203), 29–32.

Bateson, G. 1972. *Steps to an ecology of mind: collected essays in anthropology, psychiatry, evolution, and epistemology*. University of Chicago Press.

Copley, S. D., Smith, E., and Morowitz, H. J. 2010. How life began: the emergence of sparse metabolic networks. *Journal of Cosmology*, **10**, 3345–3361.

Csete, M., and Doyle, J. 2004. Bow ties, metabolism and disease. *Trends in Biotechnology*, **22**(9), 446–450.

Czárán, T., Könnyű, B., and Szathmáry, E. 2015. Metabolically coupled replicator systems: overview of an RNA-world model concept of prebiotic evolution on mineral surfaces. *Journal of Theoretical Biology*, **381**, 39–54.

Deacon, T., and Sherman, J. 2008. The pattern which connects pleroma to creatura: the autocell bridge from physics to life. Pages 59–76 of *A legacy for living systems*. Springer.

Deacon, T. W. 2006. Reciprocal linkage between self-organizing processes is sufficient for self-reproduction and evolvability. *Biological Theory*, **1**(2), 136–149.

Eigen, M., and Schuster, P. 1978. The hypercycle. *Naturwissenschaften*, **65**(1), 7–41.

Friedlander, T., Mayo, A. E., Tlusty, T., and Alon, U. 2015. Evolution of bow-tie architectures in biology. *PLOS Comput Biol*, **11**(3), e1004055.

Gánti, T. 2003. *Chemoton theory: theory of living systems*. Springer Science & Business Media.

Gause, G. F., and Witt, A. A. 1935. Behavior of mixed populations and the problem of natural selection. *The American Naturalist*, **69**(725), 596–609.

Halley, J. D., and Winkler, D. A. 2008. Critical-like self-organization and natural selection: two facets of a single evolutionary process? *Biosystems*, **92**(?), 148–158.

Halley, J., Winkler, D. A., et al. 2008. Consistent concepts of self-organization and self-assembly. *Complexity*, **14**(2), 10–17.

Hoelzer, G. A., Smith, E., and Pepper, J. W. 2006. On the logical relationship between natural selection and self-organization. *Journal of Evolutionary Biology*, **19**(6), 1785–1794.

Hordijk, W. 2013. Autocatalytic sets: from the origin of life to the economy. *BioScience*, **63**(11), 877–881.

Hordijk, W., and Steel, M. 2004. Detecting autocatalytic, self-sustaining sets in chemical reaction systems. *Journal of Theoretical Biology*, **227**(4), 451–461.

Kauffman, S., and Clayton, P. 2006. On emergence, agency, and organization. *Biology and Philosophy*, **21**(4), 501–521.

Kauffman, S. A. 1986. Autocatalytic sets of proteins. *Journal of Theoretical Biology*, **119**(1), 1–24.

Kauffman, S. A. 2002. *Investigations*. Oxford University Press.

Koonin, E. V., and Novozhilov, A. S. 2009. Origin and evolution of the genetic code: the universal enigma. *IUBMB Life*, **61**(2), 99–111.

Korzbyski, A. 1933. *Science and sanity: an introduction to non-Aristotelian systems and general semantics*. Institute of General Semantics.

Kun, A., Szilágyi, A., Könnyű, B., Boza, G., Zachar, I., and Szathmáry, E. 2015. The dynamics of the RNA world: insights and challenges. *Annals of the New York Academy of Sciences*, **1341**(1), 75–95.

Landauer, R. 1961. Irreversibility and heat generation in the computing process. *IBM Journal of Research and Development*, **5**(3), 183–191.

Lincoln, T. A., and Joyce, G. F. 2009. Self-sustained replication of an RNA enzyme. *Science*, **323**(5918), 1229–1232.

Maynard Smith, J., et al. 1995. The major transitions of evolution. *Evolution*, **49**(6), 1302–1306.

Morowitz, H., and Smith, E. 2007. Energy flow and the organization of life. *Complexity*, **13**(1), 51–59.

Mossel, E., and Steel, M. 2005. Random biochemical networks: the probability of self-sustaining autocatalysis. *Journal of Theoretical Biology*, **233**(3), 327–336.

Pascal, R., Pross, A., and Sutherland, J. D. 2013. Towards an evolutionary theory of the origin of life based on kinetics and thermodynamics. *Open Biology*, **3**(11), 130156.

Ridley, M. 2000. *Mendel's demon: gene justice and the complexity of life*. Weidenfeld & Nicolson.

Shannon, C. E. 1948. A mathematical theory of communication. *Bell System technical Journal* 27: 379-423 and 623–656. *Mathematical Reviews (MathSciNet): MR10, 133e*.

Sievers, D., and Von Kiedrowski, G. 1994. Self-replication of complementary nucleotide-based oligomers. *Nature*, **369**(6477), 221–224.

Simon, H. A. 1991. *The architecture of complexity*. Springer.

Toyabe, S., Sagawa, T., Ueda, M., Muneyuki, E., and Sano, M. 2010. Experimental demonstration of information-to-energy conversion and validation of the generalized Jarzynski equality. *Nature Physics*, **6**(12), 988–992.

Trefil, J., Morowitz, H. J., and Smith, E. 2009. A case is made for the descent of electrons. *Am Sci*, **97**, 206–213.

Tribus, M., and McIrvine, E. C. 1971. Energy and information. *Scientific American*, **225**(3), 179–188.

Vaidya, N., Manapat, M. L., Chen, I. A., Xulvi-Brunet, R., Hayden, E. J., and Lehman, N. 2012. Spontaneous network formation among cooperative RNA replicators. *Nature*, **491**(7422), 72–77.

Vasas, V., Fernando, C., Szilágyi, A., Zachár, I., Santos, M., and Szathmáry, E. 2015. Primordial evolvability: impasses and challenges. *Journal of Theoretical Biology*, **381**, 29–38.

Vetsigian, K., Woese, C., and Goldenfeld, N. 2006. Collective evolution and the genetic code. *Proceedings of the National Academy of Sciences*, **103**(28), 10696–10701.

Von Weizsäcker, C. F. F. 1980. *The unity of nature*. Farrar Straus Giroux.

Walker, S. I., and Davies, P. C. W. 2013. The algorithmic origins of life. *Journal of the Royal Society Interface*, **10**(79), 20120869.

5 On the Emerging Codes for Chemical Evolution

Jillian E. Smith-Carpenter, Sha Li, Jay T. Goodwin, Anil K. Mehta, and David G. Lynn

Life has been described as information flowing in molecular streams (Dawkins, 1996). Our growing understanding of the impact of horizontal gene transfer on evolutionary dynamics reinforces this fluid-like flow of molecular information (Joyce, 2002). The diversity of nucleic acid sequences, those known and yet to be characterized across Earth's varied environments, along with the vast repertoire of catalytic and structural proteins, presents as more of a dynamic molecular river than a tree of life. These informational biopolymers function as a mutualistic union so universal as to have been termed the Central Dogma (Crick, 1958). It is the distinct folding dynamics-the digital-like base pairing dominating nucleic acids, and the environmentally responsive and diverse range of analog-like interactions dictating protein folding (Goodwin et al., 2012)-that provides the basis for the mutualism. The intertwined functioning of these analog and digital forms of information (Goodwin et al., 2012) unified within diverse chemical networks is heralded as the Darwinian threshold of cellular life (Woese, 2002).

The discovery of prion diseases (Chien et al., 2004; Jablonka and Raz, 2009; Paravastu et al., 2008) introduced the paradigm of protein templates that propagate conformational information, suggesting a new context for Darwinian evolution. When taking both protein and nucleic acid moelcular evolution into consideration (Cairns-Smith, 1966; Joyce, 2002), the conceptual framework for chemical evolution can be generalized into three orthogonal dimensions as shown in Figure 5.1 (Goodwin et al., 2014). The 1st dimension manifests structural order through covalent polymerization reactions and includes chain length, sequence, and linkage chemistry inherent to a

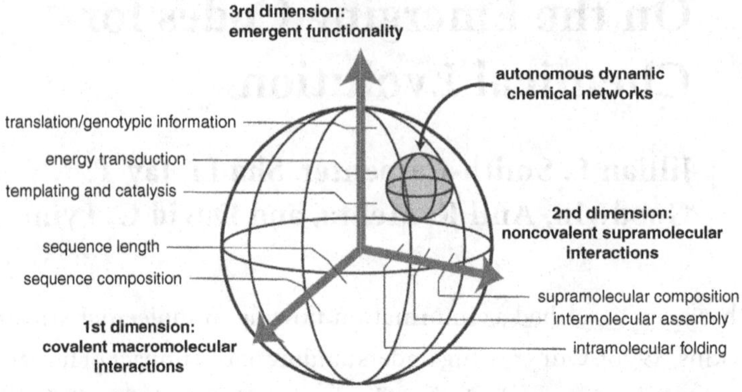

FIG. 5.1 Mapping the space for chemical evolution. 1st dimension: oligomer diversity stems from differences in sequence, length, and covalent connectivity; 2nd dimension: physical supramolecular organization extends complexity and informational content of the dynamic chemical networks; 3rd dimension: functional properties emerge from macro- and supramolecular assemblies that drive selection through feedback control of the first two dimensions. The space suitable for sustaining a coherent and progressive growth of molecular information would contain autonomous dynamic chemical networks.

dynamic chemical network. The 2nd dimension extends the order in dynamic conformational networks through noncovalent interactions of the polymers. This dimension includes intramolecular and intermolecular forces, from macromolecular folding to supramolecular assembly to multicomponent quaternary structure. Folding in this 2nd dimension certainly depends on the primary polymer sequence, and the folding/assembly diversity yields an additional set of environmentally constrained supramolecular folding codes. For example, double-stranded DNA assemblies are dominated by the rules of complementary base pairing, while the self-propagating conformations of prions are based on additional noncovalent, environmentally-dependent interactions. In the 3rd dimension, the supramolecular polymer assemblies that obtain new functions, such as templating, catalysis, energy transduction, and translating molecular information inherent to higher-order surfaces, provide constructive feedback to growth in the first two dimensions. We propose that the space mapped by these dimensions encompasses regions which access autonomous progressive growth of increasingly higher functional order.

Consider the ribosome as a mutualistic complex of more than 50 protein and RNA subunits capable of translating molecular information from one biopolymer into another (Woese, 2002). This mutualism, evolving from RNA/protein dynamic networks to exploit the digital and analog features of each polymer, traces the evolutionary trajectory for a critical Darwinian threshold of cellular life (Dawkins, 1996). Here we discuss peptide-based dynamic networks as potentially atavistic and alternative platforms for chemical evolution. Can the levels of order attained in this system be conceptually and qualitatively mapped on the three orthogonal dimensions? If so, we will have an alternative framework to explore potential spaces accessible to autonomous dynamic chemical networks for materials evolution and alternative biochemistries.

THE 1ST DIMENSION: CHEMICAL NETWORKS

Dynamic mutualisms drive ribonucleoprotein-directed information flow in the living cell. RNA is ferried at every stage, from transcription to translation, by proteins (Moore, 2005), and the ribosome, the protein/RNA organelle that serves as the molecular digital-to-analog converter of life (Goodwin et al., 2012), synthetically conjoins these two biopolymer networks. Activated nucleotides, as digitally-encoded molecular information systems, have been shown to polymerize on DNA (Bruick et al., 1997) and inorganic surfaces (Ferris and Hagan, 1986; Ferris et al., 1988); however, maintaining a dynamic system has largely remained inaccessible in the absence of sophisticated protein catalysts. The first dynamic polymerization network developed to take advantage of the complementary base pairing of nucleoside scaffolds (Figure 5.2A) altered the nature of the ligation reaction to overcome the kinetic barriers for phosphodiester polymerization (Goodwin and Lynn, 1992). Imine condensation (Figure 5.2B) was directed by complementary nucleoside base-paring via reductive amidation in a novel step-growth mechanism (Li et al., 2002,

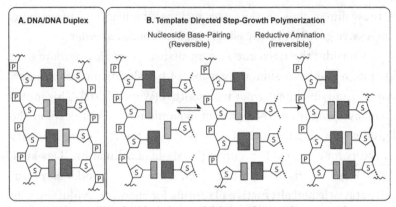

FIG. 5.2 DNA double helix model (A) with complementary base pairs, the backbone sugar (labeled S) and the phosphodiester (labeled P) and (B) the nucleoside amine polymers where the ligating amine can be reduced through reversible template-directed step-growth polymerization (Li et al., 2011).

2011). While this reductive polymerization transfers information to synthetic polymers to achieve both chain length and sequence-specific products with high fidelity, the reaction did not maintain the reversible dynamic chemical network, and therefore it limited further progressive growth.

Coiled-coil peptide networks, representing analog molecular information networks, have also been explored as templates (Rubinov et al., 2012) for mutualistic dynamic networks and have been extended to cross-coupling processes (Ashkenasy et al., 2004). Chemical systems generally harness energy input from surrounding environment in order to create order through far-from-equilibrium processes (Leunissen et al., 2005; Mattia and Otto, 2015; Tagliazucchi et al., 2014; Whitesides and Grzybowski, 2002). One of the more effective dynamic chemical networks is the aryl disulfide peptides (Carnall et al., 2010) for which template-directed assembly processes are responsive to shear force. These examples establish alternative strategies to carry and replicate molecular information (Goodwin et al., 2014). The challenge is to enrich the complexity of these chemical polymerization methods to shepherd molecular information capable of supporting progressive open-ended growth.

THE 2ND DIMENSION: MACROMOLECULAR AND
SUPRAMOLECULAR ORGANIZATION

The molecular recognition code that can direct macromolecular
growth in the 1st dimension, can also contribute to folding and
supramolecular assembly in the 2nd dimension. Cross-β assemblies
(Astbury et al., 1935; Eanes and Glenner, 1968; Geddes et al., 1968;
Parker and Rudall, 1957) are a supramolecular phase in which β-
sheets, an array of amide backbone hydrogen-bonded peptides, are
stacked and stabilized through amino acid side-chain interactions
(Geddes et al., 1968). In self-propagating cross-β assemblies, short
self-templating cross-β peptides (Childers et al., 2012; Goodwin et al.,
2012, 2014) are being used to uncover the amino acid diversity and
conformational flexibility that preferentially direct the accessible
folding landscapes. This understanding largely comes about from
elucidating peptide self-assembling pathways for amyloid diseases.
For example, both computational models (Buell et al., 2013; Tarus
et al., 2005; Yun et al., 2007) and experimental studies (Konno, 2001;
Raman et al., 2005; Shammas et al., 2011) support the importance
of nonspecific electrostatic interactions during the initial peptide
aggregation and ordering. These interactions between charged amino
acids, although nonspecific, underpin the molecular code emerging
along the 2nd dimension.

At low pH, the functionalized peptide amphiphile Ac-KLVFFAE-
NH_2, the nucleating core of the Aβ- peptide of Alzheimer's disease
with an acetyl (Ac-) and amide (-NH_2) capped N- and C-termini,
respectively, undergoes initial hydrophobic collapse into disordered
aggregate phases (Figure 5.3) (Childers et al., 2012). The paracrystalline peptide ordering within the particle can be visualized by electron microscopy and fluorescent probes, and is consistent with the
particle being an intermediate along the assembly pathway (Anthony
et al., 2012, 2014; Liang et al., 2010; Liang et al., 2014). In these
low dialectic dehydrated particles where electrostatic interactions
should dominate, antiparallel strands position the charged lysine of
adjacent peptides on opposite sides of the β-sheet. This idea was
tested with an E to Q substitution, Ac-KLVFFAQ-NH_2, capitalizing

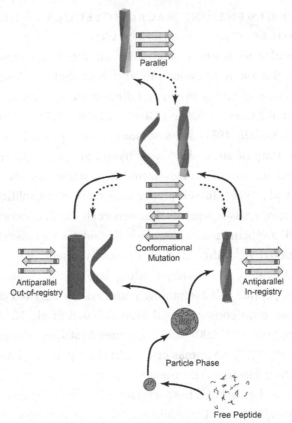

FIG. 5.3 Progressive peptide self-assembly. Intermolecular molten glob-ule particles assemble immediately on peptide dissolution. Antiparallel stranded sheets emerge from the particle phase, and during propagation, a conformational transition to the more thermodynamically stable parallel strand orientation occurs with the peptide Ac-KLVFFAQ-NH$_2$, making possible a stabilizing side-chain H-bonded array known as glutamine Q-tracks.

on the hydrogen bond donor and acceptor functionalities of the glutamine (Q) amide side chain. The backbone amide normal stretch-ing modes of β-sheet (Li et al., 2014; Smith et al., 2015) indicated that Ac-KLVFFAQ-NH$_2$ assemblies emerge from the particle phase with antiparallel relative strand orientations (Smith et al., 2015). However, during this propagation out from the particle phase, the assemblies undergo an autocatalytic transformation to a parallel

β-sheet fiber (Li et al., 2014; Smith et al., 2015) (Figure 5.3). Antiparallel strand orientation in the particle is selected kinetically, while the formation of an extended H-bonded array of glutamine side chains, Q-tracks (Smith et al., 2015), thermodynamically guides supramolecular assembly (Figures 5.3 and 5.5C). This transition, thought to proceed via conformational mutation during strand addition at the ends of the growing fiber (Li et al., 2014; Smith et al., 2015), highlights the context-dependence of the assembly pathway and the potential for staged growth. This example reveals some of the elements of progressive growth of molecular order, like a biological phylogeny, where the formation of new morphologies depends on open-ended transitions through antecedent structures (Figure 5.3).

Aside from temporal dependence on peptide self-assembly along the 2nd dimension axis, the environment – e.g., pH, temperature, solvent, and small molecule cofactors (Childers et al., 2012; Dong et al., 2007; Mehta et al., 2008) – also impacts the folding landscape. Ac-KLVFFAE-NH$_2$ assembles as homogenous nanotubes in acidic 40% acetonitrile and 20–40% methanol – stronger H-bond acceptor/donor but with similar dielectric (Figure 5.4, organic solvent phase). When the methanol is increased to 40%, spherical particles dominate the assemblies. In 60% methanol, only spherical particles are seen (Childers et al., 2012) and arrest of paracrystalline growth is easily detected in the particles (Childers et al., 2012). More dramatically, stoichiometric Zn(II) added to the extended cross-β segment H-HHQALVFFA-NH$_2$ prior to assembly reduces the lag phase and propagates a morphological transition from fibers to ribbons and nanotubes (Figure 5.4, metal association phase) (Dong et al., 2006b, 2007). These peptides assemble as antiparallel β-strands with the HH-dyad bridging the β-sheets. Both Zn(II) and Cu(II) alter assembly kinetics and morphology (Dong et al., 2006b, 2007). It may now be possible to extend such responsive networks synthetically, like a fluctuating metabolism, and use the network to select for new functional materials. Success in the creation of such dynamic networks will depend on an understanding of assembly energetics.

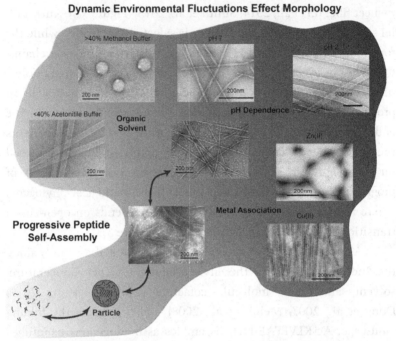

FIG. 5.4 Mapping the cross-β conformational space available to peptide congeners of the cross-β nucleating core, H-LVFFA-NH$_2$. Transmission electron microscopy images of Ac-KLVFFAQ-NH$_2$ reinforce the progressive peptide self-assembly pathway described in Figure 5.3 (Li et al., 2014; Smith et al., 2015). The shading highlights the dynamic environmentally driven transitions between the reversible conformational forms of the assemblies. Specific examples of metal associations to H-HHQALVFFA-NH$_2$ Dong et al., 2006b, 2007 and the effect of organic solvent composition and pH on the final supramolecular peptide assembly of Ac-KLVFFAE-NH$_2$ Liang et al., 2008a show the diversity of structures accessible to peptides and can be an experimental platform to further explore environmental constraints on the emerging code.

We now have a predictable folding code, analogous to base-pair associations of nucleic acid assembly, illustrated by several enticing examples. The hepta-peptide Ac-KLVFFAE-NH$_2$, at neutral pH where the glutamic acid (E) residue on the C-terminus is negatively charged, form complementary K-E salt bridges to stabilize antiparallel in-register β-sheet assemblies with positively charged lysine (K) residues (Liang et al., 2008a; Mehta et al., 2008) (Figure 5.4, pH-dependent

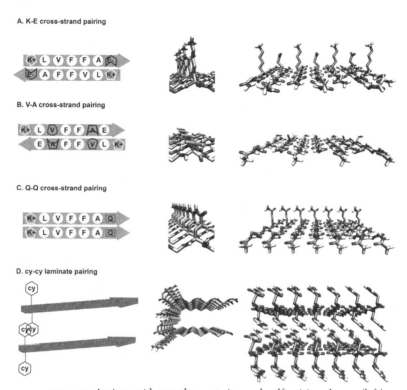

A. K-E cross-strand pairing

B. V-A cross-strand pairing

C. Q-Q cross-strand pairing

D. cy-cy laminate pairing

FIG. 5.5 Amino acid complementarity and self-pairing drawn (left) as cartoons with each arrow representing a β-strand and (middle, right) models highlighting the H-bonded peptide backbone and sidechains. (A) Salt bridge between lysine and glutamic acid (K-E) (Mehta et al., 2008). (B) Packing of bulky β-branched valine (V) against the smaller alanine (A) residue (Liu et al., 2008). (C) Side-chain hydrogen bonding via glutamine (Q) as parallel in-register strands (Liang et al., 2014). (D) Cytosine–cytosine laminate pairing (Liu et al., 2008) highlighting the cross-β architecture.

phases, and Figure 5.5A). At acidic pH the salt bridge is weakened and complementary steric pairing between valine and alanine (V-A) dominate through steric packing interactions to give antiparallel, out-of-register register β-sheet assemblies (Figure 5.5B) (Mehta et al., 2008). Consistent with this proposal, E to leucine (L), Ac-KLVFFAL-NH$_2$, removes the salt bridge (Liang et al., 2008b) and allows steric cross-strand pairing between the valine and the alanine to direct the out-of-register orientation independent of pH (Mehta et al., 2008). When the valine is substituted with tert-leucine (tL), the more sterically demanding t-butyl side chain in Ac-KL(tL)FFAE-NH$_2$ overwhelms the K/E salt bridge at neutral pH to give out-of-register nanotubes (Liu et al., 2008). Hydrogen-bonded cross-strand pairing interactions offer directional control between strands, and the E22Q substitution, Ac-KLVFFAQ-NH$_2$, mentioned above, assembles as parallel strands stabilized by H-bonded Q-tracks stretching along the sheet (Figure 5.5C) (Liang et al., 2014). Cross-sheet pairing interactions have been constructed with the nucleic acid derivative β-(cytosine-1-yl)-alanine to construct single-walled peptide nanotubes (Liu et al., 2008) (Figure 5.5D).

This expanding toolkit for cross-β assemblies has been exploited for materials synthesis, including lipid/peptide (Ni et al., 2012), fluorophore/peptide (Liang et al., 2008b), and biotin/peptide chimeric materials (Scarfi et al., 1997), and the subtle information encoded in the sequence space can now be used to create the functional feedbacks for the progressive growth of higher supramolecular order.

THE 3RD DIMENSION: CONVERGENT FUNCTIONS

A living cell maintains a dynamic tension between supramolecular assembly and covalent chemical synthesis to achieve convergent growth of new function. We have represented this tension as the confluence of a dynamic chemical network of polymer length and composition (1st dimension) with supramolecular assembly (2nd dimension) in Figure 5.1. The 3rd dimension is the convergence of these dimensions to achieve new functional order. Given that neurodegenerative protein-misfolding diseases, which emerge from self-propagating cross-β assemblies, propagate functional assemblies that compromise the actions of neurons, we sought to extend the surfaces of these assemblies to create new functions.

Biophysical characterization of the Ac-KLVFFAL-NH$_2$ nanotubes reveals a peptide bilayer (Childers et al., 2010) with

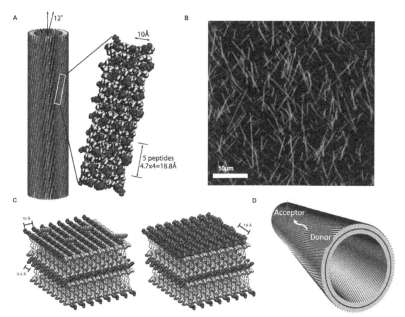

FIG. 5.6 (A) Model of Congo red (CR) bound to Ac-KLVFFAL-NH₂ nanotube surface with charged lysines and hydrophobic leucines (Childers et al., 2010). Cross-β assemblies have 4.7 Å in the H-bonding direction and 10 Å in the β-sheet stacking direction, resulting in a 1 nm × 1 nm lysine grid on the solvent-exposed surface. (B) Fluorescence imaging of Rh-LVFFAE-NH₂/Ac-KLVFFAE-NH₂ ratios at 1:75.39. (C) Molecular models of Rh-LVFFAE-NH₂ tubes and Alexa 555 binding to Ac-KLVFFAE-NH₂ tube surfaces (Childers et al., 2009a; Liang et al., 2008b). Lysine residues, rhodamine, and Alexa molecules are drawn with space filling representation. Peptide backbone drawn as sticks and all other side chains are hidden for clarity. (D) Cartoon describing the observed FRET across the surface.

4-nm-thick walls that helically coil (Childers et al., 2010; Dong et al., 2006a) to give a tube diameter of 32 ± 5 nm. The peptide strands pack in antiparallel out-of-register sheets (Mehta et al., 2008) with positively charged lysine residues exposed on the surface, as shown in Figure 5.6A. The leucine-lined hydrophobic groove binds the amyloid histochemical dye Congo red (CR) between the cross-β sheets. Electron diffraction and linear dichroism analyses place CR oriented parallel to the amyloid long axis and packed in both J- and

H-aggregates along the laminate grooves, as shown in Figure 5.6A (Childers et al., 2009b) with the peptide N-terminal lysine residues interacting with the negative charged sulfates of CR.

These lysine residues, which are ordered and densely packed on the surface at a two-dimensional concentration of almost 3 M, bind citrate-coated negatively charged gold nanoparticles (Li et al., 2014). When the lysine residue is replaced with phosphotyrosine (pY), the negatively charged Ac-pYLVFFAL-NH$_2$ assemblies are morphologically similar and bind positively charged gold nanoparticles (Li et al., 2014). These supramolecular surfaces also bind proteins as seen with the strongly basic histone H1. This basic protein, which accumulates in the cytoplasm of neurons and astrocytes in areas impacted by neurodegenerative disease, binds selectively to the negatively charged nanotubes (Li et al., 2014). Functional extensions of these assemblies also include cross-β amyloid assemblies that have been loaded with cytochrome C, which show a 10^2 increase in peroxidase activity (Kapil et al., 2015).

These extended surfaces have been used to organize functional small molecules. Fluorescence lifetime image microscopy visualized nanotube growth with the rhodamine 110–containing probe peptide, Rh-LVFFAE-NH$_2$ (Figure 5.6B). These probes provided direct mechanistic insight into not only nanotube assembly (Anthony et al., 2011, 2014; Liang et al., 2010) but also the interconversion of physical and chemical energy. Based on the insights gained from CR binding, the fluorescent nanotubes were shown to readily bind the sulfonated Alexa 555 dye (Figure 5.6C). Fluorescence resonance energy transfer (FRET) (Figure 5.6D) across the nanotube surfaces from rhodamine and Alexa 555 demonstrates an early first step in functional light energy capture and transfer (Liang et al., 2008b).

COHERENCE AND FEEDBACK

Much of our understanding of the peptide self-assembly, propagation, and conformational evolution has emerged from the study of disease-relevant peptides. The concept of a dynamic range of accessible cross-β folding phases, as shown in Figure 5.4, is presented more as a pond than a flowing river of information, and the implied dynamics may be critical to our general understanding of protein misfolding diseases. The entire conformational space is expected to be dynamically responsive to environmental input. As the cellular environment of an aging tissue changes, e.g., metal ions, metabolites, and other cellular debris in deposited plaques (Bai et al., 2013; Butterfield and Lashuel, 2010; Duce et al., 2006), the conformational network has the potential to continuously adjust, adapt, and evolve with disease progression.

In a more general sense, the proposed framework is the culmination of dynamic macromolecular chemical networks with supramolecular folding and assembly to yield new functional forms. While the axes of this dimensional system are not yet quantitatively defined, they do allow for focused experiments that address the progressive growth of molecular order for chemical evolution. As coherent behaviour emerges from designed polymer networks that generate functional feedback (De Duve, 1995), it should be possible to achieve progressive growth of information within these dynamic chemical and physical networks. An evaluation of the values for the increasing complexity in all three dimensions could become predictive for the progressive growth of functional order and the creation of new evolvable materials.

REFERENCES

Anthony, N. R., Lynn, D. G., and Berland, K. 2011. Amyloid nucleation: evidence for nucleating cores within unstructured protein aggregates. *Biophysical Journal*, **100**(3), 201a–202a.

Anthony, N. R., Lynn, D. G., and Berland, K. M. 2012. The role of interfaces in the nucleation of amyloid nanotubes. *Biophysical Journal*, **102**(3), 442a.

Anthony, N. R., Mehta, A. K., Lynn, D. G., and Berland, K. M. 2014. Mapping amyloid-β (16-22) nucleation pathways using fluorescence lifetime imaging microscopy. *Soft Matter*, **10**(23), 4162–4172.

Ashkenasy, G., Jagasia, R., Yadav, M., and Ghadiri, R. M. 2004. Design of a directed molecular network. *Proceedings of the National Academy of Sciences of the United States of America*, **101**(30), 10872–10877.

Astbury, W. T., Dickinson, S., and Bailey, K. 1935. The X-ray interpretation of denaturation and the structure of the seed globulins. *Biochemical Journal*, **29**(10), 2351.

Bai, B., Hales, C. M., Chen, P., Gozal, Y., Dammer, E. B., Fritz, J. J., Wang, X., Xia, Q., Duong, D. M., Street, C., et al. 2013. U1 small nuclear ribonucleoprotein complex and RNA splicing alterations in Alzheimer's disease. *Proceedings of the National Academy of Sciences*, **110**(41), 16562–16567.

Bruick, R. K., Koppitz, M., Joyce, G. F., and Orgel, L. E. 1997. A simple procedure for constructing 5-amino-terminated oligodeoxynucleotides in aqueous solution. *Nucleic Acids Research*, **25**(6), 1309–1310.

Buell, A. K., Hung, P., Salvatella, X., Welland, M. E., Dobson, C. M., and Knowles, T. P. J. 2013. Electrostatic effects in filamentous protein aggregation. *Biophysical Journal*, **104**(5), 1116–1126.

Butterfield, S. M., and Lashuel, H. A. 2010. Amyloidogenic protein–membrane interactions: mechanistic insight from model systems. *Angewandte Chemie International Edition*, **49**(33), 5628–5654.

Cairns-Smith, G. A. 1966. The origin of life and the nature of the primitive gene. *Journal of Theoretical Biology*, **10**(1), 53–88.

Carnall, J. M. A, Waudby, C. A., Belenguer, A. M., Stuart, M. C. A., Peyralans, J. J. P., and Otto, S. 2010. Mechanosensitive self-replication driven by self-organization. *Science*, **327**(5972), 1502–1506.

Chien, P., Weissman, J. S., and DePace, A. H. 2004. Emerging principles of conformation-based prion inheritance. *Annual Review of Biochemistry*, **73**(1), 617–656.

Childers, S. W., Ni, R., Mehta, A. K., and Lynn, D. G. 2009a. Peptide membranes in chemical evolution. *Current Opinion in Chemical Biology*, **13**(5), 652–659.

Childers, S. W., Mehta, A. K., Lu, K., and Lynn, D. G. 2009b. Templating molecular arrays in amyloids cross-β grooves. *Journal of the American Chemical Society*, **131**(29), 10165–10172.

Childers, S. W., Mehta, A. K., Ni, R., Taylor, J. V., and Lynn, D. G. 2010. Peptides organized as bilayer membranes. *Angewandte Chemie International Edition*, **49**(24), 4104–4107.

Childers, S. W., Anthony, N. R., Mehta, A. K., Berland, K. M., and Lynn, D. G. 2012. Phase networks of cross-β peptide assemblies. *Langmuir*, **28**(15), 6386–6395.

Crick, F. H. C. 1958. On protein synthesis. In *The biological replication of macromolecules*, Symposia of the Society for Experimental Biology, vol. 12. Cambridge University Press, pp. 138–163.

Dawkins, R. 1996. *River out of Eden: A Darwinian view of life*. Basic Books.

De Duve, C. 1995. *Vital dust: life as a cosmic imperative*. Basic Books.

Dong, J., Lu, K., Lakdawala, A.and Mehta, A. K., and Lynn, D. G. 2006a. Controlling amyloid growth in multiple dimensions. *Amyloid*, **13**(4), 206–215.

Dong, J., Shokes, J. E., Scott, R. A., and Lynn, D. G. 2006b. Modulating amyloid self-assembly and fibril morphology with Zn (II). *Journal of the American Chemical Society*, **128**(11), 3540–3542.

Dong, J., Canfield, J. M., Mehta, A. K., Shokes, J. E., Tian, B., Childers, S. W., Simmons, J. A., Mao, Z., Scott, R. A., Warncke, K., et al. 2007. Engineering metal ion coordination to regulate amyloid fibril assembly and toxicity. *Proceedings of the National Academy of Sciences*, **104**(33), 13313–13318.

Duce, J. A., Smith, D. P., Blake, R. E., Crouch, P. J., Li, Q., Masters, C. L., and Trounce, I.A. 2006. Linker histone H1 binds to disease associated amyloid-like fibrils. *Journal of Molecular Biology*, **361**(3), 493–505.

Eanes, E. D., and Glenner, G. G. 1968. X-ray diffraction studies on amyloid filaments. *Journal of Histochemistry & Cytochemistry*, **16**(11), 673–677.

Ferris, J. P., and Hagan, W. J. Jr. 1986. The adsorption and reaction of adenine nucleotides on montmorillonite. *Origins of Life and Evolution of the Biosphere*, **17**(1), 69–84.

Ferris, J. P., Huang, C., and Hagan, W. J. Jr. 1988. Montmorillonite: a multifunctional mineral catalyst for the prebiological formation of phosphate esters. *Origins of Life and Evolution of the Biosphere*, **18**(1–2), 121–133.

Geddes, A. J., Parker, K. D., Atkins, E. D. T., and Beighton, E. 1968. Cross-$\beta\check{z}$ conformation in proteins. *Journal of Molecular Biology*, **32**(2), 343–358.

Goodwin, J. T., and Lynn, D. G. 1992. Template-directed synthesis: use of a reversible reaction. *Journal of the American Chemical Society*, **114**(23), 9197–9198.

Goodwin, J. T., Mehta, A. K., and Lynn, D. G. 2012. Digital and analog chemical evolution. *Accounts of Chemical Research*, **45**(12), 2189–2199.

Goodwin, J. T. W., Walker, S. I., Amin, S., Armbrust, G., Burrows, C. J., and Lynn, D.G. 2014. *Alternative chemistries of life: empirical approaches.*

Jablonka, E., and Raz, G. 2009. Transgenerational epigenetic inheritance: prevalence, mechanisms, and implications for the study of heredity and evolution. *Quarterly Review of Biology*, **84**(2), 131–176.

Joyce, G. F. 2002. The antiquity of RNA-based evolution. *Nature*, **418**(6894), 214–221.

Kapil, N., Singh, A., and Das, D. 2015. Cross-β amyloid nanohybrids loaded with cytochrome C exhibit superactivity in organic solvents. *Angewandte Chemie International Edition*, **54**(22), 6492–6495.

Konno, T. 2001. Amyloid-induced aggregation and precipitation of soluble proteins: an electrostatic contribution of the Alzheimer's β (25-35) amyloid fibril. *Biochemistry*, **40**(7), 2148–2154.

Leunissen, M. E., Christova, C. G., Hynninen, A., Royall, P. C., Campbell, A. I., Imhof, A., Dijkstra, M., Van Roij, R., and Van Blaaderen, A. 2005. Ionic colloidal crystals of oppositely charged particles. *Nature*, **437**(7056), 235–240.

Li, S., Sidorov, A. N., Mehta, A. K., Bisignano, A. J., Das, D., Childers, S. W., Schuler, E., Jiang, Z., Orlando, T. M., Berland, K., et al. 2014. Neurofibrillar tangle surrogates: histone H1 binding to patterned phosphotyrosine peptide nanotubes. *Biochemistry*, **53**(26), 4225–4227.

Li, X., Zhan, Z. J., Knipe, R., and Lynn, D. G. 2002. DNA-catalyzed polymerization. *Journal of the American Chemical Society*, **124**(5), 746–747.

Li, X., Hernandez, A. F., Grover, M. A., Hud, N. V., and Lynn, D. G. 2011. Step-growth control in template-directed polymerization. *Heterocycles*, **82**(2), 1477.

Liang, C., Ni, R., Smith, J. E., Childers, S. W., Mehta, A. K., and Lynn, D. G. 2014. Kinetic intermediates in amyloid assembly. *Journal of the American Chemical Society*, **136**(43), 15146–15149.

Liang, Y., Pingali, S. V., Jogalekar, A. S., Snyder, J. P., Thiyagarajan, P., and Lynn, D. G. 2008a. Cross-strand pairing and amyloid assembly. *Biochemistry*, **47**(38), 10018–10026.

Liang, Y., Guo, P., Pingali, S. V., Pabit, S., Thiyagarajan, P., Berland, K. M., and Lynn, D. G. 2008b. Light harvesting antenna on an amyloid scaffold. *Chemical Communications*, 6522–6524.

Liang, Y., Lynn, D. G., and Berland, K. M. 2010. Direct observation of nucleation and growth in amyloid self-assembly. *Journal of the American Chemical Society*, **132**(18), 6306–6308.

Liu, P., Ni, R., Mehta, A. K., Childers, S. W., Lakdawala, A., Pingali, S. V., Thiyagarajan, P., and Lynn, D. G. 2008. Nucleobase-directed amyloid nanotube assembly. *Journal of the American Chemical Society*, **130**(50), 16867–16869.

Mattia, E., and Otto, S. 2015. Supramolecular systems chemistry. *Nature Nanotechnology*, **10**(2), 111–119.

Mehta, A. K., Lu, K., Childers, S. W., Liang, Y., Dublin, S. N., Dong, J., Snyder, J. P., Pingali, S. V., Thiyagarajan, P., and Lynn, D. G. 2008. Facial symmetry in protein self-assembly. *Journal of the American Chemical Society*, **130**(30), 9829–9835.

Moore, M. J. 2005. From birth to death: the complex lives of eukaryotic mRNAs. *Science*, **309**(5740), 1514–1518.

Ni, R., Childers, S. W., Hardcastle, K. I., Mehta, A.K., and Lynn, D. G. 2012. Remodeling cross-β nanotube surfaces with peptide/lipid chimeras. *Angewandte Chemie International Edition*, **51**(27), 6635–6638.

Paravastu, A. K., Leapman, R. D., Yau, W., and Tycko, R. 2008. Molecular structural basis for polymorphism in Alzheimer's β-amyloid fibrils. *Proceedings of the National Academy of Sciences*, **105**(47), 18349–18354.

Parker, K. D., and Rudall, K. M. 1957. The silk of the egg-stalk of the green lace-wing fly: structure of the silk of *Chrysopa* egg-stalks. *Nature*, **179**, 905–906.

Raman, B., Chatani, E., Kihara, M., Ban, T., Sakai, M., Hasegawa, K., Naiki, H., Rao, M. C. H., and Goto, Y. 2005. Critical balance of electrostatic and hydrophobic interactions is required for β2-microglobulin amyloid fibril growth and stability. *Biochemistry*, **44**(4), 1288–1299.

Rubinov, B., Wagner, N., Matmor, M., Regev, O., Ashkenasy, N., and Ashkenasy, G. 2012. Transient fibril structures facilitating nonenzymatic self-replication. *ACS Nano*, **6**(9), 7893–7901.

Scarfi, S., Gasparini, A., Damonte, G., and Benatti, U. 1997. Synthesis, uptake, and intracellular metabolism of a hydrophobic tetrapeptide-peptide nucleic acid (PNA)–biotin molecule. *Biochemical and Biophysical Research Communications*, **236**(2), 323–326.

Shammas, S. L., Knowles, T. P. J., Baldwin, A. J., MacPhee, C. E., Welland, M. E., Dobson, C. M., and Devlin, G. L. 2011. Perturbation of the stability of amyloid fibrils through alteration of electrostatic interactions. *Biophysical Journal*, **100**(11), 2783–2791.

Smith, J. E., Liang, C., Tseng, M., Li, N., Li, S., Mowles, A. K., Mehta, A. K., and Lynn, D. G. 2015. Defining the dynamic conformational networks of cross-β peptide assembly. *Israel Journal of Chemistry*, **55**(6–7), 763–769.

Tagliazucchi, M., Weiss, E. A., and Szleifer, I. 2014. Dissipative self-assembly of particles interacting through time-oscillatory potentials. *Proceedings of the National Academy of Sciences*, **111**(27), 9751–9756.

Tarus, B., Straub, J. E., and Thirumalai, D. 2005. Probing the initial stage of aggregation of the Aβ 10-35-protein: assessing the propensity for peptide dimerization. *Journal of Molecular Biology*, **345**(5), 1141–1156.

Whitesides, G. M., and Grzybowski, B. 2002. Self-assembly at all scales. *Science*, **295**(5564), 2418–2421.

Woese, C. R. 2002. On the evolution of cells. *Proceedings of the National Academy of Sciences*, **99**(13), 8742–8747.

Yun, S., Urbanc, B., Cruz, L., Bitan, G., Teplow, D. B., and Stanley, E. H. 2007. Role of electrostatic interactions in amyloid β-protein (Aβ) oligomer formation: a discrete molecular dynamics study. *Biophysical Journal*, **92**(11), 4064–4077.

6 Digital and Analogue Information in Organisms

Denis Noble

Organisms reproduce themselves, which means that their structures and processes must be copied to pass on to later generations. The question addressed in this chapter is whether they do this by encoding everything in digital format or whether analogue information is also important. The complete genome sequence of an organism is often compared to the digital information in a computer program or data file. Strong versions of biological reductionism also assume that this information is sufficient to define the organism and its development. This idea, popularised today in the idea of a genetics program, has its origins in the mechanistic philosophy of René Descartes. In his treatise on the formation of the fetus in 1664, he wrote:

> If one had a proper knowledge of all the parts of the semen of some
> species of animal in particular, for example of man, one might be
> able to deduce the whole form and configuration of each of its
> members from this alone, by means of entirely mathematical and
> certain arguments, the complete figure and the conformation of
> its members. (*De la formation du fetus*, 1664, para LXVI, p. 146)

This statement can be seen to foreshadow the Central Dogma of molecular biology (Crick, 1970), which in turn can be seen to echo the idea of the Weismann Barrier (Weismann, 1893): that the genetic material is isolated from both the rest of the organism and its environment and that it is sufficient in itself to specify the development of an organism.

THE QUESTION

The question addressed in this chapter is: Are organisms encoded as purely digital molecular descriptions in their gene sequences? By analysing the genome alone, could we then solve the forward problem of computing the behaviour of the system from this information? I argue that the first is incorrect and that the second is impossible. We therefore need to replace the gene-centric digital view of the relation between genotype and phenotype with an integrative view that also recognises the importance of analogue information in organisms and its contribution to inheritance across generations. Nature and nurture must interact. Either on its own can do nothing.

It is clearly incorrect to suppose that all biological systems are encoded in DNA alone, because an organism does not inherit only its DNA. It also inherits the complete fertilised egg cell and any non-DNA components that come via sperm. We now know that RNAs, for example, are transmitted through sperm and are responsible for non-DNA paternal effects on inheritance. Moreover, with the DNA alone, the development process cannot even get started, as DNA itself is inert until triggered by transcription factors formed by various proteins and RNAs. These factors initially come from the mother, through whom many maternal effects are transmitted (Gluckman and Hanson, 2004; Gluckman et al., 2007). It is only through an interaction between DNA and its environment, mediated by these triggering molecules, that development can begin. The centriole also is inherited via sperm, while maternal transfer of antibodies and other factors, including the microbiota (Moon et al., 2015), have been identified as a major sources of transgenerational phenotype plasticity. The non-DNA information includes three-dimensional organ, tissue, cellular, and subcellular structures that are never converted into digital format. Instead, these structures replicate via self-templating. The structures grow as the organism grows and they are divided up amongst the daughter cells during cell division. The patterns of functional biochemical and physiological networks, which in turn create the patterns of transcription factors and epigenetic effects, also replicate by self-templating. Every cell division passes these starting patterns on to the progeny. It is clear therefore that organisms must also be defined by all this analogue information as well as by the digital DNA and RNA sequences.

COMPARISON OF DIGITAL DNA INFORMATION WITH ANALOGUE INFORMATION IN PATTERNS AND STRUCTURES

How does non-DNA inheritance compare with that through DNA? So much emphasis has been placed on how large the genome information is that, perhaps surprisingly, the answer is that the non-DNA information is at least as great as that in the genome. The eukaryotic cell is an unbelievably complex structure. It is not simply a bag formed by a cell membrane enclosing a protein soup. Even prokaryotes, which were formerly thought to fit that description, are structured, and some are also compartmentalised to some degree. But the eukaryotic cell is divided up into many more compartments formed by the membranous organelles and other structures. The nucleus is also highly structured. It is not simply a container for naked DNA, which is why nuclear transfer experiments are not strict tests for excluding non-DNA inheritance. If we wished to represent these structures as digital information to enable computation, we would need to convert the three-dimensional images of the cell at a level of resolution that would capture the way in which these structures restrict the molecular interactions. This would require a resolution of around 10 nm to give at least 10 image points across an organelle of around 100 nm diameter. To represent the three-dimensional structure of a cell around 100 μm across would require a grid of 10,000 image points across. Each grid point (or group of points forming a compartment) would need data on the proteins and other molecules that could be present and at what level. Assuming the cell has a similar size in all directions (i.e., is approximately a cube), we would require 10^{12} grid points, i.e., 1000 billion points! Even a cell as small as 10 μm across would require a billion grid points. Recall that the genome is about three billion base pairs. It is therefore easy to represent the three-dimensional image structure of a cell as containing as much information as the genome, or even more since there are only four possible nucleotides at each position in the genome sequence, whereas each grid point of the cellular structure representation is associated with digital or analogue pattern information on a large number of molecular items that are present or absent locally.

There are many qualifications to be put on these calculations and comparisons (Noble, 2010). Many of the cell structures are repetitive. This is what enables cell modellers to lump together compartments like mitochondria, endoplasmic reticulum, ribosomes, filaments, and other organelles and structures, though we are also beginning to understand that, sometimes, this is an oversimplification. A good example is the calcium signalling system in muscles, where the tiny spaces in which calcium signalling occurs, which couples excitation to contraction, have to be represented at ever finer detail to capture what the experimental information tells us. Current estimates of the number of calcium ions in a single dyad (the space across which calcium signalling occurs) is only between 10 and 100, too small for the laws of mass action to be valid. And signalling to the nucleus can arise from calcium levels in microspaces near ion channels that monitor the channel activity (Ma et al., 2014).

Nevertheless, there is extensive repetition. One mitochondrion is basically similar to another, as are ribosomes and all the other organelles. But note that extensive repetition is also characteristic of the genome. A large fraction of the three billion base pairs forms repetitive sequences. Protein template regions of the human genome are estimated to be less than 1.5%. Even if 99% of the structural information from a cell image were to be redundant because of repetition, we would still arrive at figures comparable to the effective information content of the genome. And for the arguments in this chapter to be valid, it does not really matter whether the information is strictly comparable, nor whether one is greater than the other in terms of the standard definitions of information content. Significance of information matters as much as its quantity. All I need to establish at this point is that, in a bottom-up reconstruction – or indeed in any other kind of reconstruction – it would be courting failure to ignore the structural detail. That is precisely what restricts the combinations of interactions (a protein in one compartment cannot interact directly with one in another, and proteins floating in lipid bilayer membranes have their parts exposed to different sets of molecules) and may therefore make the computations possible.

DIFFERENTIAL AND INTEGRAL VIEWS OF GENETICS

These points are so obvious, and have been so ever since electron microscopes first revealed the fine details of those intricate sub-cellular structures around 50 years ago, that one has to ask how mainstream genetics came to ignore the problem. The answer lies in what I will call the differential view of genetics. At this point, a little history of genetics is relevant. The original concept of a gene was very different from the modern molecular biological definition. When Wilhelm Johannsen (1909) introduced the word, he defined it as whatever is the inheritable cause of a particular characteristic in the phenotype, such as eye colour, number of limbs/ digits, and so on. For each identifiable phenotype characteristic, there would be a gene (actually an allele – a particular variant of a gene) responsible for that characteristic. A gene could be defined therefore as something whose presence or absence makes a difference to the phenotype. It did not matter what the physical structure of the gene was. Johannsen referred to it as 'anything' ('*ein etwas*') that could be responsible. Neo-Darwinists today often resort to the same idea, as Richard Dawkins did in a seminal debate with Lynn Margulis in 2009.[1]

When genetics was combined with natural selection to produce the Modern Synthesis, which is usually called neo-Darwinism, the idea took hold that only those differences were relevant to evolutionary success and all that mattered in relating genetics to phenotypes was to identify the genetic causes of those differences. Since each phenotype must have such a cause (on this view at least) then selection

[1] http://www.voicesfromoxford.org/news/margulisdawkins-debate/158. The response was that such an inherited characteristic could be regarded as an honorary gene.

of phenotypes amounts, in effect, to selection of individual genes. It does not really matter which way one looks at it. They are effectively equivalent. The gene's eye view then relegates the organism itself to the role of disposable carrier of its genes. To quote Dawkins' famous metaphor (Dawkins, 1976, 2006), the organism is simply a 'mortal' vehicle to carry its 'immortal' genes from one generation to another. To this view we can add the idea that, in any case, only differences of genetic make-up can be observed. The procedure is simply to alter the genes, by mutation, deletion, and addition, and observe the effect on the phenotype. Meanwhile, the definition of a gene switched quite quickly after 1953 to become a DNA sequence. The change is fundamental. On the original definition, a gene was necessarily the cause of the phenotype. That is how it was defined. On the molecular biological definition, it is a matter of empirical discovery to determine the causal role of a DNA sequence. Some produce a change in phenotype, others do not. A systematic investigation of the 6,000 genes (as DNA sequences) in yeast found that 80% of knockouts are actually silent at the phenotype level in normal physiological conditions (Hillenmeyer et al., 2008).

I will call this gene-centric approach the 'differential view' of genetics to distinguish it from the 'integral view' I will propose later. To the differential view, we must add an implicit assumption. Since, on this view, no differences in the phenotype that are not caused by a genetic difference can be inherited, the fertilised egg cell (or just the cell itself in the case of unicellular organisms) does not evolve other than by mutations and other forms of evolution of its genes. The inherited information in the rest of the egg cell is ignored because (1) it is thought to be equivalent in different species (the prediction being that a cross-species clone will always show the phenotype of whichever species provides the genes), and (2) it does not evolve, or, if it does through the acquisition of new characteristics, these differences are not passed on to subsequent generations, which amounts to the same thing. Evolution requires inheritance. A temporary change does not matter.

At this stage in the argument, I will divide the holders of the differential view into two categories. The 'strong' version is that, while it is correct to say that the intricate structure of the egg cell is inherited as well as the genes, in principle that structure can be deduced from the genome information. On this view, a complete bottom-up reconstruction might still be possible even without the nongenetic information. This is a version of the old idea, referred to in the introduction to this chapter, that the complete organism is somehow represented in the genetic information. It just needs to be unfolded during development, like a building emerging from its blueprint. The 'weak' version is one that does not make this assumption but still supposes that the genetic information carries all the differences that make one species different from another.

WEAK VERSION OF THE DIFFERENTIAL VIEW

The weak version is easier to deal with, so I will start with that. In fact, it is remarkably easy to deal with. Only by restricting ourselves to the differential view of genetics could it be possible to ignore the nongenetic structural information. But nature does not play just with differences when it develops an organism. The organism develops only because the nongenetic analogue structural information is also inherited and is used to develop the organism. When we try to solve the forward problem, we will be compelled to take that analogue information into account even if it were to be identical in different species. To use a computer analogy, we not only need the 'program' of life, we also need the 'computer' of life, the interpreter of the genome, i.e., the highly complex egg cell. In other words, we have to take the context of the cell into account, not only its genome. There is a question remaining, which is whether the weak version is correct in assuming the identity of egg cell information between species. I will deal with that question later. The important point at this stage is that, even with that assumption, the forward problem cannot be solved on the basis of genetic information alone. Recall that genes need to be activated to do anything at all.

THE STRONG VERSION OF THE DIFFERENTIAL VIEW

Proponents of the strong version would probably also take this route in solving the forward problem, but only as a temporary measure. They would argue that, when we have gained sufficient experience in solving this problem, we will come to see how the structural information is somehow also encoded in the genetic information. This is an article of faith, not a proven hypothesis. As I have argued elsewhere (Noble, 2011a, 2015), following arguments also presented by Jacob (1982), Wilkins (1986), and Coen (1999), the DNA sequences do not form a 'program' that could be described as complete in the sense that it can be parsed and analysed to reveal its logic. Decades of genome sequencing have failed to find the coding that could be said to be such a program. What we have found in the genome is better described as a database of templates to enable a cell to make proteins and RNA. Unless that complete 'program' can be found (which I would now regard as highly implausible given what we already know of the structure of the genome), I do not think the strong version is worth considering further. It is also implausible from an evolutionary viewpoint. Cells must have evolved before genomes. Why on Earth would nature bother to 'code' for detail that is inherited anyway in the complete cell? This would be as unnecessary as attempting to 'code for' the properties of water or of lipids. Those properties are essential for life (they are what allow cells to form), but they do not require genes. Mother Nature would have learnt fairly quickly how to be parsimonious in creating genetic information: do not code for what happens naturally in the physico-chemical universe. Many wonderful things can be constructed on the basis of relatively little transmitted information, relying simply on physico-chemical processes, and these include what seem at first sight to be highly complex structures like spiral galaxies, tornados, and the spiral waves of cardiac arrhythmia. Even a process as important to inheritance as cell division can occur in the absence of a nucleus (Lorch, 1952), which strongly suggests that cells, and some of their key functions, developed before the evolution of DNA. Genes therefore do not need to code for everything. Nature can, as it were, get 'free rides' from the dynamic physics of structure: the attractors towards which systems move naturally. Such physical structures do not require detailed templates in the DNA sequences; they appear as the natural expression of the underlying physics. The structures can then act as templates for the self-organisation of the protein networks, thus making self-organisation a process depending on both the genome and the inherited structure.

EXPERIMENTAL EVIDENCE

Is non-DNA information inherited? Transmission of changes in structural information between generations has been observed in unicellular animals, and this phenomenon has been known for many years. Surgical modifications of the direction of cilia patterns in paramecium, produced by cutting a pole of the animal and reinserting it the wrong way round, are robustly inherited by the daughter cells down several generations (Sonneborn, 1970). Interest in this kind of phenomenon has returned, perhaps in the wake of discoveries in epigenetics that make the phenomena explicable. A good example in a multicellular organism is the work of Sun et al. on cross-species cloning of fish from different genera (Sun and Zhu, 2014; Sun et al., 2005). They enucleated fertilised goldfish eggs and then inserted a carp nucleus. The overall body structure of the resulting adult fish is intermediate. Some features are clearly inherited from the goldfish egg. The goldfish has fewer vertebrae than the carp. So does the cross-species clone. Results such as these that show non-DNA forms of inheritance are often dismissed by those favouring a strongly gene-centric interpretation of biology as rare exceptions, perhaps not representative of living processes in general. But the situation may actually be the other way round. The gene-centric view implies that the analogue structural information in organisms is unimportant, however large it may be in terms of data, because it is essentially no more than a vehicle for the digital DNA information. If that were really true, then it would not matter much which vehicle is used. There would be a kind of universal egg cell. But it does matter! It matters so much that cross-species clones generally do not work. There is incompatibility between the host egg and the guest nucleus, which ensures that embryonic development freezes at some point so that adult organisms do not result. The universal egg cell idea is therefore a nonstarter. The cross-species fish clone is actually an exception in developing as far as an adult organism. But even there the incompatibility eventually expresses itself. The cross-species clones are not fertile. The influence of the organism on DNA to steer development towards a different vertebral number also resembles the results of embryonic transfer performed in 1957 by McLaren and Michie (1958). Implantation of mice embryos between varieties having different numbers of tail vertebrae showed that the offspring displayed the vertebral number of the host, not of the donor mouse variety. The host is clearly capable of canalising the development in a direction different from that of the transferred DNA. Maternal effects as they are now known form an important and growing area of biomedical research (Gluckman et al., 2007).

MUST EVOLUTIONARY CHANGE ALWAYS BE BY DNA MUTATIONS?

The standard theory of evolution, the neo-Darwinian Modern Synthesis, postulates both that genetic change must be random and that it involves the accumulation of small changes in DNA called mutations. Yet it has been shown for more than a century that genetic change need not be restricted to mutations. The first person to propose this idea was James Baldwin, the originator of the 'Baldwin effect' (Baldwin, 1896). Baldwin realised that a subgroup of a population might choose a new niche, i.e., a change in environment. By making such a choice, the individuals will not be typical of the population as a whole. If they and their progeny flourish in the new niche, then a new variety, conceivably even a new species, could arise without it being necessary to suppose that any genetic mutations were responsible. In modern terms we would say that the subgroup's complement of alleles, i.e., existing patterns of DNA variations, had branched off from the main group. This process is of course similar to that supposed by Darwin to have occurred in the branchings of the various species of finches and tortoises on the Galapagos Islands. The important point here is that whether the branching was initiated by mutations or by self-selection in the form of the choice of the leading 'adventurous' group is in question.

In this connection it is interesting to note that genetic and epigenetic analysis has now been done on the Galapagos finches. The results show that at least as many epigenetic as genetic changes underlie the differences between the various species and that the number of epigenetic changes correlates rather better with evolution- ary distance between the species than do the genetic changes (Skinner et al., 2014). At the least this result puts the standard explanation for speciation in doubt in this iconic example. From the experimental information alone it would be impossible to say whether epigenetic changes led the speciation with subsequent assimilation into the genetic material or vice versa. Both would be possible. Even more likely, the two naturally go together.

This point raises a fundamental question about digital informa- tion and the genome. The raw digital information on each gene is not sufficient to characterise what is happening. The overall pattern of the gene alleles in a population or subgroup is also important. The genome is not just a bunch of individual genes.

This point was also established in the experiments of the developmental biologist Conrad Waddington, who is rightly viewed as the father of epigenetics. He realised that the way to test for the inheritance of acquired characteristics is first to discover what forms of developmental plasticity already exist in a population. He was then able to exploit this plasticity, and so was more likely to mimic a path that evolution itself could have taken (Waddington, 1956, 1959, 2014).

He used the word 'canalised' for this kind of persuasion since he represented the developmental process as a series of 'decisions' that could be represented as 'valleys' and 'forks' in a developmental

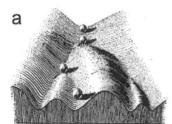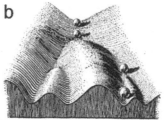

FIG. 6.1 Waddington's developmental landscape diagram. The landscape itself and the ball at the top are from his original diagram. The subsequent positions of the ball have been added to illustrate his point that development can be canalised to follow different routes (a and b). But the plasticity to enable this to happen already exists in the wild population of organisms. (Modified from www.gen.tcd.ie/mitchell/.)

landscape (Figure 6.1). He knew from his developmental studies that embryo fruit flies could be persuaded to show different thorax and wing structures simply by changing the environmental temperature or by a chemical stimulus. In his landscape diagram this can be represented as a small manipulation in slope that would lead to favouring one channel in the landscape rather than another, so that the adult could show a different phenotype starting from the same genotype. The landscape concept is of course an analogue representation of an extremely complex set of interactions between gene products (proteins, RNAs, and their network) and the environment. Like the three-dimensional structure of the organism, and unlike the DNA sequences, it is not one-dimensional digital information. Both the Baldwin effect and the work of Waddington led the great logician of science, Karl Popper, to propose a thought experiment in a lecture that he delivered to the Royal Society in 1986 (Niemann, 2014; Noble, 2014). The title of Popper's lecture was 'A New Interpretation of Darwinism'. It was given in the presence of Sir Peter Medawar, Max Perutz, and other key figures and it must have shocked his audience. He proposed a completely radical interpretation of Darwinism (but I am sure he meant neo-Darwinism), essentially rejecting the Modern Synthesis by proposing that organisms themselves are the source of the creative processes in evolution, not random mutations in DNA.

Darwinism, he said, was not so much wrong as seriously incomplete. To illustrate his interpretation of the Baldwin effect he invented an imaginary world in which there was no competition for survival, no 'selfish genes'. The organisms would still evolve. Of course, the world in which such evolution could occur would have to be effectively infinite in size to accommodate all the organisms that have ever lived. But this was just a thought experiment. Conrad Waddington understood the same point. Why then do selfish gene theorists ignore it? They do so by taking an atomistic, gene-centred view. As Popper saw, it is the insistence on just one atomistic approach that is the problem. It is combinations of genes, or rather combinatorial interactions between large numbers of their products, RNAs and proteins, that are important functionally. Most single genes contribute very little to complex functions, which is why the correlations between genes and complex diseases has been found to be a matter of large numbers of very small effects, still summing up to a small overall fraction of causation. The important point in the context of this chapter is that patterns are analogue since there can be continuous variation in expression levels of the individual components. Escaping from the straightjacket of neo-Darwinism is important for many reasons, one of the most important being that the mechanisms that lie outside the neo-Darwinist framework would contribute to the speed of evolution. Multiple mechanisms in interaction can lead to faster evolutionary change. The extent to which this could be true has been highlighted in a recent book by Andreas Wagner (2014), which shows how organisms have discovered new molecules and mechanisms in a fraction of the time that random variation would take. Evolution itself has evolved (Noble et al., 2014).

The Integrative View of the Relations between Genotypes and Phenotypes
Restricting ourselves to the differential view of genetics is rather like working only at the level of differential equations in mathematics, as though the integral sign had never been invented. This is a good

analogy, since the constants of integration, the initial and boundary conditions, restrain the solutions possible in a way comparable to that by which the cell and tissue structures restrain whatever molecular interactions are possible. Modelling of biological functions should follow the lead of modellers in the engineering sciences. Engineering models are constructed to represent the integrative activity of all the components in the system. Good models of this kind in biology can even succeed in explaining the buffering process and why particular knockouts and other interventions at the DNA level do not reveal the function (Noble, 2011a, 2011b).

The integrative view also incorporates the theory of biological relativity, which is the principle that there is, a priori, no privileged level of causation (Noble, 2012). There is no reason to privilege DNA (digital) information over structural (analogue) information in the development and evolution of organisms.

Multilevel interactions are important both in development and in evolutionary change. They are the means by which the environment interacts with these. Such interactions are also analogue in nature because they depend on constraints of lower (e.g., molecular) levels by higher-level processes that are formed as dynamic patterns. Those patterns represent continuous variation in expression levels of genes and many other factors.

ACKNOWLEDGEMENTS

Parts of this chapter are based on material from Noble (2011a, 2015).

REFERENCES

Baldwin, J. M. 1896. A new factor in evolution. *The American Naturalist*, **30**(354), 441–451.

Coen, E. 1999. *The art of genes: how organisms make themselves*. Oxford University Press

Crick, F. H. 1970. Central dogma of molecular biology. *Nature*, **227**(5258), 561–563.

Dawkins, R. 1976. *The selfish gene*. Oxford University Press.

Dawkins, R. 2006. *The selfish gene, 30th anniversary edition*. Oxford University Press.

Gluckman, P. D., and Hanson, M. A. 2004. *The fetal matrix: evolution, development and disease.* Cambridge University Press.

Gluckman, P. D., Hanson, M. A., and Beedle, A. S. 2007. Non-genomic transgenerational inheritance of disease risk. *Bioessays*, **29**(2), 145–154.

Hillenmeyer, M. E., Fung, E., Wildenhain, J., Pierce, S. E., Hoon, S., Lee, W., Proctor, M., Onge, R. P. St., Tyers, M., Koller, D., et al. 2008. The chemical genomic portrait of yeast: uncovering a phenotype for all genes. *Science*, **320**(5874), 362–365.

Jacob, F. 1982. *The possible and the actual*. Pantheon.

Johannsen, W. 1909. *Elemente der exakten Erblichkeitslehre*. Gustav Fischer.

Lorch, J. 1952. Enucleation of sea-urchin blastomeres with or without removal of asters. *Quarterly Journal of Microscopical Science*, **93**(24), 475–486.

Ma, H., Groth, R. D., Cohen, S. M., Emery, J. F., Li, B., Hoedt, E., Zhang, G., Neubert, T. A., and Tsien, R. W. 2014. γCaMKII shuttles Ca^{2+}/CaM to the nucleus to trigger CREB phosphorylation and gene expression. *Cell*, **159**(2), 281–294.

McLaren, A., and Michie, D. 1958. An effect of the uterine environment upon skeletal morphology in the mouse. *Nature*, **181**, 1147–1148.

Moon, C., Baldridge, M. T., Wallace, M. A., Burnham, C. D., Virgin, H. W., and Stappenbeck, T. S. 2015. Vertically transmitted faecal IgA levels determine extra-chromosomal phenotypic variation. *Nature*, **521**(7550), 90–93.

Niemann, H. 2014. *Karl Popper and the two new secrets of life: including Karl Popper's Medawar lecture 1986 and three related texts*. Mohr Siebeck.

Noble, D. 2010. Differential and integral views of genetics in computational systems biology. *Interface Focus*, rsfs20100444.

Noble, D. 2011a. Neo-Darwinism, the Modern Synthesis and selfish genes: are they of use in physiology? *Journal of Physiology*, **589**(5), 1007–1015.

Noble, D. 2011b. Successes and failures in modeling heart cell electrophysiology. *Heart Rhythm*, **8**(11), 1798–1803.

Noble, D. 2012. A theory of biological relativity: no privileged level of causation. *Interface Focus*, **2**(1), 55–64.

Noble, D. 2014. Secrets of life from beyond the grave. *Physiology News*, **97**, 34–35.

Noble, D. 2015. Evolution beyond neo-Darwinism: a new conceptual framework. *Journal of Experimental Biology*, **218**(1), 7–13.

Noble, D., Jablonka, E., Joyner, M. J., Mueller, G. B., and Omholt, S. W. 2014. Evolution evolves: physiology returns to centre stage. *Journal of Physiology*, **592**(11), 2237–2244.

Skinner, M. K., Gurerrero-Bosagna, C., Haque, M. M., Nilsson, E. E., Koop, J. A. H., Knutie, S. A., and Clayton, D. H. 2014. Epigenetics and the evolution of Darwin's finches. *Genome Biology and Evolution*, **6**(8), 1972–1989.

Sonneborn, T. M. 1970. Gene action in development. *Proceedings of the Royal Society of London B: Biological Sciences*, **176**(1044), 347–366.

Sun, Y., and Zhu, Z. 2014. Cross-species cloning: influence of cytoplasmic factors on development. *Journal of Physiology*, **592**(11), 2375–2379.

Sun, Y., Chen, S., Wang, Y., Hu, W., and Zhu, Z. 2005. Cytoplasmic impact on cross-genus cloned fish derived from transgenic common carp (*Cyprinus carpio*) nuclei and goldfish (*Carassius auratus*) enucleated eggs. *Biology of Reproduction*, **72**(3), 510–515.

Waddington, C. H. 1956. Genetic assimilation of the bithorax phenotype. *Evolution*, 1–13.

Waddington, C. H. 1959. Canalization of development and genetic assimilation of acquired characters. *Nature*, **183**(4676), 1654–1655.

Waddington, C. H. 2014. *The strategy of the genes*. Vol. 20. Routledge.

Wagner, A. 2014. *Arrival of the fittest: solving evolution's greatest puzzle*. Penguin.

Weismann, A. 1893. *The germ-plasm: a theory of heredity*. C. Scribner's Sons.

Wilkins, A. S. 1986. *Genetic analysis of animal development*. Wiley-Liss.

7   From Entropy to Information

Biased Typewriters and the Origin of Life

Christoph Adami and Thomas LaBar

So much has been written about the possible origins of life on Earth (see, e.g., the popular books by Deamer, 1994; deDuve, 1995; Koonin, 2011; Morowitz, 2004) that it sometimes seems that – barring an extraordinary breakthrough in experimental biochemistry (e.g., Patel et al., 2015), or the discovery of the remnants of an ancient biochemistry (Davies et al., 2009) – nothing new can be said about the problem. But such a point of view does not take into account that perhaps not all the tools of scientific inquiry have been fully utilized in this endeavor to unravel our ultimate origin on this planet. Indeed, origin-of-life research has historically been confined to a fairly narrow range of disciplines, such as biochemistry and geochemistry. Today, a much broader set of tools is being unleashed on this problem, including mathematical (England, 2013; Smith, 2008; Vetsigian et al., 2006) and computational approaches (Mathis et al., 2015; Nowak and Ohtsuki, 2008; Segre et al., 2000; Vasas et al., 2012; Walker et al., 2012). Computational approaches to the study of possible origins of life are often derided because they lack a particular feature of biochemistry, or "because they do not take into account the specific properties of individual organic compounds and polymers" (Lazcano and Miller, 1996). Such a point of view ignores the possibility that life may be not a feature that is dependent on a particular biochemistry (Benner et al., 2004), but could instead be a feature of *any* chemistry that is capable of encoding information.

If the one invariant in life is information (information about how to replicate, that is), it then becomes imperative to understand the general principles by which information could arise by chance. It is generally understood that evolution, viewed as a computational process (Adami, 1998; Mayfield, 2013), leads to an increase in information on average. The amount of information that evolution has accumulated to date differs from organism to organism, of course, and precise numbers are not known. A rough estimate of the amount of information stored in an organism's genome can be obtained by calculating the amount of functional DNA in an organism.[1] The general idea here is that only functional DNA can be under selection, as after all information is that which guarantees survival (Adami, 2002, 2012). For humans (assuming a functional percentage of about 8%; Rands et al., 2014), this means that our DNA codes for about half a billion bits.[2]

Almost all of the information contained in our genome (and any other organism's) owes its existence to the evolutionary process. But the algorithm that is evolution cannot be at work in the absence of replication, and therefore cannot explain the origin of life. It is in principle possible that the first replicator did not originate on Earth but rather arrived on Earth from extraterrestrial sources (Arrhenius, 1908; Hoyle and Wickramasinghe, 1981; Wickramasinghe, 2011). Even if that were the case, such an origin story does not obviate the need for emergence *somewhere*, so we may ask generally: "What is the likelihood of spontaneous emergence of information?" The question in itself is not new, of course. Howard Pattee asked as early as 1961, shortly after the discovery of the structure of DNA (but before the discovery of the genetic code) (Pattee, 1961):

(1) How did a disordered collection of elements which forms sequences with no restrictions produce, with reasonable probability, enough initial order to result in the general property

---

[1] It is not necessary to consider epigenetic variation in the estimate of information content, as all epigenetic changes are performed by enzymes whose information is already stored within DNA.

[2] This number is (given the functional percentage of 8%) an upper limit on the information content, as protein-coding regions display considerable variation and redundancy, which lowers information. However, as open reading frames account for only 1% of the human genome and regulatory sequences (the other 7%) are much less redundant, the true information content of human DNA is likely not much lower.

of self-replication? (2) Assuming the general property of self-replication for all sequences, how did those particular sequences which now exist arise, with reasonable probability, from the set of all possible sequences?

In order to estimate the likelihood of spontaneous emergence of a self-replicator, it is necessary to estimate the *minimal information* necessary to replicate, because the length of the sequence is not a good indicator of fitness. A quick Gedankenexperiment can clarify this. Imagine that a symbolic sequence (written using ASCII characters) can replicate if and only if anywhere on the string the exact sequence origins appears. This is a 7-letter sequence, and the total number of possible sequences of length 7 is 26^7, or about 8 billion. The likelihood to find this sequence by chance if a billion sequences are tried is, obviously, about 1 in 8. But suppose we try sequences of length 1,000. If we ask only that the word appears *anywhere* in the sequence, increasing sequence length obviously increases both the number of possible sequences and the number of self-replicators. Thus, the likelihood to find a self-replicator scales exponentially not with the length of the sequence (it does not become $26^{-1,000}$), but rather with the *information content* of the sequence (as we will see momentarily). In the present example, the information content is clearly 7 letters. But how do you measure the information content of biomolecules?

INFORMATION CONTENT OF BIOMOLECULES

Generally speaking, the information content of a symbolic sequence is equal to the amount of uncertainty (about a particular ensemble) it can reduce. This information can be written mathematically in terms of the entropy of the ensemble (described by the random variable X that can take on states $x_1, \ldots, x_n$ with probabilities $p_1, \ldots, p_n$

$$H(X) = -\sum_{i=1}^{n} p_i \log p_i \tag{7.1}$$

and the conditional entropy $H(X|s)$, where s is the sequence whose information content we would like to measure, as

$$I(s) = H(X) - H(X|s).$$ (7.2)

The latter entropy is given by the conditional entropy distribution $p_{i|s}$ instead. So, for example, the sequence Colonel Mustard reduces the uncertainty about the identity of the murderer in a popular board game from $\log_2 6 \approx 2.83$ bits to zero (as there are a priori six suspects, and the sequence fingers the perpetrator), so the information content is 2.83 bits. The sequence length, on the contrary, is 15 (counting the space as a symbol), which translates to $15 \log_2(27) \approx 71.3$ bits. Thus, sequence length and information content can be very different: information is about something, while sequence length is just entropy.

Unfortunately, we cannot measure the information content of biomolecules in the same manner, because we do not know the entropy of the ensemble that the biomolecular sequence is information about. Let us call this random variable E (for "environment"), as it represents the environment within which the sequence is functional, in the same sense that X above was the environment within which the sequence Colonel Mustard is functional. However, an information-theoretical "trick" allows us to make progress. Let s be a functional biomolecule (a polymer of length L), and its information content (per the formula above)

$$I(s) = H(E) - H(E|s),$$ (7.3)

that is, it is the entropy of the "world" minus the entropy of the world given that we know s. We can also define the average information content as

$$\langle I \rangle = \sum_s p(s)I(s) = H(E) - H(E|S) = H(E : S),$$ (7.4)

where $H(E : S)$ is the shared entropy between environment and sequences, but again that formula is not useful because we do not know $H(E)$. However, the formula can also be written as

$$\langle I \rangle = H(S) - H(S|E) \tag{7.5}$$

in terms of the entropy of sequences $H(S)$ and the conditional entropy of the sequences given an average environment. This is also not useful, as the world is not an average of environments, but one very particular one $E = e$. Could we write this in terms of a difference of entropies as in Eq. (7.3)? We then would guess that

$$I(s) = H(S) - H(S|e), \tag{7.6}$$

but Eq. (7.6) is not mathematically identical to Eq. (7.3), as the identity holds only for the averages. However, Eq. (7.6) can be derived from an approach embedded in Kolmogorov complexity theory (Adami, 1998, 2002; Adami and Cerf, 2000), where that equation represents the "physical complexity" of the sequence. Furthermore, Eq. (7.6) is practical to the extent that its value can be estimated. For example, as S is the ensemble of sequences, its entropy is simply given by $\log N$, where N is the total number of sequences of that length (it is possible to extend this formalism to sequences of varying length). Sequences with an arbitrary function in environment $E = e$ have an entropy smaller than $\log N$. Let us imagine that the number of polymers with that function (in $e \in E$) is N_e (with $N_e \ll N$). Then (here we specify the base of the logarithm by the number of possible monomers D)

$$I(s) = -\log_D \frac{N_e}{N} \tag{7.7}$$

which, it turns out, is identical to Szostak's "functional complexity" measure (Szostak, 2003). It allows us to quantify the information content of a biomolecular sequence if the "density" of functional sequences N_e/N is known, and makes it possible to calculate the likelihood of emergence (by chance), of a molecule with information content I. As the likelihood must be given by the density of molecules of that type within the set of all molecules of that length, we find

$$P = \frac{N_e}{N} = D^{-I}, \tag{7.8}$$

where the relationship to information content follows directly from Eq. (7.7). Thus we see (as advertised earlier) that this likelihood depends *only* on the information content of the sequence, but not on its length. Below, we will test this prediction using the digital life system Avida and find it violated. However, the origin of this apparent violation is easily tracked down, and we are confident that the equality holds exactly in principle.

TESTING THE LIKELIHOOD OF EMERGENCE BY CHANCE

We first tested the likelihood to find the sequence origins by creating random ASCII polymers of length 7 using an alphabet of $D = 26$ (no spaces or other punctuation), and where each symbol was drawn from a uniform distribution over the letters a to z. When testing a billion sequences we did not find origins, which is in accord with the probability $P = 26^{-7}$ calculated above. Note that for ASCII strings (unlike the biomolecules) there is never any redundancy, so that $N_e = 1$ always. We then randomly searched for self-replicating sequences within the digital chemistry of the Avida Artificial Life system (Adami, 1998; Adami and Brown, 1994; Ofria and Wilke, 2004; Ofria et al., 2009). In Avida, ASCII sequences *can self-replicate*, but only because these sequences are translated to instructions that are executed on virtual CPUs (see Figure 7.1). In this sense, the sequences are really self-replicating computer programs, and because these sequences can mutate as they are copied, they evolve in a strictly Darwinian manner (see Table 7.1 for the arbitrary assignment of ASCII letters to Avidian instructions). The Avida system has been used for more than 20 years to test evolutionary dynamics (see, e.g. the review by Adami, 2006 covering mostly the first 10 years), and the likelihood of emergence of functional information (but not self-replication) has been studied in this system before (Hazen et al., 2007). (See also Pargellis, 2003 for an investigation of spontaneous emergence of digital life in a related digital system.)

TABLE 7.1 Instruction set of the Avidian programming language used in this study. The notation ?BX? implies that the command operates on a register specified by the subsequent nop instruction (for example, nop-A specifies the AX register, and so forth). If no nop instruction follows, use the register BX as a default. More details about this instruction set can be found in Ofria et al. (2009).

Instruction	Description	Symbol
nop-A	no operation (type A)	a
nop-B	no operation (type B)	b
nop-C	no operation (type C)	c
if-n-equ	Execute next instruction only if ?BX? does not equal complement	d
if-less	Execute next instruction only if ?BX? is less than its complement	e
if-label	Execute next instruction only if template complement was just copied	f
mov-head	Move instruction pointer to same position as flow-head	g
jmp-head	Move instruction pointer by fixed amount found in register CX	h
get-head	Write position of instruction pointer into register CX	i
set-flow	Move the flow-head to the memory position specified by ?CX?	j
shift-r	Shift all the bits in ?BX? one to the right	k
shift-l	Shift all the bits in ?BX? one to the left	l
inc	Increment ?BX?	m
dec	Decrement ?BX?	n
push	Copy value of ?BX? onto top of current stack	o
pop	Remove number from current stack and place in ?BX?	p
swap-stk	Toggle the active stack	q
swap	Swap the contents of ?BX? with its complement	r
add	Calculate sum of BX and CX; put result in ?BX?	s
sub	Calculate BX minus CX; put result in ?BX?	t
nand	Perform bitwise NAND on BX and CX; put result in ?BX?	u
h-copy	Copy instruction from read-head to write-head and advance both	v
h-alloc	Allocate memory for offspring	w
h-divide	Divide off an offspring located between read-head and write-head	x
IO	Output value ?BX? and replace with new input	y
h-search	Find complement template and place flow-head after it	z

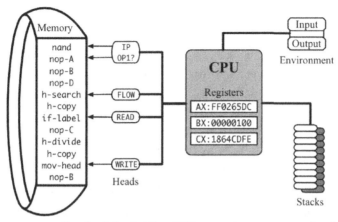

FIG. 7.1 Sketch of the Avidian CPU, executing a segment of code. The CPU uses three registers (AX, BX, CX) and an instruction pointer (IP) that reads the program into the CPU. A read-head, a write-head, and a flow-head are used to specify positions in the CPU's memory. For example, the "copy" command reads from the read-head and writes to the write-head, while "jump"- type statements move the instruction pointer to the flow-head. The CPU uses two stacks to simulate an "infinite Turing tape" and input/output buffers to communicate with its environment (reproduced from Ofria et al. (2009), with permission).

The likelihood that any particular sequence coded within 26 instructions can replicate depends strongly on the meaning of each instruction. If a single letter (monomer) were to be interpreted as "replicate the entire sequence it is in," then self-replicators would be very easy to find. Over the years of development of Avida, the meaning of each symbol has changed as the instruction set itself has changed over time, so the absolute values for the information content of self-replicators may also change in the future. We are here interested only in the rate at which self-replicators can be found in relationship to the information content, and how this rate depends on other factors in the environment that can be modified. Translated to a search for the origins of life, we are interested in how local (environmental) conditions can favorably increase the likelihood to find a self-replicator with information content I purely by chance.

We first focused on Avidian sequences constrained to length $L = 15$, as there already is a hand written standard replicator of that length in Avida, given by the string wzcagczvfcaxgab. If every instruction in this replicator were information, the likelihood of finding it by chance would be $26^{-15} \approx 6 \times 10^{-22}$. Even if we tested a million sequences per second per CPU, on 1000 CPUs running in parallel, we only would expect to find a single self-replicator in about 50,000 years of continuous search. We tested one billion sequences of $L = 15$ and found 58 self-replicators (all of them unique) by chance, indicating that the information content of self-replicators is vastly smaller than 15 mers. Indeed, we can estimate the information content as

$$I(15) = -\log_D(58 \times 10^{-9}) \approx 5.11 \pm 0.04 \text{ mers}, \qquad (7.9)$$

with a one-σ error. Here, the "mer" is a unit of information obtained by taking logarithms to the base of the alphabet size, so that a single monomer has up to one mer of entropy (Adami, 2002, 2012). This means that, within the replicating 15-mers, only about 5 of those 15 mers are information.

We next tested the information content of sequences constrained to several different lengths. Among a billion random sequences of $L = 30$, we found 106 replicators, which translates to

$$I(30) = -\log_D(106 \times 10^{-9}) \approx 4.93 \pm 0.03 \text{ mers}, \qquad (7.10)$$

which is significantly different from $I(15)$. In fact, the calculated information content suggests that perhaps replicators of length five or six might exist, but an exhaustive search of all 11,881,376 $L = 5$ sequences and all 308,915,776 $L = 6$ sequences reveals this not to be the case. When searching a billion sequences of $L = 8$ we found 6 unique self-replicators, implying an information content

$$I(8) = -\log_D(6 \times 10^{-9}) \approx 5.81 \pm 0.13 \text{ mers}. \qquad (7.11)$$

The six sequences we found are qxrchcwv, vxfgwjgb, wxvxfggb, vhfgxwgb, wxrchcvz, and wvfgjxgb.

We can understand this trend of decreasing information content with increasing length (violating Eq. 7.8) as a consequence of the way we treat Avidian sequences, namely as having a beginning and an end. Indeed, while the genome itself is circular, execution always begins at a marked instruction. We can see this effect at work using the example origins sequence that we used before. If we add a single letter to the 7-mer origins, the number of sequences that spell the word increases by 52 (adding the letter to the beginning or the end of the word), while the total number of possible sequences increases only by 26. Thus, the density of self-replicators increases with length, leading to a decrease of information.

We tested whether this decrease of information with increasing sequence length would continue by testing 300 million sequences of length 100. We found 17 self-replicators among this set, which translates to $I(100) = 5.10 \pm 0.09$ mers and suggests that the trend not only does not continue (which of course would have been absurd) but may reverse itself. There is a subtle information-theoretic reason for an increase in information with increasing sequence length. Suppose that there is a single instruction that could abrogate self-replication if it is to be found anywhere within the sequence, when in its absence the sequence replicates (a "kill" instruction, so to speak). Even though such an instruction is obviously not information about how to self-replicate, its needed absence actually *is* information. When the sequence length increases, the presence of such a "kill" instruction becomes more and more likely, and therefore the absence of the instruction over the increasing sequence length represents an increase in information. This is the trend suggested in Figure 7.2.

BIASED TYPEWRITERS

In a sense, the random search for self-replicators is very inefficient: it is known that functional molecular sequences cluster in genetic space, while vast regions of that space are devoid of function. Yet the random generation of sequences searches all of genetic space evenly. Is there a way to focus the search more on sequences that are likely to

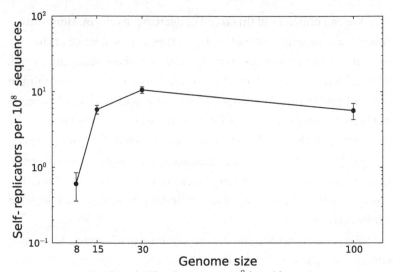

FIG. 7.2 Number of self-replicators per 10^8 found for various genome sizes using an unbiased (uniform) probability distribution of monomers. The number of self-replicators per 10^8 for $L = 100$ is estimated from sampling 300 million sequences only (all others used samples of 10^9). Error bars are standard deviations.

be functional? It turns out there is, and this method requires only the generation of monomers using a biased probability density function that more resembles that generated by functional sequences (Adami, 2015). We first present a simple example (the biased typewriter), and then outline the theory behind the enhanced search.

Words in the English language have a very characteristic letter-frequency distribution that makes it possible to distinguish English text from random sequences of letters, and even text written in different languages. Figure 7.3 (using data from Lewand, 2000) shows the frequency distribution of letters in English text, showing that "e" appears more frequently than "t," which itself is more frequent than "a" and so on. As this is the expected frequency of letters in English, a focused search should generate words with these expected frequencies; that is, the " monomers" of English words should be generated with the frequency distribution shown in Figure 7.3, rather than uniformly. When we did this for 1 billion sequences of 7 letters,

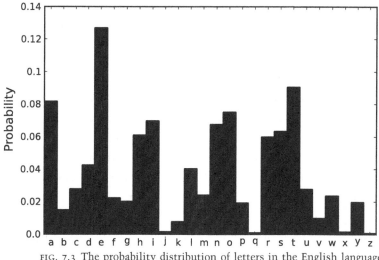

FIG. 7.3 The probability distribution of letters in the English language. Data from Lewand (2000).

we found origins twice. How large is the expected increase in likelihood?

We saw earlier that the information content of sequence s can be written as

$$I(s) = -\log \frac{N_e}{N}, \tag{7.12}$$

which itself is an approximation of the form

$$I(s) = H(S) - H(S|e), \tag{7.13}$$

assuming that the distribution of functional sequences in genetic space is uniform.[3] The remaining entropy (given the current environment $E = e$) $H(S|e)$ is not known a priori, but we can estimate it. This entropy of the polymer $s \in S$ can be written in terms of the entropy of monomers, the shared entropy of all monomer pairs, triplets, and so on, using a formula that was first derived by Fano in a very different context (Fano, 1961, p. 58):

---

[3] The distinction between the entropy written as $\log N_e$ or as $-\sum_s p(s|e) \log p(s|e)$ can viewed as the same distinction that is made in thermodynamics, where the former is known as the entropy in the "micro-canonical ensemble," whereas the latter entropy pertains to a "canonical ensemble" if $p(s|e)$ is the canonical distribution; see, e.g., Reif (1965).

$$H = \sum_{i=1}^{L} H(i) - \sum_{i>j}^{L} H(i:j) + \sum_{i>j>k}^{L} H(i:j:k) - \cdots \qquad (7.14)$$

where $H(i)$ is the entropy of the ith monomer, $H(i:j)$ is the shared entropy between the ith and jth monomer, and so on. The sum in Eq. (7.14) has alternating signs of correlation entropies, culminating with a term $(-1)^{L-1}H(1:2:3:\cdots:L)$. The per-site entropies $H(i)$ can easily be obtained if ensembles of functional molecular sequences are known, as multiple alignment of these sequences can give us the probability distribution $p(i)$ at each site. The pairwise entropies $H(i:j)$ are important too, in particular if the monomers in the polymer interact functionally, as is often the case if the sequence folds into a structure (Gupta and Adami, 2016). Here we will use only the first term in Eq. (7.14) to discuss the likelihood of information emergence by chance, but we will discuss the effect of neglecting the other terms below.

In the following, we will use the symbol I_0 for the information content of a self-replicator measured using only the first term in Eq. (7.14), given by

$$I_0 = L - \sum_{i=1}^{L} H(i). \qquad (7.15)$$

The first term in Eq. (7.15) is, of course, the first term in Eq. (7.13) if $H(S) = \log(N)$ and we agree to take logarithms to the base of the size of the alphabet. In that case, $\log_D N = \log_D D^L = L$. Using this expression, the likelihood to find self-replicators by chance is approximated as

$$P_0 = D^{-I_0} = D^{-L+\sum_{i=1}^{L} H(i)}. \qquad (7.16)$$

Let us define the "average biotic entropy" H_b as the average entropy per site for functional sequences (hence the name "biotic")

$$H_b = \frac{1}{L} \sum_{i}^{L} H(i). \qquad (7.17)$$

We distinguish this biotic entropy from the "abiotic" entropy $H_\star$, which is the entropy per site within a sequence assembled at random.

If each monomer appears with uniform probability, then the abiotic entropy is maximal: $H_\star = 1$. Using this definition, we can write Eq. (7.16) as

$$P_0 = D^{-L(1-H_b)}. \tag{7.18}$$

If we were to generate ASCII sequences with a probability distribution obtained from English words (the equivalent of the biotic sample; see Figure 7.3), the abiotic entropy would be smaller than 1 (namely, $H_\star \approx 0.89$, the entropy of the distribution in Figure 7.3) while the biotic entropy must be zero, as there is only a single origins among 7-mers. Using the probability distribution of letters in English rather than the uniform distribution raises the probability to find the 7-mer origins to

$$P_\star = 26^{-7(0.89)}. \tag{7.19}$$

This seems like a small change, but the mean number of successes out of 10^9 tries is increased from about 1 in 8 billion to 1.53 per billion. And indeed, we found the word twice when searching a billion sequences with the biased distribution shown in Figure 7.3. Note, however, that the entropy of English is equal to the entropy $\frac{1}{L}\sum_i^L H(i)$ only if sequences cannot be aligned, and therefore that all $H(i) \approx H_\star$.

Can searching with a biased probability distribution increase the chance of finding a self-replicator in Avida? We first took the 58 self-replicators we found when searching $L = 15$ sequences and created a monomer-probability distribution $p_\star$ out of them. This distribution shown in Figure 7.4 shows that within these randomly created replicating sequences, the 26 instructions appear far from uniformly in the sequence (as of course is expected), in the same way as English (because it conveys information) has a nonuniform letter distribution. The entropy of the distribution shown in Figure 7.4 is $H_\star = 0.91$ mers. According to the approximation we made above, biasing the monomer creation process using this particular probability distribution should lead to an enhancement E of the likelihood of finding a self-replicator

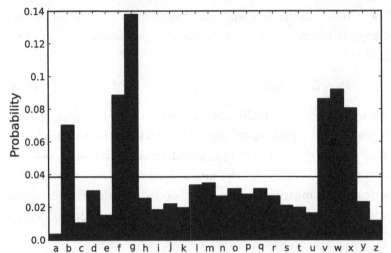

FIG. 7.4 The biased probability distribution p_* of Avida instructions obtained from the genomes of 58 randomly generated $L = 15$ replicators (the meaning of each letter is described in Table 7.1). The black line represents the probability for a uniform distribution.

$$E = \frac{P_*}{P_0} \approx \frac{D^{-L(H_* - H_b)}}{D^{-L(1 - H_b)}} = D^{L(1 - H_*)}. \qquad (7.20)$$

Eq. (7.20) suggests that the enhancement factor E depends only on the bias in the distribution and the length of sequence. However, we should not be fooled into believing that any reduced entropy H_* will lead to an enhancement in the probability to find self-replicators by chance: the distribution p_* needs to be close to the distribution of actual replicators. For example, omitting the instruction "x" (the h-divide instruction that splits off a completed copy; see Table 7.1) certainly leads to an entropy less than one, but using such a biased distribution cannot net a self-replicator, as h-divide is required for replication.

We proceeded to test Eq. (7.20), by searching for self-replicators using the biased distribution p_* (see the section on methods). Among a billion sequences of $L = 15$ generated in this manner, we found 14,495 self-replicators, an enhancement of $E = 14,495/58 \approx 250$, while Eq. (7.20) predicted an enhancement of $E = 81.3$. We also

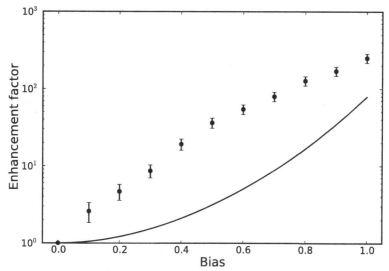

FIG. 7.5 The enhancement factor E to find entropies for genomes of 15 instructions as a function of the bias, using an interpolated probability distribution $p(i, b)$. Here, $b = 0$ means unbiased, and $b = 1$ uses a fully biased distribution $p_\star$. Black dots represent estimates (calculated as the number of self-replicators per 10^8 for a biased distribution divided by the number of self-replicators per 10^8 for a uniform distribution), while error bars are standard deviations. The solid line is the naive prediction given by Eq. (7.20).

tested whether changing the probability distribution from uniform gradually toward $p_\star$ leads to a gradual increase in the E. The empirical enhancement factor shown in Figure 7.5 indeed increases with the bias and is larger than the one predicted from the simple approxima-tion Eq. (7.20). This difference is likely due to a number of effects. On the one hand, we are neglecting any higher-order correlations in Eq. (7.14). On the other hand, we are assuming that $H_\star \approx H(i)$ for all i, that is, that the entropy at each site is the same. This is not at all true for functional sequences that can be aligned (see, e.g., Adami and Cerf, 2000; Adami et al., 2000; Gupta and Adami, 2016). Sequences that are obtained from a random procedure (rather than from an evolutionary process) are likely difficult to align, and therefore $H_\star \approx H(i)$ may hold.

The enhancement works for sequences of any length, but depends on how well the biased distribution represents actual

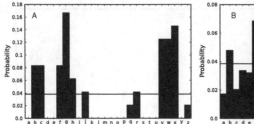
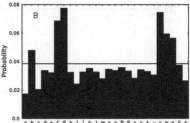

FIG. 7.6 Probability distribution of instructions. (A) $p_\star(8)$ obtained from the replicators of length $L = 8$, giving rise to an entropy $H_\star(8) = 0.71$ mers. (B) $p_\star(30)$ obtained from the replicators of length $L = 30$, giving rise to an entropy $H_\star(30) = 0.98$ mers. The solid horizontal line denotes the uniform probability distribution $1/26$ in both panels.

functional replicators. For example, as we found only 6 self-replicators of length 8, the distribution $p_\star(8)$ is fairly coarse (see Figure 7.6A), while the distribution we obtained from the 106 $L = 30$ replicators has a significant uniform contribution (Figure 7.6B), because among the 30 instructions only a handful need to carry information in order for the sequence to be able to replicate. We show in Figure 7.7 the enhancement achieved by biasing the search for each of the three length classes $L = 8, 15$, and 30.

Could we use the probability distribution for sequences obtained in one length group to bias the search in another length group? Such a procedure might be useful if the statistics of monomer usage is poor (as for the case $L = 8$) or if the distribution was obtained from a sequence with too much entropy (as for the case $L = 30$). It turns out that this is not the case: biasing the $L = 30$ search using $p_\star(15)$ does not work well (144.3 replicators found per 10^8) compared with biasing with the "native" $p_\star(30)$ (297 per 10^8). In the same manner, biasing the $L = 8$ search works best with the "native" bias $p_\star(8)$, yielding 230 per 10^8, as opposed to only 15.8 per 10^8 biasing with $p_\star(15)$.

Finally we asked whether taking the self-replicators obtained from a biased search (and that consequently nets many more replicators) gives rise to a more accurate probability distribution $p_\star$, which then could be used for a more "targeted" biased search. By "rebiasing"

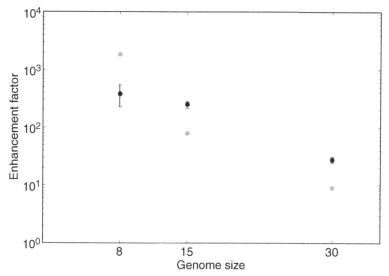

FIG. 7.7 Empirical enhancement factor (black dots, with 1σ counting error), along with the predicted enhancement factor using the entropy of the distribution based on Eq. (7.20) (gray dots) for $L = 8, 15$, and 30.

successively (see the section on methods), we did indeed obtain more and more replicators, albeit with diminishing returns (see Figure 7.8).

DISCUSSION

One of the defining characteristics of life (perhaps *the* defining characteristic) is that life encodes information, and information leaves a trace in the monomer abundance distribution (a nonrandom frequency distribution) (Dorn and Adami, 2011; Dorn et al., 2011) of the informational polymers. As life evolves, the information contained in it increases on average (Adami et al., 2000), but evolution cannot explain where the first bits came from. Information can in principle arise by chance, just as an English word can appear by chance within an ASCII string that is created randomly, as per the "dactylographic monkeys" metaphor. The "infinite monkey theorem" posits that a million monkeys typing on a million keyboards, if given enough time (and typing randomly) could ultimately type out all of Shakespeare's works. However, the theorem is misleading, as even correctly typing

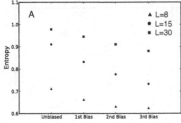
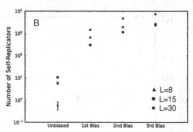

FIG. 7.8 (A) Average per-site entropy H_* for replicators in different length classes, at various stages of biasing. "Unbiased" reports the average per-site entropy obtained from the self-replicators that were found in an unbiased search, and whose biased distribution was used to find the self-replicators whose average per-site entropy is shown in "1st Bias." Those in turn were used for a biased search that gave rise to replicators with bias shown in "2nd Bias," and so on. (B) Number of self-replicators (per billion) found at each biasing stage. Biasing the distribution with more "focused" probability distributions p_* leads to an increasing yield of self-replicators, albeit with a diminishing return. In rebiasing with $L = 8$, some duplicate sequences were obtained, and those are not included in the count.

out the first 30 characters of Hamlet's soliloquy ("To be or not to be ...") cannot occur during the time our universe has been around (about 4.36×10^{17} seconds), as Hamlet's 30-mer is one in about 3×10^{42}. Using biased typewriters will not allow the monkeys to finish either, as it is only accelerating the search by a factor $E \approx 46{,}700$.

We can ask whether more sophisticated methods of biasing exist. One look at Eq. (7.14) suffices to answer this question in the positive. We could begin by generating sequences biased in such a way that the more common 2-mers are generated with increased likelihood. In English text, for example, the "dimers" "th", "he", and "in" appear with frequencies 3.56%, 3.07%, and 2.43%, respectively, which are significantly larger than the random dimer expectation $\approx 0.15\%$. Indeed, as the frequency of "or" is 1.28%, while "ig" appears at 0.255%, our 7-mer origins would be found fairly fast. Likewise, in our 6 replicators of length $L = 8$ the dimer gb appears significantly more often than expected by the product of the likelihood of g and b.

Such biased search procedures can also accelerate the search for functional biomolecules where the target is a function other than self-replication. For example, when designing random peptide libraries (either for screening purposes or to perform directed evolution), researchers often bias the codons in such a way that the stop codon is rare (so-called NNB or NNS/NNK libraries (Barbas et al., 1992)). Hackel et al. (2010) went beyond such simple biases and constructed a protein library to screen for binding to a set of 7 targets. To bias the random sequences, they mimicked the amino-acid distribution in human and mouse CDR-H3 loops (complementarity determining regions, which are found in antibodies) and found that such a library outcompetes even NNB libraries significantly: of the 20 binders that they found, 18 were traced back to the CDR-biased library.

The implications of the present theoretical and computational analysis of the emergence of informational "molecules" by chance for the problem of understanding the origin of life are straightforward. It is well known that monomers do not form spontaneously at the same rate. The abiotic distribution of amino acids is heavily skewed both in spark synthesis experiments and in meteorites (Dorn et al., 2011), and the same is true for other monomers such as carboxylic acids and many other candidate alphabets in biochemistry. In many cases, the abiotic skew (often due to thermodynamic considerations) will work against the probability of spontaneous emergence of information, but in some cases it may work in its favor. In particular, we might imagine that in complex geochemical environments the abiotic distributions can be significantly different in one environment compared with another, raising the chance of abiogenesis in one environment and lowering it in another.

We also immediately note that in chemistries where molecules do not self-replicate but catalyze the formation of other molecules, the abundance distribution of monomers would change in each catalysis step. If these monomers are recycled via reversible polymerization (Walker et al., 2012), then the activity of the molecules can change the entropy of monomers, which in turn changes the likelihood of spontaneous discovery. Should this process "run in the right direction," it is possible that self-replicators are the inevitable outcome. This hypothesis seems testable in digital life systems such as Avida.

Methods

In order to explore the spontaneous emergence of self-replicators in Avida, we generated random genomes of length L. These genome sequences were generated with different probability distributions for the Avidian instructions (we used Avida version 2.14, which can be downloaded from https://github.com/devosoft/Avida). First, we generated 10^9 random genomes for lengths $L = \{8, 15, 30\}$ and 3×10^8 sequences for $L = 100$ with an unbiased (i.e., uniform) instruction distribution 1/26 (because there are 26 possible instructions). In order to decide whether a genome could successfully self-replicate, we performed two tests. First, we checked whether the organism would successfully divide within its lifespan. Here, we used the traditional Avida parameters for an organism's lifespan: it must divide before it executes $20 \times L$ instructions. While this indicates that an Avidian could successfully reproduce, it does not imply that the Avidian's descendants could also reproduce. In our search we found many viable Avidians that would successfully divide into two nonviable organisms. Therefore, we only counted Avidians that could self-replicate and produce offspring that could also self-replicate as true self-replicators (in other words, they are " colony-forming"). This does not mean that every self-replicator would produce a perfect copy of itself in the absence of mutation; in fact, most of these replicators undergo implicit mutations solely due to their genome sequence, and their offspring differ in length from the parent (LaBar et al., 2015). In analyzing a genome's ability to self-replicate, we used the default Avida settings, described for example by Ofria et al. (2009).

Next, we generated random genome sequences with a biased instruction distribution. These biased distributions were calculated by altering the probability that each instruction was generated by our random search. The probability of an instruction i being generated for a biased search was set at

$$p(i, b) = (1 - b)(1/26) + bp_\star(i), \tag{7.21}$$

where b is the bias, $0 \leq b \leq 1$, and $p_\star(i)$ is the probability that instruction i appears in the set of all genomes that were classified as self-replicators in the unbiased search. When $b = 0$, the distribution is the uniform distribution, and when $b = 1$, the distribution is the frequency distribution for the instructions in the set of self-replicators $p_\star$ found with the unbiased search for a given length. The parameter b allows us to set the bias, and thus the entropy, of the distribution to detect the role of the instruction entropy in determining the likelihood of spontaneous self-replicator emergence. For genomes of $L = 15$, we generated 10^9 random sequences with $b = 1$ and 10^8 random sequences with $b = \{0.1, 0.2, 0.3, 0.4, 0.5, 0.6, 0.7, 0.8, 0.9\}$.

Finally, we performed searches where we iteratively biased the distribution of instructions. First, we generated self-replicators with an unbiased instruction distribution. We then created another set of self-replicators with a biased distribution of instructions using the above equation with $b = 1$ (referred to as "1st bias"). However, as opposed to stopping the self-replicator generation process, we then searched for self-replicators two more times (referred to as "2nd bias" and "3rd bias"). Each time, we used the set of self-replicators from the previous bias: the distribution of instructions for the 2nd bias was derived from the set of self-replicators obtained from the 1st bias, and the distribution of instructions for the 3rd bias was derived from the set of self-replicators from the 2nd bias (in both of these we set $b = 1$). We generated 10^8 random genomes using the 1st bias for $L = \{8, 30\}$ and 10^8 random genomes using the 2nd and 3rd bias for $L = \{8, 15, 30\}$ with a biased instruction distribution. For $L = 15$, we used the 10^9 random genomes described above to obtain the 1st bias.

Acknowledgments

We thank Arend Hintze and Charles Ofria for extensive discussions, and Piet Hut and Jim Cleaves for the suggestion to carry out the kind of computational work presented here. This work was supported by the National Science Foundation's BEACON Institute for the Study of Evolution in Action under contract No. DBI-0939454. We gratefully acknowledge the support of the Michigan State University High Performance Computing Center and the Institute for Cyber Enabled Research (iCER).

REFERENCES

Adami, C. 1998. *Introduction to artificial life*. Springer Verlag.

Adami, C. 2002. What is complexity? *BioEssays*, **24**(12), 1085–1094.

Adami, C. 2006. Digital genetics: unravelling the genetic basis of evolution. *Nature Reviews Genetics*, **7**(2), 109–118.

Adami, C. 2012. The use of information theory in evolutionary biology. *Annals NY Acad. Sci.*, **1256**, 49–65.

Adami, C. 2015. Information-theoretic considerations concerning the origin of life. *Origins of Live and Evolution of the Biospheres*, **45**, 9439.

Adami, C., and Brown, C. T. 1994. Evolutionary learning in the 2D artificial life system Avida. Pages 377–381 of Brooks, R., and Maes, P. (eds.), *Proceedings of the 4th International Conference on the Synthesis and Simulation of Living Systems (Artificial Life 4)*. MIT Press.

Adami, C., and Cerf, N. J. 2000. Physical complexity of symbolic sequences. *Physica D*, **137**(1), 62–69.

Adami, C., Ofria, C., and Collier, T. C. 2000. Evolution of biological complexity. *Proc. Natl. Acad. Sci. U.S.A.*, **97**(9), 4463–4468.

Arrhenius, S. 1908. *Worlds in the making: The evolution of the universe*. Harper & Row.

Barbas, C. F. III, Bain, J.D., Hoekstra, D. M., and Lerner, R. A. 1992. Semisynthetic combinatorial antibody libraries: a chemical solution to the diversity problem. *Proc Natl Acad Sci U S A*, **89**(10), 4457–4461.

Benner, S. A., Ricardo, A., and Corrigan, M. A. 2004. Is there a common chemical model for life in the universe? *Curr. Opin. Chem. Biol.*, **8**, 672–689.

Davies, P. C. W, Benner, S. A., Cleland, C. E., Lineweaver, C. H., McKay, C. P., and Wolfe-Simon, F. 2009. Signatures of a shadow biosphere. *Astrobiology*, **9**(2), 241–249.

Deamer, D. 1994. *Origins of life: the central concepts*. Jones & Bartlett.

DeDuve, C. 1995. *Vital dust: life as a cosmic imperative*. Basic Books.

Dorn, E. D., and Adami, C. 2011. Robust monomer-distribution biosignatures in evolving digital biota. *Astrobiology*, **11**(10), 959–968.

Dorn, E. D., Nealson, K. H., and Adami, C. 2011. Monomer abundance distribution patterns as a universal biosignature: examples from terrestrial and digital life. *J Mol Evol*, **72**(3), 283–295.

England, J. L. 2013. Statistical physics of self-replication. *J Chem Phys*, **139**(12), 121923.

Fano, R. M. 1961. *Transmission of information: a statistical theory of communication*. MIT Press and John Wiley.

Gupta, A., and Adami, C. 2016. Strong selection significantly increases epistatic interactions in the long-term evolution of a protein. *PLOS Genetics*, **12**, e1005960.

Hackel, B. J., Ackerman, M. E., Howland, S. W., and Wittrup, K. D. 2010. Stability and CDR composition biases enrich binder functionality landscapes. *J Mol Biol*, **401**(1), 84–96.

Hazen, R. M., Griffin, P. L., Carothers, J. M., and Szostak, J. W. 2007. Functional information and the emergence of biocomplexity. *Proc Natl Acad Sci U S A*, **104 Suppl 1**(May), 8574–81.

Hoyle, F., and Wickramasinghe, N. C. 1981. *Evolution from space*. Simon & Schuster.

Koonin, E. V. 2011. *The logic of chance: the nature and origin of biological evolution*. FT Press.

LaBar, T., Adami, C., and Hintze, A. 2015. Does self-replication imply evolvability? Pages 596–602 of Andrews, P., Caves, L., Doursat, R., Hickinbotham, S., Polack, F., Stepney, S., Taylor, T., and Timmis, J. (eds.), *Proc. of European Conference on Artificial Life 2015*. MIT Press.

Lazcano, A., and Miller, S. L. 1996. The origin and early evolution of life: prebiotic chemistry, the pre-RNA world, and time. *Cell*, **85**(6), 793–798.

Lewand, R. E. 2000. *Cryptological mathematics*. Mathematical Association of America.

Mathis, C., Bhattacharya, T., and Walker, S. I. 2015. *The emergence of life as a first order phase transition*. arXiv:1503.02777.

Mayfield, J. E. 2013. *The engine of complexity: evolution as computation*. Columbia University Press.

Morowitz, H. J. 2004. *Beginnings of cellular life: metabolism recapitulates biogenesis*. Yale University Press.

Nowak, M. A., and Ohtsuki, H. 2008. Prevolutionary dynamics and the origin of evolution. *Proc Natl Acad Sci U S A*, **105**(39), 14924–14927.

Ofria, C., Bryson, D. M., and Wilke, C. O. 2009. Avida: a software platform for research in computational evolutionary biology. Pages 3–35 of Adamatzky, A., and Komosinski, M. (eds), *Artificial life models in software*, 2nd ed. Springer Verlag.

Ofria, C., and Wilke, C. O. 2004. Avida: a software platform for research in computational evolutionary biology. *Artificial Life*, **10**(2), 191–229.

Pargellis, A. N. 2003. Self-organizing genetic codes and the emergence of digital life. *Complexity (Wiley)*, **8**, 69.

Patel, B. H., Percivalle, C., Ritson, D. J., Duffy, C. D., and Sutherland, J. D. 2015. Common origins of RNA, protein and lipid precursors in a cyanosulfidic protometabolism. *Nature Chemistry*, **7**, 301–307.

Pattee, H. H. 1961. On the origin of macromolecular sequences. *Biophys J*, **1**(Nov), 683–710.

Rands, C. M., Meader, S., Ponting, C. P., and Lunter, G. 2014. 8.2% of the human genome is constrained: variation in rates of turnover across functional element classes in the human lineage. *PLOS Genet*, **10**(7), e1004525.

Reif, F. 1965. *Fundamentals of statistical and thermal physics*. McGraw-Hill.

Segre, D., Dafna, B., and Lancet, D. 2000. Compositional genomes: prebiotic information transfer in mutually catalytic noncovalent assemblies. *Proc. Natl. Acad. Sci. USA*, **97**, 4112–4117.

Smith, E. 2008. Thermodynamics of natural selection I: energy flow and the limits on organization. *J Theor Biol*, **252**(2), 185–197.

Szostak, J. W. 2003. Functional information: molecular messages. *Nature*, **423**, 689.

Vasas, V., Fernando, C., Santos, M., Kauffman, S., and Szathmary, E. 2012. Evolution before genes. *Biol Direct*, **7**, 1.

Vetsigian, K., Woese, C., and Goldenfeld, N. 2006. Collective evolution and the genetic code. *Proc Natl Acad Sci U S A*, **103**(28), 10696–10701.

Walker, S. I., Grover, M. A., and Hud, N. V. 2012. Universal sequence replication, reversible polymerization and early functional biopolymers: a model for the initiation of prebiotic sequence evolution. *PLOS ONE*, **7**(4), e34166.

Wickramasinghe, C. 2011. Bacterial morphologies supporting cometary panspermia: a reappraisal. *International Journal of Astrobiology*, **10**, 25–30.

Part III Life's Hidden Information

8   Cryptographic Nature

David Krakauer

Nature loves to hide.

– Heraclitus

SECRET LIFE

The principle of energy conservation underpins much of physics and chemistry, to include mechanics, thermodynamics, and relativity (Mach, 1910). By contrast, the idea of statistical information provides a common language for genetics, development, and behavior, where information encodes "coordinates" for free energy landscapes – memory required by search strategies directed at appropriating metabolic resources (Krakauer, 2011). This memory is often encoded in genetic sequences that express enzymes, signaling molecules, and receptors, all of which can orient toward or process metabolic free energy. Whereas our understanding of energy can be reduced ultimately to symmetry principles (Brading and Castellani, 2003), information is derived from its antithesis, symmetry breaking (Gell-Mann, 1995). Stored information records which of several alternative adaptive configurations have been selected historically.

Adaptive matter requires metabolic energy for biological work – growth, reproduction, and repair. These requirements extend upward to populations, ecosystems, and even cities. The patchy distribution of free energy in space and time, combined with the fact that energy is transformed and thermalized when metabolized, requires efficient and accurate procedures for energy seeking. This scarcity problem is the reason why life can be thought of as a suite of evolved inferential mechanisms dependent on both memory storage – information – and "computation" for adaptive search.

Evolution by natural selection – a population-based search mechanics – produces outcomes in which information is restricted and concentrated ensuring that energy is distributed nonuniformly and anticompetitively: sequestered within cells, bodies, and communities, where it can be preferentially utilized and monopolized (Krakauer et al., 2009).

Organizations – cells, organisms, and populations – with accurate information about the whereabouts of metabolic energy sources endeavor to keep this information to themselves and share informative signals only with those with whom they have found a means to cooperate. These preferred signals include those generating the immunogenic self versus nonself, mating types, restriction systems, species isolating mechanisms, and, in some cases, languages.

The best way to restrict the flow of information is to protect it or to encrypt it (Shannon, 1949). This is the subject of this chapter: the many ways in which evolved information flows are restricted and metabolic resources protected and hidden, the thesis of living phenomena as evolutionary cryptosystems.

IMITATIVE ENTROPY

What does it mean for life to be a secret? If it is a secret, it seems to be a secret like Edgar Allan Poe's purloined letter, a secret that is hiding in plain sight (Poe et al., 1976).

One way to think about adaptive secrecy is that it is a necessary step in the evolution of life – protolife and ongoing – ensuring that replicative and metabolic mechanisms and products are concentrated and protected. Fidelity and privacy are required for continued replicative kinetics.

The replicator–mutator equation (Nowak, 2006) describes the frequency of strains x_i in an evolving population of n strains subject to mutation Q_{ij} and selection k_i:

$$\dot{x}_i = \sum_j Q_{ij} k_j x_j - x_i \bar{k}(\mathbf{x})$$

The operator Q is a bistochastic matrix of transition probabilities. The dynamics are restricted to the simplex S_n, $\sum_i x(t)_i = 1$, and the population mean fitness is defined as $\bar{k}(x(t))$. Under the operation of Q, trajectories flow from outside the positive orthant into S_n with new strains emerging through mutation that are not present in the initial distribution of $x_i(0)$. It is a classical result of evolutionary dynamics that the elements of the Q matrix select between ergodic and nonergodic equilibrium states; if the off-diagonal elements of Q sum to a value greater than a critical threshold value ("error threshold"), then the distribution of $x_i(t) \rightarrow 1/n$ (Eigen, 2000).

The vector k_i encodes the information that a strain X_i possesses about the environment in which it lives. This information is assumed to translate directly into its instantaneous rate of growth $k_i x_i$ and instantaneous relative fitness $x_i(k_i - \bar{k}(\mathbf{x}))$. Such selection equations tell us nothing about the adaptive mechanisms through which the information in $\mathbf{k}$ is acquired. This requires an adaptive dynamics for $\mathbf{k}$. Under a purely positive imitative dynamics in which $k_i > 0$ is information and $k_i = 0$ is zero information (ignoring the original source of the information), we can include learning,

$$\dot{k}_i = x_i \sum_j c_{ij} x_j |k_i - k_j| - e k_i$$

where c_{ij} is a matrix of imitation rate constants and e is an entropic parameter that introduces the loss of information. When information is barely lost $e \approx 0$ and $Q_{ii} = Q_{jj}$ for $\forall_j \in n$ (all strains are equally mutable) and $c_{ij} = c$, this learning rule ensures that all strains evolve to a constant quantity of information and the population will converge on $x_i \rightarrow 1/n$. Thus without some form of secrecy, evolution becomes "neutral" as a result of imitative entropy much the same way that evolution becomes a pure drift process above a dissipative error threshold. For evolution to proceed we need to place restrictions – encryption – on the imitation rate matrix $\mathbf{c}$. In order to understand constraints on $\mathbf{c}$ we consider in more detail the structure of individual strains in terms of combinatorics and information theory.

THE COMMUNICATION THEORY OF SECRECY SYSTEMS

The essential elements of a secrecy system are a message (sometimes called the plaintext) and an encrypter that makes use of a key to transform the message into a cryptogram or ciphertext. The cryptogram is transmitted – whereupon it can be intercepted – to a decrypter that yields the original message using prior knowledge of the key (Shannon, 1949).

For example, a "one-time pad" is a method of encryption where each bit of a plaintext is encrypted by combining it with the corresponding bit or character from a pad/key using modular addition. The key is random and of the same length as the message. Since every cryptogram is associated with a unique random key, the cryptogram cannot be deciphered; it is information-theoretically secure.

Shannon proved that information-theoretic security is a general property of any effective probabilistic cryptogram. Define the entropy of a message source as $H(M)$, the entropy of keys as $H(K)$, and the entropy of cryptograms as $H(C)$. The desirable properties of an encrypted message can be stated in terms of a triplet of inequalities using mutual information:

(1) There is no information in the messages about the cryptograms, $I(M, C) = 0$.
(2) The information in the cryptogram about the key is greater than or equal to zero, $I(C, K) \geq 0$.
(3) The entropy/uncertainty in the key must be greater than or equal to the entropy/uncertainty in the message, $H(K) \geq H(M)$.

This is the requirement for "theoretical security," meaning the cryptogram can never be decoded even with unlimited computing resources. Another way to present this idea is in terms of combinatorics (Massey, 1986).

Consider a set of messages M. This set consists of 2^l binary sequences of length n, where $n \geq l$. The one-time pad has a key with length n where all 2^n keys are equally probable. We calculate the cryptogram through modular addition,

$$C = K \oplus M$$

where $\oplus$ is addition modulo 2. This can be shown to ensure that $P(C = c|M = m) = 2^{-n}$. The probability of discerning a given cryptogram for a given message is strictly chance, and the message and the cryptogram are statistically independent. Hence the key must be of the same dimension as the message when a unique key is used to send a single message.

When a potential attacker has prior information, the demands on the key grow exponentially. For example, if we assume that the attacker has almost perfect knowledge of the cryptosystem with a memory of $2^l - 2$ message–cryptogram pairs, then one requires 2^n bits of key information per individual bit in the message. This requirement was not met in the case of the Enigma machine and lead to its eventual deciphering.

The take-home message is that secrecy requires that the information content of a key exceed the information content of a message. In terms of imitative entropy, this implies that any adaptive system vulnerable to an imitation-exploit requires for ongoing evolution an equal quantity of randomizing information as functional information.

FUNCTIONAL REQUIREMENTS FOR EVOLUTIONARY ENCRYPTION

The only sure way an evolving lineage can overcome imitation-exploits is to set to zero its vulnerable imitation coefficients in the matrix **c**. In order to do this, any visible component of an adaptive strategy needs to be mathematically composed with an effectively random key in order to generate an adaptive "one-way-function." A one-way function is a function that is computationally hard to invert given a random input (Katz and Lindell, 2007). For example, for most pairs of primes their integer product cannot be inverted. For a key-encrypted message, such a function, $C = F_K(M)$, has the property that finding M by inverting F_K requires an exhaustive search without the key. Note that once this inversion is found it will be of immediate selective value to an imitator.

Selective one-way functions can be generated a number of different ways. Pseudorandom generators provide algorithms for generating sequences that are indistinguishable from random. Pseudorandom permutations build on pseudorandom functions to generate permutations of sequences in memory with uniform probability. A common example of a pseudorandom permutation would be a block cipher that involves an encryption function, $F_K(M)$, and an inverse decryption function, $F_K^{-1}(C)$. The encryption function breaks information into two blocks of length k and m for key and message and generates a cryptogram of length n:

$$F_K(M) := F(K, M) : \{0, 1\}^k \times \{0, 1\}^m \to \{0, 1\}^n$$

The decryption achieves the inverse,

$$F_K^{-1}(C) := F^{-1}(K, C) : \{0, 1\}^k \times \{0, 1\}^m \to \{0, 1\}^n,$$

such that

$$\forall_K : F_K^{-1} F_K(M) = M.$$

This typically implies that for every unique choice of key one permutation from a set of size $(2^n)!$ is selected.

There are a variety of algorithms in computer science that implement block ciphers; of interest here are those that plausibly map onto candidate adaptive one-way functions.

CANDIDATE ADAPTIVE ONE-WAY-FUNCTIONS

In this section I review a small number of naturally occurring "eco-algorithms" (Valiant, 2013) that generate a significant amount of randomness in biological sequences that prevents some form of natural "signal detection." In a few cases, the function of randomization is not known; in others, there is more or less direct evidence for a form of encryption that achieves protection through reduced competition. It is evident, however, that randomization in the service of crypsis is ubiquitous in evolved organisms even if the encryption is approximate and far from the limit of "information-theoretic" security.

Shuffled Parasites

The best-studied mechanism for generating random variation in order to confuse would-be surveillants is antigenic variation in endoparasites. Parasites that live within long-lived hosts expose their surface proteins to the innate and adaptive immune system. Once recognized, parasites can be neutralized by antibodies that target exposed markers (antigens) and induce a panoply of clearance mechanisms, including agglutination and phagocytosis. It is in the interest of parasites to present "noise" to their hosts and thereby evade surveillance. The same is true of cancer cells but to a far lesser degree, as somatic cells have evolved not over many epochs of Earth's history to evade detection but only over the ontogenetic time scale of the cancerous organism.

The protozoan *Trypanosoma brucei* causes sleeping sickness and is covered in a protective "variant surface glycoprotein" (VSG) that acts somewhat like a cloak of invisibility from the immune systems (Hutchinson et al., 2007). Within a single clone of a single infection it is not unusual for more than 100 VSG strains to be cocirculating. In order for the parasite to be transmitted to a new host, at least one of these strains needs to evade detection. The VSG "key" is composed of one protein from a reservoir of between 1,000 and 2,000 "mosaic gene" sequences carried on about 100 mini-chromosomes (estimated to be around 10% of the total genetic information). For every cell generation, a new gene from which the VSG is translated is activated stochastically with a low probability ($p \approx 0.01$). Moreover, the mosaic genes – as the name suggests – are generated through stochastic sequence recombination. Thereby trypanosomes give the appearance of randomness to each immune system that they encounter and reduce the ability of immune memory to mount an effective response to new infections. Comparable randomizing mechanisms are found broadly across the protozoans and the fungi (Verstrepen and Fink, 2009).

Combinatorial Ciliates

The ciliates are an extremely diverse class of unicellular organisms that includes the the freshwater and free-living *Paramecia* and *Tetrahymena*, and the giant *Stentor*, or trumpet, animalcule. Ciliates have two defining characteristics: they are fringed by small hair-like cilia, giving them under the microscope the appearance of independent-minded eyelids fringed by beautiful eyelashes, and they contain two genomes – a small micronucleus (MIC) and a large macronucleus (MAC).

The MIC acts as the reproductive germline and is passed from parent to progeny, whereas the macronucleus (MAC) performs the somatic functions of transcription and translation required for growth and asexual cell division. The MAC is generated by the MIC at the start of each new generation through an incredibly complicated sequence of genetic rearrangements, concatenations, and amplifications (Prescott, 2000). If the MAC is Melville's *Moby Dick* (around 200,000 words), then the MIC is T. S. Eliot's *The Waste Land* (around 3,000 words). We further require that *Moby Dick* is written through an ingenious shuffling and deletion(!) of content from *The Waste Land*.

The ability to construct the MAC requires unshuffling the apparently random MIC into highly ordered genetic sequences using a variety of epigenetic mechanisms, including RNA interference (RNAi) and RNA-binding proteins. In some ciliates, around 90% of the spacer DNA in the MIC is destroyed in preparing the MAC, and this seems to play the role of a "cipher key" in allowing the adaptive sequence required by the MAC to be decoded in each new generation (Bracht et al., 2013).

Unlike the scrambling of the immunogenic surface proteins by protists, it is not known why the MIC of the ciliates is so highly randomized. The discoverer of these mechanisms, David Prescott, has suggested that adaptive variability supporting evolvability could play a role (Prescott, 2000). The hypothesis that I favor is that of encryption that keeps critical information required for replication away from parasites such as viruses that might otherwise appropriate MIC genes for their own translational objectives. For a virus to to steal from the MIC it would need to discover and encode the full ciliate decryption function, $F_K^{-1}(C)$, which given the size constraints placed on viruses, is effectively impossible.

The Genetics of Speciation

Species are defined in many different ways, but one feature shared by all definitions is a reduced rate of genetic recombination between species over that within a species. The theory that best accounts for this empirical regularity was provided by Dobzhansky (1936) and then Muller (1939) (Orr, 1996). Their idea was that two genetic loci shared by two diploid organisms (aa, bb) diverge into two incompatible genomes, (Aa, bb) and (aa, Bb). The new genomes that can arise from selfing among these mutants (AA, bb) and (aa, BB) remain viable, whereas mixtures (Aa, Bb) do not.

The argument is that the alleles A and B have not previously met and coadapted, hence are likely to interact deleteriously. It is now known that numerous so-called complementary loci play a role in promoting incompatibility between organisms and that many of the alleles that confer incompatibility do little else but promote species isolation. In other words "speciation genes" or, perhaps more accurately, speciation networks, behave like cryptographic keys ensuring that genetic information conferring locally adaptive information cannot be decoded on recombination. As Orr has shown, these complementary loci ("cryptographic loci") are expected to accumulate at an exponential rate through a mutation-selection process, and so, in time, genomes will house a very significant quantity of encryption (Orr, 1995).

CRYPTOGRAPHIC COMPLEXITY

An immediate implication of "cryptographic nature" is that a considerable quantity of stored information is required to generate effective keys. The more historical knowledge competitors have stored about the correlation of messages to cryptograms, the more key-related resources are required.

By analogy with engineered cryptosystems, it is anticipated that key-related, random information will grow and possibly overshadow functional information. This has the implication of creating a rather perverse scaling law. As the quantity of pragmatic adaptive information (information of inferential and metabolic value to an organism) increases, the quantity of key-related information increases exponentially. Thus apparently "junk" sequences could grow to exceed coding sequences, and this effect will be most pronounced in those lineages with the most functional information. Note that this junk need not be true junk in the sense of random genetic material but regulatory sequences implicated in the production of functional proteins.

The "c-Value" Paradox

It has been recognized for several decades that variation in genome size does not correlate in an obvious way with variation in organismal phenotypes. For example, some flowering plants contain up to 10^{11} base pairs, whereas some insects have as few as 10^8 base pairs, and mammals have not been measured to exceed 10^{10} base pairs. This variation seems to be at odds with the feature sets of these lineages – why should small, sedentary plant genomes exceed the genome size of large and behaviorally complicated mammals? The "characteristic-value" ("c-value") paradox describes the surprise biologists feel when confronted with these facts (Eddy, 2012).

The dominant explanation for the c-value anomaly is that genomes contain large quantities of "junk DNA," DNA that has not been hitherto implicated in coding for functional proteins. This has long been thought to reflect the replicative efforts of selfish transposons – sequences that endlessly "copy and paste" themselves throughout a genome with impunity. However, the recent project ENCODE finds that more than 80% of junk is actively transcribed, and this at least challenges the most naive selfish replicator model.

Of significant interest has been the discovery of a veritable zoo of regulatory RNA molecules (Dandekar and Sharma, 1998) and conserved noncoding sequences (Dermitzakis et al., 2005). Both of these classes of genetic sequence are known to play a crucial role in regulating their less numerous "coding" cousins.

The c-value hypothesis suggests that this noncoding sequence could play the role of a "key" to obscure adaptive information that might otherwise be appropriated by eavesdroppers such as parasites. This is of course one of the key hypotheses for the evolution of recombinant sexuality (Hamilton et al., 1990). Comparative studies on bacterial genome variation suggest a possible role for parasitism in inducing an expansion in genome size (Wernegreen, 2005) and correspondingly a reduction in genome size attendant on adopting mutualistic relationships (Krakauer, 2002).

COMPLEXITY SCIENCE AS CRYPTOGRAPHIC INQUIRY

Theoretical science seeks to achieve a compression of empirical regularities. Variability in replicated observations is accounted for in terms of minimal models that are (1) explanatory, to imply that they are interoperable with models for nominally distinct observations often at more microscopic levels; (2) predictive, capable of producing verifiable outputs for inputs not in the domain of model construction, namely out of sample fidelity; and (3) comprehensible, revealing of causal relationships between observables. The phrase "mechanistic model" stresses properties 1 and 3. Statistical models stress property 2. Phenomenological models stress properties 2 and 3. The more "fundamental" an observation, the more successful we have been at achieving all three goals through minimal models. The theory of electromagnetism would be a prototypical example.

In the domain of "complex" phenomena we struggle to obtain all of these properties in a minimal model. All physical systems can be decomposed into regular and noisy quotients, but complex systems are typified by an unusually high contribution from randomness. And this randomness can be of a special kind that introduces new forms of regularity. Forms of "fixed" randomness that produce new regularities allow for a direct connection to encryption.

Frozen Accidents

To clarify the role of "regularized randomness" or "frozen accidents" (Gell-Mann, 1995), consider the world of one-dimensional cellular automata (CA). CA are characterized by a 5-tuple,

$$C = (S, s_0, N, d, f) \tag{8.1}$$

The value S is the set of states and s_0 is the initial state of the CA. The neighborhood is N, the dimension d, and the local update rule is $f : S^i \to S$. At each point in time the global system is updated, $F : C_t \to C_{t+1}$. We shall consider $d = 1$ and a binary state space $S \in \{0, 1\}$. The neighborhood is defined by a symmetric radius, $r = 1$ – nearest neighbors, in which case $N = \{-r, 0, r\}$, which implies a neighborhood size of 3.

For this kind of CA there are 256 update rules given by the state of three binary random variables, p, q, r. For example, rule 240 depends only on p. Hence this rule is indifferent to the values of q and r. An initial state $s_0 = (p = 1, q = 0, r = 0)$ generates a global pattern that resembles a diagonal line oriented at 45% to the right of vertical. As does any initial state defined by $s_0 = (1**)$. At some time τ the system will possess a state S_τ.

Now assume a stochastic dynamic that operates on the space of update rules, $D(i, j) : f^{(i)} \to f^{(j)}$, where i indexes a rule $f^{(i)}$ in the set of 256 possible rules. We might choose that rules can "mutate" into another as a function of their absolute distance in "rule space" with some probability $p << 1$. Hence,

$$D(i, j) = p^{|i-j|} \tag{8.2}$$

Each new mutation is itself unpredictable, but the consequence of each mutation is to change the deterministic laws of motion. The observed system $C^{(i)}$, where i indexes the current update rule, will randomly walk in a discrete lattice of highly regular dynamics.

For example, if we start with rule 240, we might observe one realization of this random walk to be 241, 243, and return to 242. This corresponds to a motion in the space of boolean functions (adopting the conventional symbols $\vee$ for OR, $\neg$ for NOT, and $\wedge$ for AND):

$$p \to p \vee (\neg(q \vee r)) \to p \vee (\neg q) \to p \vee (r \wedge ((\neg q))) \qquad (8.3)$$

This illustrates the role that an evolutionary process can have on the prediction of dynamics from a given initial condition, s_0. Mutations break the symmetry of sites q and r. Without knowledge of the precise transitions, this message has been effectively encrypted through the function $D(i, j)$. In order to predict the evolution of the system from the initial condition, we would need to know both the time at which mutation took place and the selected rule.

Historical Encryption

Complexity science can be likened to efforts at decoding a message that has been combined with a variety of historical sources of noise. Note that the noise needs to remain low in order that the original signal persists. The degree of minimalism that any denoised model can achieve will then by captured through concepts such as "effective complexity" (Gell-Mann and Lloyd, 1996).

The effective complexity of a complexity theory, to the extent that it seeks to fit observed regularities, is expected to be rather large. Even a parsimonious, predictive model will need to take into account a known "historical key" that accounts for each branching event in which a symmetry has been broken. Unlike human cryptosystems, evolved keys have not been transmitted in any deliberative way, and the encryption is not "designed" to conceal the nature of reality; this arises as an accident of history promoting a switch between different dynamical scenarios – hence "frozen accidents."

One way in which we might decode dynamics in the scientific search space is to consider complex systems at the level of universality or "equivalence classes" under coarse-grained measurements.

Compressed Historical Cryptographic Keys

One means of reducing the quantity of historical keys is to recognize that conservation laws impose significant constraints on the space of realizable forms and functions. All adaptive solutions need to be consistent with what we know of physics and chemistry.

For the CA examples described previously,

$$p \to p \vee (\neg(q \vee r)) \to p \vee (\neg q) \to p \vee (r \wedge ((\neg q))) \qquad (8.4)$$

We establish a correspondence to the random walk in the space of "coarse-grained" Wolfram universality classes (numbered 1–4) (Wolfram, 1984):

$$2 \to 2 \to 2 \to 2 \qquad (8.5)$$

In other words, this random sequence of deterministic cases is invariant with respect to its statistical evolution, which preserves those pattern categories in which we find a set of separated simple stable or periodic structures (Class 2). This need not be true; for different rules, we could observe at the "macroscopic" level transitions between universality classes. However, through a higher level of description we can often significantly increase our predictive capability. The same logic is pursued in natural science as is described in Chapter 10.

The theory of allometric scaling and regular features of evolved networks provide a means of " compressing" contingent information by exploiting constraints on metabolic energy optimization. It is well known that all multicellular organisms conform to scaling laws with quarter power exponents (West et al., 1999).

At the genetic level it is observed that mutation rate scales with mass according to $p = km^{-1/4}$. If we assume that organisms seek to maximize the rate of mutation subject to the upper limit of the error threshold, then we can derive a maximum genome size, $L = k'm^{1/4}$ (Krakauer, 2011) (note that this is the requirement for indefinite replication of a lineage and does not apply to the patholgoical case of cancer, which has a very short genealogy by any evolutionary standards and is free to grow entropically and without bound for a brief period only). Hence without any knowledge of history we can place an upper abound on the dimensions of the genome. This does not tell us which sequence an organism is using, but it does define a metabolic equivalence class for all organisms with a given quantity of genetic information. This kind of prior information makes decrypting evolutionary history somewhat easier, but a significant challenge remains.

The same kind of argument, albeit somewhat weaker, comes from network science. There is evidence for empirical laws that describe the distributions of network connectivity (Newman, 2006). The degree distributions of many evolved networks to include genetic regulatory networks, neural networks, and ecological networks fall into a rather small number of functional forms to include power laws. Higher-order properties of these networks include the logarithmic scaling of geodesics with network size, so-called small world networks (Watts and Strogatz, 1998). As in the case of allometry, these regularities provide us with prior information that we can use to restrict the search space of historical keys – in this case network growth rules or developmental processes that are capable of producing the observed statistical features of the networks.

CONCLUSION: FROM INVARIANCE TO COMPLEXITY

Eugene Wigner, in his influential paper, "Events, Laws of Nature, and Invariance Principles" (1964), maintains that it is the role of physics to account for those regularities that are called "laws of nature." The elements of the behavior that are not specified by the laws of nature are called "initial conditions." It is by the definition of initial conditions that they are arbitrary with respect to physical laws, but as I have suggested, complex phenomena accumulate an alarming quantity of initial conditions and they are far from functionally arbitrary. This places us in an pickle of a problem.

One way to reduce the magnitude of this challenge is to restrict the volume of "initial conditions" by using our knowledge of physics and biophysics. Allometry, network regularities, and error thresholds are all principled approaches to placing bounds on free parameters, using geometric invariance principles, to limit the size of a theory.

Another approach suggested in this chapter is to begin to formalize the complexity science of adaptive systems in terms of mathematically suggestive concepts from the field of cryptography. This resembles "metamathematics" – the mathematical study of mathematical objects – in that we are endeavoring to formalize generic informational properties that result from any evolutionary, communicative process. In order to explain specific cases, we have no choice but to significantly increase the number of measured parameters, and our theories and models are anticipated to grow accordingly.

REFERENCES

Bracht, J. R., Fang, W., Goldman, A. D., Dolzhenko, E., Stein, E. M., and Landweber, L. F. 2013. Genomes on the edge: programmed genome instability in ciliates. *Cell*, **152**(3), 406–416.

Brading, K., and Castellani, E. 2003. *Symmetries in physics:philosophical reflections.* Cambridge University Press.

Dandekar, T., and Sharma, K. 1998. *Regulatory RNA.* R. G. Landes.

Dermitzakis, E. T., Reymond, A., and Antonarakis, S. E. 2005. Conserved non-genic sequences: an unexpected feature of mammalian genomes. *Nature Reviews Genetics*, **6**(2), 151–157.

Eddy, S. R. 2012. The C-value paradox, junk DNA and ENCODE. *Current Biology*, **22**(21), R898–R899.

Eigen, M. 2000. Natural selection: a phase transition? *Biophysical Chemistry*, **85**, 101–123.

Gell-Mann, M. 1995. *The quark and the jaguar: adventures in the simple and the complex.* St. Martin's Griffin.

Gell-Mann, M., and Lloyd, S. 1996. Information measures, effective complexity, and total information. *Complexity*, 44–52.

Hamilton, W. D., Axelrod, R., and Tanese, R. 1990. Sexual reproduction as an adaptation to resist parasites (a review). *Proceedings of the National Academy of Sciences*, **87**(9), 3566–3573.

Hutchinson, C. O., Picozzi, K., Jones, N. G., Mott, H., Sharma, R., Welburn, S. C., and Carrington, M. 2007. Variant surface glycoprotein gene repertoires in *Trypanosoma brucei* have diverged to become strain-specific. *BMC Genomics*, **8**(1), 234.

Katz, J., and Lindell, Y. 2007. *Introduction to modern cryptography: principles and protocols.* CRC Press.

Krakauer, D. C. 2002. Evolutionary principles of genomic compression. *Comments on Theoretical Biology*, **7**, 215–236.

Krakauer, D. C. 2011. Darwinian demons, evolutionary complexity, and information maximization. *Chaos*, **21**(3), 037110.

Krakauer, D. C., Page, K. M., and Erwin, D. H. 2009. Diversity, dilemmas, and monopolies of niche construction. *American Naturalist*, **173**(1), 26–40.

Mach, E. 1910. *History and root of the principle of the conservation of energy*. Open Court Publishing.

Massey, J. L. 1986. Cryptography: a selective survey. In E. Biglieri and G. Prati (eds.), *Digital Communications*. North-Holland, pp. 3–21.

Newman, M. E. J. 2006. The structure and function of complex networks. *SIAM Review*, **45**(2), 167–256.

Nowak, M. A. 2006. *Evolutionary dynamics: exploring the equations of life*. Belknap Press.

Orr, H. A. 1995. The population genetics of speciation: the evolution of hybrid incompatibilities. *Genetics*, **139**(4), 1805–1813.

Orr, H. A. 1996. Dobzhansky, Bateson, and the genetics of speciation. *Genetics*, **144**(4), 1331–1547.

Poe, E. A., Levine, S., and Levine, S. 1976. *The short fiction of Edgar Allan Poe: an annotated edition*. University of Illinois Press.

Prescott, D. M. 2000. Genome gymnastics: unique modes of DNA evolution and processing in ciliates. *Nature Reviews Genetics*, **1**(3), 191–198.

Shannon, C. E. 1949. Communication theory of secrecy systems. *Bell System Technical Journal*, **28**(4), 656–715.

Valiant, L. 2013. *Probably approximately correct: nature's algorithms for learning and prospering in a complex world*. Basic Books.

Verstrepen, K. J., and Fink, G. R. 2009. Genetic and epigenetic mechanisms underlying cell-surface variability in protozoa and fungi. *Annu Rev Genet.*, **43**(1), 1–24.

Watts, D. J., and Strogatz, S. H. 1998. Collective dynamics of small-world networks. *Nature*, **393**(6684), 440–442.

Wernegreen, J. J. 2005. For better or worse: genomic consequences of intracellular mutualism and parasitism. *Current Opinion in Genetics & Development*, **15**(6), 572–583.

West, G., Brown, J., and Enquist, B. 1999. The fourth dimension of life: fractal geometry and allometric scaling of organisms. *Science*, 1677–1679.

Wigner, E. P. 1964. Events, laws of nature, and invariance principles. *Science*, **145**(3636), 995–999.

Wolfram, S. 1984. Universality and complexity in cellular automata. *Physica D: Nonlinear Phenomena*, **10**(1–2), 1–35.

9   Noise and Function

Steven Weinstein and Theodore P. Pavlic

We are surrounded by noise. The controlled explosions of internal combustion engines combine with the roar of rubber on asphalt to create the drone of road and highway traffic. Weed-whackers, lawn mowers, and leaf blowers can turn sunny suburban summer days into a buzzing confusion. Indeed, the entire universe is filled with the faint din that is the cosmic microwave background (CMB) radiation, leftover electromagnetic radiation from the Big Bang. The CMB was discovered when Penzias and Wilson (1965) went looking for the source of annoying hiss plaguing their new radio telescope, hiss that threatened to obscure signals from distant stars and galaxies. Noise seems to be entirely destructive, thus something to be eliminated, if possible.

Noise can be beneficial, however, in at least two ways. One is familiar, the other paradoxical and far less well known. The familiar way is simply as a source of variety. For example, genes undergo random mutation from processes both external to the organism (e.g., cosmic rays) and internal (Dobrindt and Hacker, 2001). This genotypic variation is the source of heritable variation in phenotype, which is of course essential for the process of natural selection (Wagner, 2014). Phenotype can even vary within isogenic populations due to variation in gene expression (Fraser and Kaern, 2009). This *phenotypic noise* is thought to provide an evolutionary advantage for some microorganisms, as it increases the chance that some will survive under stressful conditions. The CMB noise, though destructive from the standpoint of the users of radio telescopes, plays a constructive role in generating the tiny variations in energy density in the early universe that are the seeds of structure formation. The random fluctuations we call noise gave rise to stars and galaxies and galactic clusters. This much is familiar, at least to those working in the relevant areas of biology or astrophysics.

Considerably less familiar is the role that noise can play in nonlinear systems, in particular systems with one or more thresholds, points at which small differences in input give rise to disproportionate differences in output. Converting an analog signal into a digital signal involves sampling the signal at regular intervals and writing down a digital approximation of the amplitude at each point. For example, if one has a one-bit digital system with two possible values to represent the interval from 0 to 1, then there will be a threshold at the analog value 0.5, below which any value will be digitally recorded as 0, and above which any value will be recorded as 1. More bits simply mean more thresholds, more ways to cut up the interval into discrete chunks. Neurons behave like single-threshold devices, firing when and only when the voltage across the cell membrane reaches a certain activation threshold. What noise can do in a threshold system is push the signal over the threshold, but in a way very much unlike an amplifier. Amplifiers multiply the signal, whereas noise is additive. The implications and applications of this nonstandard amplification are both deep and wide. Here we will lay out as simply as possible the principles behind this sort of noise benefit and then illustrate its application.

SHADES OF GRAY: NOISE IN IMAGE PROCESSING

Photographs are never veridical. The information coming through the lens is inevitably greater than the information stored on the recording medium. A digital camera sensor has a finite spatial resolution; the camera's sensor consists of a matrix of smaller individual sensors corresponding to a single "pixel" of the image. Any features of the image smaller than an individual pixel will be lost. Associated with each pixel is a color. The color spectrum in the real world is continuous, but the digital encoding of color is discrete so that, in general, the color stored will only be an approximation of the actual color. In other words, color information must be rounded off in order to be stored as a number on a digital computer. The number of binary places available for each number is referred to as the *bit depth*.

Suppose we have a digital image consisting of 7 megapixels. Let's consider a "black and white" camera for simplicity, so that the colors are shades of gray. Each pixel has an 8-bit number attached to it indicating the shade of gray, with 0 (00000000) corresponding (by convention) to black, and 255 (11111111) corresponding to white. There are a total of $2^8 = 256$ shades of gray. Figure 9.1A shows the palette, alongside an image of a woman known as Lena (Hutchison, 2001) rendered using this palette (Figure 9.1B).

Now, suppose we want to print these images. Indeed, you may well be reading a printed version of this page, printed on a laser printer capable of printing 300 dots per inch (dpi). That resolution gives us around 7 million evenly spaced points on a typical sheet of paper, so if we were to use up an entire page to print the 7-megapixel image, we would have a one-to-one correspondence between pixels and dots. If the printer could print 256 shades of gray at any given point, then we'd have perfect reproduction of the stored image. But the printer is not nearly that flexible. At each dot location, most printers can print either a black dot or nothing. Because the number of pixels and the number of dots are approximately the same (in our example), we are effectively reducing an 8-bit (per pixel) image to a 1-bit (per dot) image; 256 shades of gray at each point get mapped to either black or white.

The obvious way to map the shades of gray is to impose a threshold as before, whereby we print a black dot at a point if the corresponding pixel is more than 50% gray (numbers between 0 and 127), and we otherwise leave it blank (white) (numbers between 128 and 255). For an image that has an equal distribution of lighter and darker grays, this might appear to be as good as one can do. But a quick glance at Figure 9.1C and D shows the limitations of this simple thresholding method; an enormous amount of detail is lost. Increasing the number of thresholds to create 8 shades of gray, as in Figure 9.1E and F, yields a noticeable improvement, but a printer that can print only in black and white is limited to the performance of the 2-shade case.

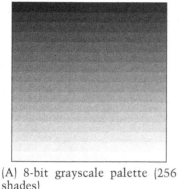

(A) 8-bit grayscale palette (256 shades)

(B) 8-bit Lena (256 shades)

(C) 1-bit grayscale (2 shades)

(D) 1-bit Lena (2 shades)

(E) 3-bit grayscale (8 shades)

(F) 3-bit Lena (8 shades)

FIG. 9.1 Shades of gray.

However, there are clever methods to improve the fidelity of black-and-white image reproductions. Traditional printed newspapers use varying dot size to represent darker and lighter portions of an image. Applying this *halftone* concept to a device like a laser printer

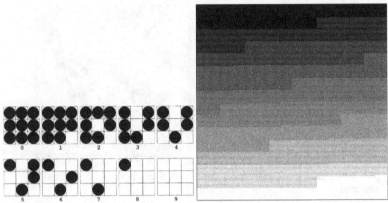

(A) Halftone array   (B) Halftone grayscale palette

FIG. 9.2 Grayscale using halftone.

or an LCD display with *fixed* dot or pixel size involves representing gray by varying the density of the distribution of black dots in an array. Using 3×3 arrays of dots, we can represent 10 different shades of gray shown in Figure 9.2A, which allows for a grayscale palette like that shown in Figure 9.2B. A 300-dpi printer can print 100×100 patches of gray per inch, where each patch has a 3×3 pixel area. Thus, with a three-fold reduction in the effective spatial resolution of the image, the number of shades that can be represented is increased from two shades to 10. The same technique could be applied to 4×4 arrays of dots to achieve 17 shades of gray at the cost of further decreasing the spatial resolution.

Although deterministic methods like halftoning can be effective ways of trading spatial resolution for color resolution, they introduce noticeable artifacts. Note the abruptness of the shifts in the halftone grayscale palette in Figure 9.2B. To avoid this blockiness while still being able to trade spatial resolution for color resolution, a very different approach can be used based on the appropriate addition of random variation to pixel values. For each pixel, we take the original grayscale value and add a random number between 0 and 255. An image made up of these random values would look like visual "noise" – it is a distribution of dots in arbitrary shades of gray.

(A) 8-bit grayscale (256 shades) with 8-bit noise

(B) 8 bit Lena (256 shades) with 8-bit noise

(C) 1-bit grayscale (2 shades)

(D) 1-bit Lena (2 shades)

FIG. 9.3 Reduction from 256 shades (8 bits) to 2 shades (1 bit) using 8-bit random dither noise.

The result of adding this noise to the image is an array of pixels with values between 0 and 510. We can turn the resulting array back into an image by dividing the values by two, thereby restoring the original range of 0 (black) to 255 (white). Figure 9.3A and B shows the result: noisy versions of the original grayscale palette and the original image. This of course is not an improvement over the non-noisy 8-bit grayscale image with 256 shades of gray. The noise does what we generally expect noise to do: it degrades the image.

But recall that we added the noise not to improve the grayscale image but to get a better result when we subsequently impose a threshold at 127 and convert to a black and white (2-shade) image. Imposing the threshold, we map any pixel with value 128 or above to white (255), and any pixel 127 or below to black (0). The resulting pseudo-grayscale palette looks like Figure 9.3C, while the resulting pseudo-grayscale Lena looks like Figure 9.3D. The results are instructive. The grayscale palette looks better than the version in which noise was not imposed before thresholding (Figure 9.1C), giving the impression of a variety of shades of gray. Lena, however, does not look very good by comparison with Figure 9.1D. The problem with the Lena image is not that she has more shades of gray but that the shades tend to change over the scale of a few pixels. Images of this sort are better treated by more sophisticated techniques, such as the Floyd–Steinberg error diffusion method (Floyd and Steinberg, 1976).

However, if we avail ourselves of 8 shades of gray (3 bits) rather than just black and white (1 bit), the use of random noise is much more effective. Figure 9.4A and B shows the original images augmented with a low level of noise, spanning 1/8 of the total range (32 shades of gray). That is, the noise randomizes the 3 least significant bits of the 8 bits in use. If we now reduce to 8 shades (encodable by 3 bits) having added this noise, we get Figure 9.4C and D, which are a decided improvement on Figure 9.1E and F, which are what we get if we go from 256 shades to 8 shades without first adding noise.

The process of adding noise to an image or a signal in order to preserve information once the signal is subjected to quantization[1] (digitization) is called dithering.[2] In the example above, we took an already discrete signal (each pixel having one of 256 shades of gray) and made it more discrete, mapping the 256 shades into 2 shades (black and white). However, the initial process of moving from an

[1] "Quantization" here refers not to the physicist's process of finding a quantum-mechanical version of a classical theory, but rather to the process of discretizing the properties of an image or signal.

[2] Dithering also includes related methods that use not random noise but some other signal that is uncorrelated with the signal of interest. The ring laser gyroscope, for example, uses periodic (sinusoidal) dither to prevent its counterrotating laser beams from locking under conditions of slow rotation.

(A) 8-bit grayscale (256 shades) with 3-bit noise

(B) 8 bit Lena (256 shades) with 3-bit noise

(C) 3-bit grayscale (8 shades)   (D) 3-bit Lena (8 shades)

FIG. 9.4 Reduction from 256 shades (8 bits) to 8 shades (3 bits) using 3-bit random dither noise (randomizing the 3 least significant bits).

image with a continuum of shades to one with 256 shades is also an example of quantization. Were we dealing with digital audio, we would be working with a one-dimensional stream of samples of the waveform, each of which has a continuously valued amplitude (the volume) that must be mapped into a finite set of numbers for storage in a computer, say 24 bits (16,777,216 possible values). This can then be further reduced to 16 bits (64,436 possible values) for CD encoding. Dither is routinely used in this process.

Most digital representations involve more than one threshold. The seminal work of Roberts (1962) considered the problem of transmitting digital television signals. At the time, a 6-bit-per-pixel resolution was considered adequate. Roberts proposed a scheme whereby a 3-bit-per-pixel signal could effectively encode the necessary detail if pseudo-random noise were added to the 6-bit representation prior to rounding to 3 bits.[3] Shortly thereafter, others realized that this technique was akin to a technique called "dithering" that had been conceived several decades prior as a technique to overcome the tendency of certain mechanical systems to stick for various reasons, rendering them insensitive to small changes in operational parameters (Schuchman, 1964). Engineers designed electromotor circuits to apply dither in the form of small zero-mean oscillations that would allow devices to respond more easily to small steering signals from the operator (Farmer, 1944; Korn and Korn, 1952). In all of these cases, proper application of dither tends to linearize a nonlinear system; the dither blurs thresholds, eliminating some of the jaggedness that goes along with systems that have one or more thresholds.

But the addition of noise does something else along the way. In a physical system, adding noise adds energy to the system. If the system has one or more thresholds, this has the effect of taking subthreshold signals and boosting them, albeit stochastically. This is the phenomenon known as *stochastic resonance*. Let's take a look at a simple example, again using printed images, before moving on to the role stochastic resonance can play in natural – including biological – phenomena.

We're accustomed to the fact that there are sounds we can't hear because they're too soft, and sights we can't see because they're too faint. These are thresholds of hearing and vision, respectively. By analogy, consider an image we can't make out because it's too light: a very light shade of gray indistinguishable from its white background. Figure 9.5A shows the full grayscale palette with a line separating the grays that are dark enough to distinguish from those that, for some

---

[3] The scheme of Roberts (1962) was an early example of what is called subtractive dither, where the noise is subtracted from the image after transmission

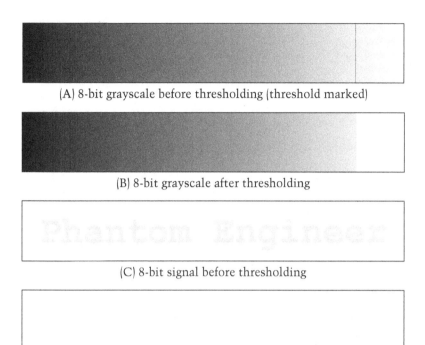

(A) 8-bit grayscale before thresholding (threshold marked)

(B) 8-bit grayscale after thresholding

(C) 8-bit signal before thresholding

(D) 8-bit signal after thresholding

FIG. 9.5 Rendering a signal imperceivable by thresholding. The signal is too light to survive the imposition of the threshold.

focal individual, are not. We can represent the indistinguishability of any grays below the threshold by rendering them as white, as in Figure 9.5B. Thus, when a faint image of the words "Phantom Engineer" (Figure 9.5C) is rendered in this very light gray, it will look like Figure 9.5D to someone for whom the threshold represents the limits of their perceptual acuity. The image will be invisible.

Now, suppose we add noise by randomly darkening each pixel, including the background. We will use low-level, 3-bit noise, as shown in Figure 9.6A. This corresponds to randomly selected shades of the very light grays lying below the threshold of perceivability. Thus, like the image itself, the noise will be invisible – see Figure 9.6B – to someone who cannot make out very light shades of gray. Adding this noise to our original image as in Figure 9.6C has the remarkable effect of bringing a noisy but very legible version of the image

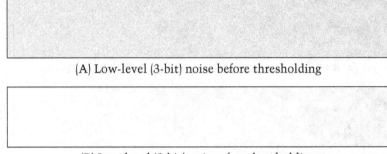

(A) Low-level (3-bit) noise before thresholding

(B) Low-level (3-bit) noise after thresholding

(C) 8-bit signal plus 3-bit noise before thresholding

(D) 8-bit signal plus 3-bit noise after thresholding

FIG. 9.6 Stochastic resonance: signal boosting with noise. Random dither noise spanning the 3 least significant bits is added. With the addition of noise, the signal becomes dark enough to remain visible even after the imposition of the threshold.

above the threshold of perceivability, as is evident in Figure 9.6D, in which the lightest, subthreshold grays are removed. The ability of noise to boost a signal above threshold is the essence of stochastic resonance.

One of the salient characteristics of stochastic resonance, indeed the feature that makes it somewhat akin to a true resonance phenomenon, is the dependence on the amplitude and specific properties of the noise. This is true of dither noise in general. Too much noise (here, too large a spectrum of grays) threatens to obliterate the signal, while too little will fail to push the signal above threshold at all and will have no effect. The relevance of this for understanding the role of stochastic resonance in nature is significant, for there is noise of all kinds and all amplitudes everywhere. Often it does what we think noise does: it interferes with the signal, the image, or the operation of a dynamical system. But when the noise level is proportional to the level of one or more significant thresholds in a system of interest, we can and should look for stochastic resonance, as it may be key to understanding the function of the system (Gammaitoni et al., 1998).

AMPLIFICATION BY NOISE IN NATURAL DYNAMICAL SYSTEMS

We will now take a look at how the stochastic resonance effect can be used in modeling a dynamical system existing in nature. The example we'll study is the one in which the term 'stochastic resonance' was originally introduced. Though there is no resonance in the ordinary physicist's sense (though see Gammaitoni et al., 1995), we are once again presented with a situation in which the addition of noise permits the system of interest – in this case the climate – to straddle a threshold.

The Earth has existed in two relatively stable climates around 10 degrees Celsius apart for millions of years. Periods in which the climate is cooler are called "ice ages." In 1978, Bhattacharya and Ghil conjectured that the two stable climates correspond to the two minima of a double-well pseudo-potential like the one shown in Figure 9.7. The horizontal axis represents the Earth's temperature, and the curve acts like a potential energy term in ordinary mechanics, with a single unstable equilibrium forming an energy barrier between two stable equilibria. In this model, the Earth's climate inevitably converges to one of the two stable equilibria. The equilibrium on the left represents the ice age, and the one on the right is a temperate period like the present.

Thus, we have a primitive model of a system with two stable states. But the stability of these states means that there is no way to transition between them, thus no way to explain how the climate shifts from one to the other. However, it was observed that the eccentricity of the Earth's orbit varies over a period roughly the

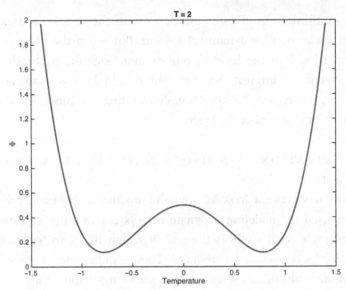

FIG. 9.7 Double-well pseudo-potential representing the Earth's climate as a dynamical system.

same as the time between ice ages – around 100,000 years (Hays et al., 1976). Because the eccentricity of the orbit is correlated with small variations in the amount of solar heating ("insolation"), it was conjectured that these small variations might be sufficient external drivers of the Earth's dynamical system to cause the observed periodic climate changes. This suggests that one augment the model by introducing a time-varying oscillation in the pseudo-potential such that the double-well shape is a transient feature separating two epochs in which the potential morphs into a single well, a single, quasi-stable equilibrium.

The problem with this idea was that the estimated insolation differences were too small to be responsible for such a change. At best, the resulting time-varying pseudo-potential takes forms like those in Figure 9.8, where the barrier between the two stable equilibria is always maintained, and where the change in temperature due to the displacement of each local minimum is of the order of only 1 degree Celsius. So a simple dynamical systems approach does not provide an

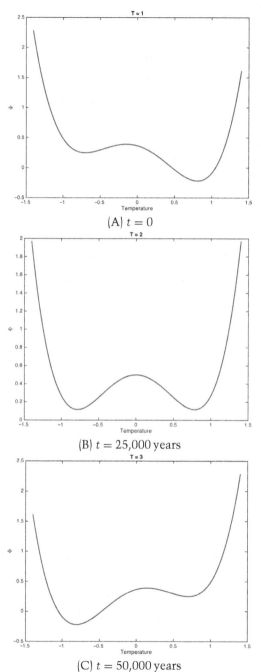

(A) $t = 0$

(B) $t = 25,000$ years

(C) $t = 50,000$ years

FIG. 9.8 Time-varying double-well pseudo-potential with period of 100,000 years.

explanation for the congruence of the period of the insolation signal and the period of Earth's climate switching.

Independently, Benzi et al. (1982) and Nicolis (1982) arrived at similar explanations for how the climate might actually shift. They proposed that including the fine-scale, shorter-time variations in heating and cooling due to various other factors might result in the climate hopping from one well to the other. In other words, factoring in the existence of a certain level of noise in the climate system might account for the ability, as it were, of the climate to surmount the otherwise insurmountable threshold, the hump between the two minima. After all, the geological record shows not only periodicity in the Earth's climate but also significant small-scale variations, which indicated that the Earth's dynamics must include some internal noise. So, following the approach of Nicolis, the deterministic double well pseudo-potential is augmented with a noise source, converting an ordinary differential equation into a stochastic differential equation. In other words, the climate is now modeled as a diffusion process – a random walk that is pulled downhill but can, on occasion, take several steps uphill. For such a diffusion process in a double well, the mean time to transition from one stable equilibrium to another is well characterized by a formula parameterized by the height of the barrier between the equilibria and the strength of the internal noise. What Nicolis realized was that the changes in the barrier height due to noise could lead to large changes in the mean residence time. If the noise is of the right amplitude, then the climate is likely to hop from one not-quite-stable minimum to the other when the barrier is low. Consequently, Benzi et al. (1982) named the phenomenon *stochastic resonance* based on its similarity to the frequency-selective properties of conventional deterministic resonance. Whereas "resonance" in the traditional sense is between the frequency of an input and the characteristic response of a system, the resonance here is between the frequency of the long-term oscillation in insolation (the input) and the amplitude of the noise (a characteristic feature of the system). If the noise is too small, nothing special will happen; the system will never transit from one climate to the other. If the noise is too great, the system will never settle in one climate or another, as the noise will dominate the oscillation. If the noise is within the correct range, however, the climate will oscillate at approximately the 100,000-year period of the subthreshold background oscillation in the ellipticity of the Earth's orbit.

NOISE IN LIVING SYSTEMS: DECISION-MAKING IN ANT COLONIES

Over the past 20 years, an intriguing body of evidence has pointed to a role for stochastic resonance in a variety of biological processes (McDonnell and Abbott, 2009). Extensive work has been done demonstrating the role of noise in general and stochastic resonance, in particular, in the neural systems that carry out sensory information processing (Moss et al., 2004), but it is important at the macroscopic level as well. We will conclude our discussion of stochastic resonance by discussing its role in the social dynamics of group decision-making in certain species of ants.

There are a wide variety of *mass-recruiting ants* that form charismatic foraging trails that concentrate all foraging effort onto a single food source for a short period of time (Holldobler and Wilson, 1990). Since many species of mass-recruiting ant have a heterogeneous foraging force, it is thought that it may be beneficial to concentrate the foraging force all in one area in order to guarantee there is an adequate representation of each worker type. So it is expected that these ants must make use of some decentralized mechanism that can drive their foraging force to a quick consensus on the best of several available foraging options. A typical feature of mass-recruiting ants is the use of pheromone trails (Holldobler and Wilson, 1990). Although details vary across different mass-recruiting ant taxa, the observed pattern is usually a variation of what follows. A focal ant leaves her nest and searches for food. When she finds food, she can choose to return to her nest and deposit some quantity of pheromone along her path back to the nest. The amount of pheromone she deposits is related to the quality of the discovered food, with higher-quality foods leading to more deposited pheromone. Although that deposited pheromone will eventually evaporate, for a short time after deposition, the pheromone near the nest will attract the attention of other foragers that would otherwise search randomly for food. They will then have an increased likelihood of finding the same food source as the focal ant and then also lay a pheromone trail on their return visit. So, initially, a set of food items will be discovered randomly. Due to the positive feedback inherent in the recruitment system, the highest quality of those food sources will eventually attract all of the foragers.

Until recently, it has been believed that such trail-laying mass-recruitment mechanisms had a flaw similar to the one described in Figure 9.9 for the early deterministic models of climate change – the ants were thought to be rigidly bistable and unable to cope with changes in food availability after a critical point in the recruitment process. In other words, the positive feedback in the recruitment would eventually become so strong that the system would become entirely insensitive to changes in relative food source quality, just as early mathematical models of climate change were not properly sensitive to the variations in insolation. This intuition was verified in early experiments with *Lasius niger* (Beckers et al., 1990). Moreover, early dynamical mathematical models of trail-laying were also shown to be insensitive to changes in relative food quality (Camazine et al., 2001; Nicolis and Deneubourg, 1999). However, several recent experiments show that many other trail-laying ants are able to dynamically reallocate their foraging forces to track changes in the environment (Dussutour et al., 2009; Latty and Beekman, 2013; Reid et al., 2011). For example, Dussutour et al. (2009) presented colonies of *Pheidole megacephala* with a laboratory dynamic environment summarized in Figure 9.9. During the 180-minute experiment, colonies were placed at the mouth of a Y-bridge with two legs of different lengths, and the experiment proceeded in three 60-minute phases:

- During the first 60 minutes, equal-quality feeders were placed at the ends of both legs. Because one leg was shorter, it eventually dominated the

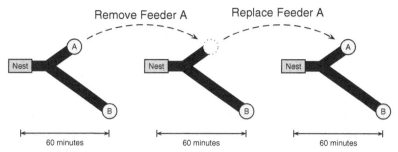

FIG. 9.9 Graphical summary of dynamic foraging experiment used by Dussutour et al. (2009) to study flexibility of trail-laying in *Pheidole megacephala* big-headed ant colonies. In the first 60 minutes of the experiment, colonies are given a choice between two feeders, A and B, that differ only in distance to the nest. During the second 60 minutes of the experiment, the nearest feeder (A) is removed. Finally, during the final 60 minutes of the experiment, the nearest feeder (A) is replaced.

collective attention of the colony, and a single trail was formed to the feeder on that leg.

• During the second 60 minutes of the experiment, the feeder on the short leg was removed. With the disappearance of food, the pheromone trail was not reinforced, and the colony was eventually able to return to random search and subsequently converge on the feeder at the end of the long leg.

• During the final 60 minutes of the experiment, a feeder was returned to the short leg of the Y-maze. The traditional model of trail-laying recruitment would predict that the new feeder would be ignored because all foragers would be latched into following the existing pheromone trail. However, contrary to those predictions, the short leg was rediscovered and ants returned to exploiting the closer feeder.

To explain the results of the experiment, Dussutour et al. propose a slight extension to the traditional mathematical model of trail-laying recruitment inspired by stochastic resonance. They observed that ants in their experiments would often make "errors" in their trail-following behavior that would lead a minority of the ants down the opposite leg of the Y-maze. In an attempt to capture this phenomenon, they augmented the traditional mathematical model of trail-laying behavior with an "error" level that would cause an individual to rarely, but measurably often, choose the leg of the

Y-maze with the smaller quantity of pheromone deposited on it. At small error levels, the theoretical system had nearly identical decision-making latency and accuracy characteristics to the deterministic system when presented with a static choice set. However, at specific nonzero error levels, the system could produce switching dynamics that matched those of experimental data from ants such as *P. megacephala* that have the ability to follow changes in relative feeder quality. Furthermore, the mathematical model predicted that different error levels would correspond to different random natural switching times between alternatives, and amplification of variations in food quality would be possible if the periodicity of those variations matched the natural error-driven switching time.

The time-scale matching argument for the switching behavior observed in some trail-laying, mass-recruiting ants is identical to the one used in the early models of stochastic resonance in climate systems that are matched to the periodicity of solar insolation. However, in the case of the ants, differences in error level across different ant taxa could be explained by natural selection. In particular, the individual error level could be tuned by natural selection so that the stochastic switching time of the colony would match the natural periodicity of changes in food quality in the natural environment. Colonies with individuals that make errors at the appropriate rate would have an advantage over colonies that make more or fewer errors. Making too many individual-level errors would mean switching too frequently from good choices to bad choices, and making too few individual-level errors would mean focusing for too long on one choice even though a better choice was now available. Consequently, the "errors" at the individual level would be better described as random variability (noise) that was itself a trait under selection, and the prediction would be that ants that evolved in more ephemeral environments would also have larger amounts of noise in individual-level response to pheromone trails. In fact, as Dussutour et al. (2009) discuss, the flexible *P. megacephala* ants in their study that are well modeled by nonzero noise do come from an environment where food quality changes more frequently than in the case of the *L. niger* ants that had previously been used to support the deterministic modeling of trail-following behavior with no noise. If location of the best-quality food source is viewed as a signal that tends to change over some characteristic time scale, then the current location of the main foraging trail can be viewed as a version of that signal amplified using stochastic resonance. This argument is identical to those made in the stochastic-resonance literature where some input-to-output measure, such as mutual information, is maximized by varying the amount of noise added to the input signal (Neiman et al., 1996). Similarly, in the earlier image-processing example, a certain amount of noise is sufficient to push the text 'Phantom Engineer' above the threshold of visibility (Figure 9.6). Too little noise will not do the trick. Adding noise will make the text more visible up to a point, after which the readability goes down as the entire image becomes dominated by noise. In the case of the ants, the signal being amplified is the relative quality of the feeders, and the output is the selection of a path by the colony.

The idea that apparent "errors" could actually be an adaptive phenotypic trait under selection is not unprecedented and goes beyond the examples of possible stochastic resonance in natural phenomena. For example, the idea that noise can be of benefit in decision-making is relatively old. In what they called "ethological cybernetics," Haldane and Spurway (1954) performed an information-theoretic analysis of the statistical distributions of honeybees responding to communicated information from so-called waggle dances. Honeybees have the ability to communicate information from one forager to another through a dance language that communicates the relative polar coordinates (i.e., distance and direction) of a discovered food source. However, after a bee communicates these coordinates to another, the bee receiving the information will often make "mistakes" and explore a location slightly different from the one discovered by the original bee. The *average* location explored by an ensemble of receiver bees will closely match the originally discovered food source, and so these variations are viewed as "noise" due to imperfect communication of the coordinates in the dance-communication channel. Haldane and Spurway determine that the bee-to-bee channel communicates roughly 4 bits of information about the direction of the target. That is, a bee can communicate only 1 of 16 different cardinal directions; any finer resolution appears to be impossible. This may seem to reflect a fundamental limitation, such as a physiological or neurological constraint, but it could also be an adaptive response to dispersed food sources in an environment. If a honeybee is dancing to communicate the location of a nectar source, such as a flower, the resulting noisy scatter of her colony mates will likely find other flowers in a similar location. Thus, the amount of error in the communication may be tied to some ecological measure of forage patchiness, and honeybees selected for environments with a different patchiness may communicate with different levels of error.

As the honeybee example does not involve dynamically changing signals in the environment, it is not an example of stochastic resonance in the strict sense, but it does reflect how the amount of noise expressed in a behavior is itself a phenotype that nature can adapt to match natural variation. However, a very similar information-theoretic analysis of fire ants does suggest additional ties to stochastic resonance and behaviors shaped by nature. In particular, Wilson (1962) described a consistent error distribution in the distance and direction information communicated by fire-ant pheromone trails leading to prey. In the trail-laying examples above, the actual food sources were static, and the experimenters could change the location of different sources at discrete instants in time. This was appropriate for the particular ant species under study. However, fire ants are a natural example of a species adapted to continuously varying food quality and location. These ants track moving food sources: living prey items that have the ability to flee. They must have the ability to dynamically adapt and follow fleeing prey until the prey is sufficiently subdued. Wilson suggests that the relatively poor ability of individual fire ants to follow trails is actually an adaptation. The result of the ensemble of error-prone trail followers is a cloud of ants in the general vicinity of the original location of the discovered prey. If this cloud is large enough, it can track the motion of the escaping prey. Too much noise will cast too large of a net and lead to too thin coverage over a prey item, and too little noise will not disperse the ants far enough to catch the escaping prey. So the dispersal of the trail followers could be matched to the escape dynamics of the typical kinds of prey. Just as in traditional stochastic resonance examples, a certain critical amount of noise helps a dynamic output (the ultimate location of the end of a fire-ant foraging trail) follow a dynamic input (the trajectory of the escaping prey).

CONCLUSION

An ant colony's ability to react to changes in the food supply and the climate's ability to react to small changes in insolation are examples of the power of noise to qualitatively change the way a system responds to its environment. If we disregard the noise – disregard the small, random variations in the properties of the system – we find the system converges to a fixed point of its dynamical equations and stays there indefinitely. The ant colony remains fixated on a single food source, insensitive to changes in food supply; the climate remains where it is, never shifting. But when the model of the system is modified to include noise, the system is able to surmount a barrier and transition to a qualitatively distinct state. The colony is able to discover and consolidate a new path. The climate reacts to the slight change in insolation over the course of millenia. Noise makes these systems more sensitive to their environment.

A promising area to look for further noise effects in biology lies at smaller scales. The fundamental process at the foundation of all life is the expression of genes as proteins. The chemical reactions involved are constrained by both the availability of the reactants and their proximity, making the process as a whole subject to fluctuations that have come to be called *gene expression noise* (Kaern et al., 2005). The sources of the noise and the role it plays in the process of development and reproduction are matters of intense contemporary investigation (Sanchez et al., 2013; Viney and Reece, 2013). For example, Fernando et al. (2009) proposed the existence of an intracellular genetic perceptron, a single-cell gene network capable of associative learning. Remarkably, Bates et al. (2014) show that the performance of the perceptron is actually enhanced by gene expression noise at a specific level.

One of the factors impacting gene expression in bacteria is the intercellular communication mechanism broadly known as *quorum sensing*, in which bacteria both emit and detect signaling molecules, allowing them to infer the concentration of other bacteria and act accordingly (Popat et al., 2015; Waters and Bassler, 2005). This, too, is a noisy process, subject to the whims of diffusion in the intercellular environment. Like the examples of stochastic resonance we have considered, it is a threshold-oriented system, whereby genes are switched on or off depending on whether a critical density of other bacteria is sensed in the neighborhood. Karig et al. (2011) have applied stochastic resonance to the development of a synthetic biological system that utilizes gene expression noise to boost a time-varying molecular signal consisting of varying concentrations of the molecule used by Gram-negative bacteria in quorum sensing.

The recent, fascinating work on the role in biological systems of stochastic resonance in particular, and noise in general, is surely the tip of an iceberg. Hoffmann (2012) advances the idea that the molecular machines such as kinesin that do physical work within the cell make use of the random noise that is the thermal motion of the water molecules in the cytoplasm. Rolls and Deco (2010) provide an extended look at the role of noise in brain function. The idea that noise can and does do work, enhancing the information processing that is essential to life, is an idea whose time has come.

REFERENCES

Bates, R., Blyuss, O., and Zaikin, A. 2014. Stochastic resonance in an intracellular genetic perceptron. *Physical Review E*, **89**(3), 032716.

Beckers, R., Deneubourg, J., Goss, S., and Pasteels, J. M. 1990. Collective decision making through food recruitment. *Insectes Sociaux*, **37**(3), 258–267.

Benzi, R., Parisi, G., Sutera, A., and Vulpiani, A. 1982. Stochastic resonance in climate change. *Tellus*, **34**, 10–16.

Bhattacharya, K., and Ghil, M. 1978. An energy-balance model with multiply-periodic and quasi-chaotic free oscillations. Pages 299–310 of *Evolution of planetary atmospheres and climatology of the earth*. Centre National d'Études Spatiales.

Camazine, S., Deneubourg, J., Franks, N., Sneyd, J., Theraulaz, G., and Bonabeau, E. 2001. *Self-organization in biological systems*. Princeton University Press.

Dobrindt, U., and Hacker, J. 2001. Whole genome plasticity in pathogenic bacteria. *Current Opinion in Microbiology*, **4**(5), 550–557.

Dussutour, A., Beekman, M., Nicolis, S. C., and Meyer, B. 2009. Noise improves collective decision-making by ants in dynamic environments. *Proc. R. Soc. B*, **276**(1677), 4353–4361.

Farmer, W. C. (ed). 1944. *Ordnance field guide: restricted*. Military Service Publishing Company.

Fernando, C. T., Liekens, Anthony M. L., Bingle, L. E. H., Beck, C., Lenser, T., Stekel, D. J., and Rowe, J. E. 2009. Molecular circuits for associative learning in single-celled organisms. *Journal of the Royal Society Interface*, **6**(34), 463–469.

Floyd, R. W., and Steinberg, L. 1976. An adaptive algorithm for spatial grey scale. *Proc. Soc. Inf. Disp.*, **17**, 75–77.

Fraser, D., and Kaern, M. 2009. A chance at survival: gene expression noise and phenotypic diversification strategies. *Molecular Microbiology*, **71**(6), 1333–1340.

Gammaitoni, L., Marchesoni, F., and Santucci, S. 1995. Stochastic resonance as a bona fide resonance. *Physical Review Letters*, **74**(7), 1052.

Gammaitoni, L., Hanggi, P., Jung, P., and Marchesoni, F. 1998. Stochastic resonance. *Reviews of Modern Physics*, **70**(1), 223.

Haldane, J. B. S, and Spurway, H. 1954. A statistical analysis of communication in "*Apis mellifera*" and a comparison with communication in other animals. *Insectes Sociaux*, **1**(3), 247–283.

Hays, J. D., Imbrie, J., and Shackleton, N. J. 1976. Variation in the earth's orbit: pacemaker of the ages. *Science*, **194**(4270), 1121–1132.

Hoffmann, P. M. 2012. *Life's ratchet: how molecular machines extract order from chaos*. Basic Books.

Holldobler, B., and Wilson, E. O. 1990. *The ants*. Harvard University Press.

Hutchison, J. 2001. Culture, communication, and an information age Madonna. *IEEE Prof. Commun. Soc. Newsl.*, **45**(3), 1, 5–7.

Kaern, M., Elston, T. C., Blake, W. J., and Collins, J. J. 2005. Stochasticity in gene expression: from genotypes to phenotypes. *Nat. Rev. Genet.*, **6**(June), 451–464.

Karig, D. K, Siuti, P., Dar, R. D., Retterer, S. T., Doktycz, M. J., and Simpson, M. L. 2011. Model for biological communication in a nanofabricated cell-mimic driven by stochastic resonance. *Nano Communication Networks*, **2**(1), 39–49.

Korn, G. A., and Korn, T. M. (eds). 1952. *Electronic analog computers: D-C analog computers*. McGraw-Hill.

Latty, T., and Beekman, M. 2013. Keeping track of changes: the performance of ant colonies in dynamic environments. *Anim. Behav.*, **85**(3), 637–643.

McDonnell, M. D., and Abbott, D. 2009. What is stochastic resonance? Definitions, misconceptions, debates, and its relevance to biology. *PLOS Comput. Biol.*, **5**(5), e1000348.

Moss, F., Ward, L. M., and Sannita, W. G. 2004. Stochastic resonance and sensory information processing: a tutorial and review of application. *Clinical Neurophysiology*, **115**(2), 267–281.

Neiman, A., Shulgin, B., Anishchenko, V., Ebeling, W., Schimansky-Geier, L., and Freund, J. 1996. Dynamical entropies applied to stochastic resonance. *Phys. Rev. Lett.*, **76**(23), 4299–4302.

Nicolis, C. 1982. Stochastic aspects of climatic transitions: response to a periodic forcing. *Tellus*, **14**, 1–9.

Nicolis, S. C., and Deneubourg, J. 1999. Emerging patterns and food recruitment in ants: an analytical study. *J. Theor. Biol.*, **198**, 575–592.

Penzias, A. A., and Wilson, R. W. 1965. A measurement of excess antenna temperature at 4080 Mc/s. *Astrophys. J.*, **142**, 419–421.

Popat, R., Cornforth, D. M., McNally, L., and Brown, S. P. 2015. Collective sensing and collective responses in quorum-sensing bacteria. *Journal of the Royal Society Interface*, **12**(103), 20140882.

Reid, C. R., Sumpter, D. J. T., and Beekman, M. 2011. Optimisation in a natural system: Argentine ants solve the towers of Hanoi. *J. Exp. Biol.*, **214**(January 1), 50–58.

Roberts, G. L. 1962. Picture coding using pseudo-random noise. *IRE Trans. Inf. Theory*, **8**(2), 145–154.

Rolls, E. T., and Deco, G. 2010. The noisy brain. *In Stochastic dynamics as a principle of brain function*. Oxford University Press.

Sanchez, A., Choubey, S., and Kondev, J. 2013. Regulation of noise in gene expression. *Annual Review of Biophysics*, **42**, 469–491.

Schuchman, L. 1964. Dither signals and their effect on quantization noise. *IEEE Trans. Commun. Technol.*, **12**(4), 162–165.

Viney, M., and Reece, S. E. 2013. Adaptive noise. *Proceedings of the Royal Society of London B: Biological Sciences*, **280**(1767).

Wagner, A. 2014. *Arrival of the fittest: solving evolution's greatest puzzle*. Penguin.

Waters, C. M., and Bassler, B. L. 2005. Quorum sensing: cell-to-cell communication in bacteria. *Annu. Rev. Cell Dev. Biol.*, **21**, 319–346.

Wilson, E. O. 1962. Chemical communication among workers of the fire ant *Solenopsis saevissima* (Fr. Smith): 2. An information analysis of the odour trail. *Anim. Behav.*, **10**(1–2), 148–158.

10 The Many Faces of State Space Compression

David Wolpert, Eric Libby,
Joshua A. Grochow, and Simon DeDeo

Historically, scientists have defined the "macrostate" of a system, and the associated "level" or "scale," in a purely informal manner, based on insight and intuition. For example, in evolutionary biology often the macrostate is an entire species, a quantification of a set of coevolving organisms that ignores within-species diversity, fundamental dependencies between organisms, and complicated subunits such as tissues and cells that can also reproduce. Similarly, in economics the macrostates of the world's socioeconomic system are often defined in terms of firms, industrial sectors, or even nation-states, neglecting the internal structure of these highly complex entities.

One might view the reliance of many sciences on such vague human "insight" into how to even describe a physical system – a reliance that has no formal justification – as troubling. How do we know that these choices for the macrostates are the best ones with which to analyze the system? How do we even quantify the quality of a choice of macrostate? Might there be alternatives that are superior to our choices? A superior choice might, for example, allow greater accuracy in our prediction of the evolution of the system and/or reduce the computational cost of making such predictions. Given the possibility that superior choices might exist, can we *solve* for the optimal macroscopic state space with which to investigate a system?

This question, of how best to compress a microstate of a system into a macrostate, is the general problem of state space compression (SSC). To address this problem, we must first decide how to quantify the quality of a proposed map $x_t \to y_t$ that compresses a dynamically evolving "fine-grained" microstate x_t into a dynamically evolving "coarse-grained" macrostate y_t.[1] Given a definition of the quality of any compression and a microstate dynamics x_t, we can try to *solve* for the best map compressing x_t into a higher-scale macrostate y_t. The dynamics of such an optimally chosen compression of a system can be viewed as defining its emergent properties. Indeed, we may be able to iterate this process, producing a hierarchy of scales and associated emergent properties, by compressing the macrostate y to a yet higher-scale macrostate y'.

Here we adopt a utilitarian approach to SSC: a proposed compression is good if it is relatively easy to measure the associated macrostate, compute its evolution, and use its future values to accurately predict a given observable of interest concerning the system. Indeed, many of the quantities that are studied in the sciences are considered interesting precisely because they are good compressions in this sense. For instance, in physics, macrostates of a fine-grained system involve thermodynamic variables, such as temperature or pressure, that can be measured and used for accurate prediction of many observables of interest, at relatively low computational cost.

EXAMPLES

In this section we present some high-level examples to help clarify how our approach to SSC could be applied in various settings.

Example 10.1 *Consider a flock of N birds with fixed mass exhibiting coordinated flight (Attanasi et al., 2014; Bialek et al., 2012; Couzin,*

---

[1] Here we will use the term "coarse-graining" interchangeably with "compression" and so will consider "coarse-graining" maps that are completely arbitrary. There is a large literature that instead interprets the term "coarse-graining" to refer to the special case where x is the spatial position of a set of N particles and we wish to map each x to the spatial position of $m < N$ "effective particles." See Saunders and Voth (2013) for a review. There is also work that interprets the term "coarse-graining" to refer to the case where x is a network structure and we wish to map each x to a smaller network (see Blagus et al., 2012; Itzkovitz et al., 2005). For discussion of the general problem of coarse-graining for computational, cognitive, biological, and social systems, see, e.g., DeDeo, 2011, 2014; Krakauer et al., 2014, 2010; and references therein.

2009). The microstate of the flock, x, is the phase space position of the flock, i.e., the positions and velocities of all N different birds; in particular, the space of microstates is then $X = (\mathbb{R}^3)^{2N}$. The stochastic dynamics $P(x_t \mid x_0)$ of the microstate is given solely by some bird-level rule whereby each bird's acceleration is determined by its position and velocity as well as the positions and velocities of other birds in the flock. If we are interested in predicting the center of mass of the flock at all future times, then our observable of interest is a point in $\Omega \equiv \mathbb{R}^3$, whose value is given by the center-of-mass function $\mathcal{O}: \mathbb{R}^{6N} \to \mathbb{R}^3$. To evaluate a candidate compression, we can assign an accuracy cost C that is the Euclidean distance between two points in $\Omega = \mathbb{R}^3$ (i.e., $C: \Omega \times \Omega \to \mathbb{R}$), where the first argument of C is given by applying $\mathcal{O}$ to x_t, while the other one arises by predicting the time-t center of mass based on evolving a compressed description of the state of the flock.

As an example, one way to successfully predict the position of the center of mass is to evolve the stochastic dynamics of the microstate. This means simply choosing the macrostate to identically equal the microstate. However, this may be computationally challenging for large N since $X = \mathbb{R}^{6N}$. As an alternative, a successful SSC of the microstate dynamics would be a map from the high-dimensional vector x to some other much smaller-dimensional vector of "sufficient statistics," y, such that we can easily and accurately compute the evolution of y into the future and at each future time t recover the center of mass of the flock from y_t.

For example, consider a macrostate y comprising the $\mathbb{R}^3 \times \mathbb{R}^3 \times (\mathbb{R}^3)^3 = \mathbb{R}^{15}$ vector

> {Position of the center of mass of the flock; Momentum of the entire flock; Components of a Gaussian ellipsoid fit to the shape of the flock}.

Since one component of this particular y is precisely the quantity we are interested in (i.e., the position of the flock's center of mass), we can recover the desired prediction from y_t with the prediction function $\rho: Y \to \Omega$, which in this case simply selects the first three coordinates (the center of mass) from the preceding description. If this macrostate can be easily and accurately evolved forward in time, without concern for any other information in x not reflected in those fifteen real numbers, then it would be a good state compression, with a small combined value of the computation cost of the dynamics of the macrospace $Y = \mathbb{R}^3 \times \mathbb{R}^3 \times (\mathbb{R}^3)^3$ and the Euclidean error of the prediction of the center of mass. If not, this would say that there is something in the dynamics of the flock that is not well captured by these macrovariables.

Example 10.2 *Consider an agent-based simulation of the behavior of N humans in a set of interacting firms in an economy evolving in real time, where each player has time-varying observational data concerning the history of the entire economy and a time-varying set of endowments (e.g., property, income). Suppose the dynamics of the system is a set of stochastic rules for each of the players, specifying what move they will make at each moment as a function of their current observation vector and endowment set. Suppose as well that we are interested in the future total gross domestic product (GDP) of the economy, and in particular with how that GDP depends on some exogenous parameter θ that an external regulator can set, e.g., a parameter in the tax code.*

In this scenario, the microstate x_t at any time t includes θ together with the observation vector, endowment set, and moves of each of the N players at t. So the total GDP at all future times is specified by the vector of future microstates, x. Suppose we take the space of observables Ω to be $\mathbb{R}_{\geq 0}$, representing the total GDP at all future times, so that the observation map $\mathcal{O}(x)$ is the vector of future total GDPs of the economy specified by x. We define the accuracy function $C: \Omega \times \Omega \to \mathbb{R}$ to be the future discounted average of the absolute values of the difference between the predicted and actual GDPs. The challenge is that given the dependence of the microscopic stochastic dynamics x_t on the parameter θ that the regulator sets, and therefore the dependence of future GDPs on θ, it may be very challenging to calculate those future GDPs for any particular value of θ.

Now it may be that we can coarse-grain X into a compressed multidimensional space Y that specifies only the current state of each of the firms – ignoring the individuals within the firms and we can both evolve any $y \in Y$ easily and accurately and infer the GDP at each time future t knowing only y_t. In this case the coarse-graining of the economy into characteristics of the firms in the economy would be a good compression of the state space of the economy.

However, in many situations this state compression will not work well, giving large expected error for the prediction of future GDP. It may be that in some of these situations, rather than agglomerate all players into their associated firm to form each component of y, better predictions result from some other compression. In particular, it may be that a compression where each component of y involves the states of players from multiple firms results in far more accurate predictions of future GDP, without any gain in the computation cost of making those predictions. In such a situation, if we could find that better compression, it would provide major insight into what drives the evolution of the economy. For certain restricted classes of macrostates and their dynamics, this is the aggregation problem of economics (Chiappori and Ekeland, 2011; Hildenbrand, 2007).

Example 10.3 *In weather prediction we want to know the temperature and precipitation several days into the future. We could make a perfect prediction, in theory, if we knew the state of every atom on the planet, together with a detailed understanding of the flux of light and other particles from the sun. However, obtaining this information to the requisite accuracy is essentially impossible. Furthermore, prediction at such a detailed level would be computationally intractable. As an alternative, current forecasts use data from weather stations and satellites to make predictions. These are* *"compressed" quantifications of the current state of the atmosphere and solar radiation, at a scale much larger than atomic. What is the optimal trade-off between the cost of building, deploying, and maintaining new measurement devices, the computational cost of making a prediction from the data (which currently uses some of the world's largest super-computers), and the accuracy with which we predict the weather? State space compression provides a framework in which to make such questions mathematically precise, by considering the trade-off between measurement cost, computation cost, and accuracy cost.*

STATE SPACE COMPRESSION

General Considerations for Formalizing SSC

As illustrated above, often we are not interested in predicting the future value of a microstate *x in toto*. Instead, there are certain observables associated with *x* that we are interested in predicting, e.g., the center of mass of a flock of birds or the GDP of an economy. Accordingly, any compression of the microstate of a system that is then used for prediction involves three quantities: a compression map $X \to Y$, dynamical laws for evolution over Y, and a map from Y to the observable of interest.

In the approach to SSC considered here, the goal is to find such a triple that minimizes the associated cost of computation while maintaining high accuracy in predicting future values of the observable of interest. So we consider an SSC $X \to Y$ to be good if:

1. Our choice of the dynamics over Y is far easier to calculate than the dynamics over X. (Often – but not always – this will imply that $|Y|$ is significantly smaller than $|X|$.)
2. The observables concerning future values of x_t that we are interested in can be accurately and easily estimated from the corresponding values $y_t \in Y$.

From the perspective of coding theory, the goal is to compress a system's fine-grained state down to an element in a (usually) smaller space, evolve that compressed state, and then use it to predict the observables of interest concerning the future fine-grained state. Stated more prosaically, we wish to simulate the dynamics of x_t, using a simulation that optimally trades off its computation and measurement costs with its accuracy in predicting some observable of interest concerning future values of x.

It is important to emphasize that just like the choice of the observable of interest, the choice of how to quantify the "accuracy" of a prediction of that observable is problem-dependent. In some scenarios the way to quantify this accuracy will be exogenously provided, e.g., as a metric over the values of the observable. There are other scenarios, though, where there is no obvious metric over the space of observables. In these scenarios it often makes sense to use information-theoretic concepts such as mutual information, conditional entropy, or Kullback-Leibler divergence to quantify the accuracy of a prediction for an observable of interest.

Note as well that we will often be concerned with predicting aspects of the entire future evolution of an observable of interest, rather than its value at a single, particular moment in the future. Phrased differently, we are often interested in future *behavior*, which is an inherently multi-time-step phenomenon. Indeed, by requiring that the compression/decompression accurately reconstructs the time series as a whole, we ensure that it captures the dynamics of the system.

Compressing the state space of a system to reduce computation and measurement costs while maintaining predictive accuracy is a core concern not only in science and engineering. It is also a guiding principle in how the human nervous system operates. Computation is very costly for a biological brain, in terms of heat generated that needs to be dissipated, calories consumed, etc. Moreover, at an abstract level, the (fitness-function-mediated) "goal" of one's senses and the associated computational processes in the brain is to produce a compressed description of the environment such that the brain then uses it to produce accurate predictions of the future state of certain fitness-relevant attributes of the environment – all at minimal caloric cost (Clark, 2013; Krakauer, 2011; Shadmehr et al., 2010).[2] So in "designing" the human brain, natural selection is faced with the precise problem of optimal SSC as we have formulated it. This suggests that we may be able to construct powerful heuristics for optimal SSC by considering how the human nervous system processes information.

There are also human-constructed artifacts whose design reflects the trade-offs considered in SSC; results concerning SSC might thus be helpful in such design. One such example occurs in CPU design and the problem of branch prediction. In modern CPUs, whenever a conditional "if-then-else" statement is encountered, the CPU first guesses – based on certain heuristics and a record of previous branches – which branch of the conditional to take. It starts speculatively computing as though that branch were taken; if it later discovers that it guessed wrong, it terminates the speculative computation and continues along the correct branch, having incurred a delay for guessing incorrectly. We see here the trade-off between accuracy and computation cost: it makes sense to do branch prediction only if there is a computationally efficient predictor that is correct most of the time. In any other case, branch prediction would end up wasting more CPU cycles than it saves. Branch prediction is enormously successful in practice, partly because, based on the prior distribution of how most people (or compilers) write and use conditional statements, relatively simple heuristics give (almost shockingly) accurate results. Interestingly, some of the best branch predictors are in fact based on (very) simple models of neurons, as suggested in the discussion above (see, e.g., Jiménez, 2005).

---

[2] Recent work has suggested that biological individuality itself – cells, organisms, species – may be understood as the emergence of new coarse-grained partitions over the biochemical world (Krakauer et al., 2014).

Basic Concepts and Notation

Formalizing the issues discussed above leads to the following model of the variables and functions in SSC.[3] The first four are specified exogenously by the problem being studied:

1. the **microstate** $x \in X$, which evolves in time according to a stochastic process $P(x_t, \ldots)$, that in particular specifies a prior distribution over the value of x at a current time t_0
2. a space of possible values of an **observable** of interest Ω, and an associated observation conditional distribution $\mathscr{O}(\omega \in \Omega \mid x)$, which may be deterministic[4]
3. a (typically) real-valued **accuracy function** $C : \Omega \times \Omega \to \mathbb{R}$ that quantifies how good a predicted value is compared with the true observable
4. a **weight function** $W(t)$ used to determine the relative importance of predictions for the observable at all moments in the future.

Note that this formulation of SSC involves a distribution over initial values of the microstate, $P(x_{t_0})$. Often we will be interested in how well a particular prediction of a future value of an observable of interest, arising from a particular $x \sim P(x_{t_0})$, matches the actual value. For example, this is the case when predicting the future center of mass of a flock of birds that starts with some particular configuration of every bird in the flock. In other cases, though, we will be interested in how well the entire distribution of predictions of the future values of the observable, based on the entire distribution over initial microstates, matches the associated distribution of actual future values. For example, this is often the case when using Monte

---

[3] We adopt the convention that, depending on the context, uppercase variables indicate either an entire space or a random variable defined over that space and lowercase variables indicate elements of a space or values taken by a random variable. Measures of the integrals will all be implicitly matched to the associated space. So when we use the integral symbol for a countable space, we are using the counting measure, turning that integral into a sum.

[4] In addition to allowing stochastic dynamics, whenever it was natural to use a function $f : A \to B$ between two spaces, we instead use the more general setting of a conditional distribution $d_f(b|a)$.

Carlo sampling to predict future evolution of a distribution. In such cases it may make sense to replace (3) with a function that gives some sort of measure of the difference between two probability distributions over Ω, rather than the difference between two points from Ω.

The remaining three objects are chosen by the scientist – possibly by SSC optimization:

5. a **macrospace** or **compressed space** Y, with **macrostates** (y_t) that evolve according to a stochastic process $\phi_{t,t_0}(y_t \mid y_{t_0})$
6. a **compression** distribution, $\pi(y \mid x)$, which compresses x to y (and may be a deterministic function)
7. a **prediction** distribution $\rho(\omega \in \Omega \mid y)$, which gives the prediction for the observable based on the compressed state y.

Three terms will contribute to our overall state compression objective function. The first is the average future value of the accuracy function, evaluated under $W(t)$. We call this the **accuracy cost**; it is formalized below. The second term is the average future **computation cost** of iterating both ϕ and ρ. The third term is the cost of evaluating π once – or otherwise performing a measurement, e.g., of the physical world – to initialize y. For example, it may reflect a process of physically measuring an initial value x, with a noisy observation apparatus producing a value $z \in Z$ according to a distribution $P(z \mid x)$, where the observed initial value z is then mapped to the starting value of our simulation via a distribution $P(y_0 \mid z)$ that we set. In this case $\pi(y \mid x) = \int dz\ P(y_0 \mid z)P(z \mid x)$, where $P(z \mid x)$ might be more accurately described as a "measurement cost." Whenever we write "computation cost," it should be understood in this more general manner.

Unless specified otherwise, we take the weight function $W(t)$ to be a probability distribution over t. A particularly important type of $W(t)$ is a simple future geometric discounting function, $W(t) \propto [1 - \gamma]^t$. This is often appropriate when there is a priori uncertainty on how far into the future the scientist will end up running the computation; the basic idea is that having an answer tomorrow is more valuable than having it a month from now.[5] As alternatives, we could consider the (undiscounted) average accuracy cost and average computation cost over some temporal window $t \in [0, T]$. Another alternative, for computations known to halt, is to consider the average of those two costs from $t = 0$ to the (dynamically determined) halting time. Formally, this means extending the definition of W to be a function of both t and y_t.

We, the scientists, are provided the original stochastic process $P(x_{t'} \mid x_t)$ and determine the observable $\mathcal{O}$ and weight function $W(t)$ that capture aspects of the system that interest us. Our task is to choose the compression function π, associated space Y, compressed state dynamics ϕ, and prediction map ρ to the space of observables, such that π, ϕ, and ρ are relatively simple to calculate, compared with the dynamics of x_t, and the resultant accuracy costs are minimal (e.g., the resultant predictions for $\mathcal{O}(x_{t'})$ for $t' > t$ are minimal distortions of the true values). The best such tuple $\{\pi, Y, \phi, \rho\}$ is the one that best trades off the (average) quality of the reconstructed time series with the costs of implementing π and calculating the dynamics of y_t, according to some real-valued objective function $K(\pi, Y, \phi, \rho)$. Intuitively, a good state compression is like a good encoding of a video movie: it produces a compact file that is simple to decode and from which we can recover the characteristics of the original movie that the human user is attuned to, with high accuracy.

In many cases, maps $x \rightarrow y$ that are good compressions of a dynamical system have the property that the dynamics of y_t is (first-order) Markovian. (Indeed, as mentioned below, obeying *exactly* Markovian dynamics is a core component of the definition of a "valid compression" considered in Görnerup and Jacobi (2008, 2010), Israeli and Goldenfeld (2004, 2006), Jacobi and Goernerup (2007), Pfante et al. (2014), and Shalizi and Moore (2003).) Even if the best compression is

---

[5] Also see the literature on decision theory, where geometric discounting with a constant γ is justified as necessary to maintain consistency across time in how a decision-maker makes a decision, and especially the literature on reinforcement learning and dynamic programming (Sutton and Barto, 1998).

not Markovian, so long as we are considering a good compression of x_t into y_t, once we set the initial value of y often we can approximate well the future values of y_t with a Markovian process, with little further information needed from later x_t to correct that dynamics. For these reasons, we often restrict attention to compressed spaces whose associated dynamics is Markovian. However, we do not require this. Furthermore, we do not require ϕ and P to be constant throughout time.

Our formulation of SSC is a framework rather than a universally applicable specification of what a "good" compression is. Its use requires that we make explicit choices for things such as how to quantify accuracy and computation costs. Rather than provide discipline-independent prescriptions, we emphasize that the problem at hand should determine the choices adopted for the various terms described.

USING SSC TO ADDRESS OTHER SCIENTIFIC CHALLENGES

Although our SSC framework is most directly concerned with simplifying computation of the future of a dynamical system, it may provide insight into several other topics in science as well. In this section we present several examples of this. We intend to investigate them in more depth in the near future.

A Form of Complexity Motivated by SSC

Say we find a particularly good compression (Y, π, ϕ, ρ), which substantially reduces the value of the objective function K compared with its value under the "null compression," which leaves the original space X unchanged. Formally, using "id" to indicate an identity map (with the spaces it operates varying and implicit), such a tuple (Y, π, ϕ, ρ) results in a value of

$$\frac{K(\pi, \phi, \rho; P)}{K(\mathrm{id}, \mathrm{id}, \mathrm{id}; P)} \tag{10.1}$$

close to its lower bound of 0 (assuming we have chosen the objective function so that it is guaranteed nonnegative). When this ratio is close to 0, it means that we have found a way to simulate the dynamics of the original system that is both computationally easy and that accurately predicts the future of the (observable of interest concerning the) underlying system. When it is large, however (i.e., only barely below 1), either the simulation is computationally difficult to run or it results in poor predictions of the future of the underlying system, or both.

Accordingly, we can define the **compression complexity** of a system x_t evolving according to $P(x_t)$ as

$$\frac{\min_{\pi,\phi,\rho} K(\pi, \phi, \rho; P)}{K(\text{id}, \text{id}, \text{id}; P)} \tag{10.2}$$

ranging from 0 for highly compressible (noncomplex) systems to 1 for highly incompressible (complex) systems. Importantly, compression complexity is low for completely random systems as well as "simple" systems. (This is in contrast to algorithmic information content, which is low for simple strings but high for random ones.)

More precisely, consider the case where $\Omega = X$, $\mathcal{O}$ is the identity map, and the fine-grained dynamics is independent and identically distributed (IID) noise, i.e., $p(x_{t+1} \mid x_t) = p(x_{t+1})$ with a high entropy $H(X_t)$. Suppose as well that accuracy cost is the negative of the time-averaged form of mutual information between the coarse-grained and fine-grained states (defined formally in Eq. (10.7) below as $-I_{\overline{\mathcal{P}}_{\pi,\phi,\rho}}(X'; X)$). For this situation, no choice of π, ϕ, and ρ results in smaller accuracy cost than when no compression is used. So if our SSC objective function were accuracy cost alone, we would conclude that compression complexity is high (just as with algorithmic information content [AIC]).

However, consider compressing X to a Y that contains only a single element. We lose nothing in accuracy cost. But computation cost is now zero. So the value of the *full* SSC objective function is greatly reduced. This illustrates that "a fine-grained dynamics that is IID noise" is assigned a small compression complexity.

When compression complexity is low, the compressed system $\{Y, \pi, \phi, \rho\}$ may provide substantial physical insight into the dynamics P of the microstate X, since the compressed system intuitively tells us "what's important about X's time-evolution." To illustrate this, return to Example 10.2, where the microstate specifies the state of all the people in an evolving economy. As discussed there, potentially the best choice for the compression π is to map the joint state of all the people in the economy to a small number of characteristics of the firms that employ those people. Intriguingly, though, it may well be that the best macrostates instead are characteristics of the relation among the firms, or characteristics of the relation among individuals at multiple firms. In such a situation, SSC would guide us to model the economy not on a firm-by-firm basis, but rather in terms of those SSC-optimal macrostates that involve the states of multiple firms and individuals simultaneously.

As defined above, compression complexity is normalized by the value $K(\mathrm{id}, \mathrm{id}, \mathrm{id}; P)$. This is not always appropriate, e.g., if there is some concrete physical meaning associated with the value of the objective function. In such cases, there are several possible modifications to the definition of compression complexity that may make sense, e.g., $\min_{\pi, \phi, \rho} K(\pi, \phi, \rho; P)$. Note as well that compression complexity is defined *with respect to a given accuracy cost and associated observation operator*. So the same fine-grained system may be characterized as "complex" or "not complex" depending on what one wants to predict concerning its future state.

Unlike many other complexity measures, compression complexity is tailored to measuring the complexity of *dynamic* systems. Indeed, in its simplest form, it does not apply to static objects, like images. There are various modifications of compression complexity that can apply to static objects, though. For example, if we have a generative process that creates images, we could measure the compression complexity of the generative process that produces that image.

On the other hand, note that as a system evolves its optimal SSC will change. So as we go from t to $t + 1$ to $t + 2$, etc., if we evaluate the associated values of compression complexity, taking each of those successive times as the "initial time," in general we would expect that complexity to undergo a dynamic process. This may provide a useful framework for analyzing informal suppositions of many fields concerning how complexity evolves in time, e.g., concerning how the complexity of large biological systems changes in time.

Finally, we emphasize that despite its name, we do not promote compression complexity as "the" way to measure complexity. Rather, we are simply highlighting that it has many aspects that match well to characteristics of complexity that have been proposed in the past and that it may lead to novel insight into physical phenomena. (Indeed, unlike many other complexity measures, compression complexity is tailored to measuring the complexity of *dynamic* systems, not static ones.)

Using SSC to Define Information Flow among Scales of a System
Many conventional measures of "information flow," such as causal information flow (Ay and Polani, 2008), transfer entropy (Prokopenko and Lizier, 2014; Prokopenko et al., 2013; Schreiber, 2000), and Granger causality (Kamiński et al., 2001), are motivated by considering the relation between systems that are physically separated from one another. The dynamic behavior of such systems can be coupled by earlier confounding variables, or not, and appropriately removing such confounders is a central concern of the usual motivations of these measures of information flow.

Arguably this makes these measures ill-suited to quantifying information flow among the scales of a single system. After all, behavior at different scales of a single system is in many respects *solely* due to such "confounders," in that the previous states of the more detailed scale often completely determine the current state of the less detailed scale, in what might be seen as a form of "bottom-up causation." As an example, the most straightforward way to use transfer entropy to measure the information flow from (the coarse-grained space) Y_t to (the fine-grained space) X_{t+1} would quantify it as the conditional mutual information $I_{\pi,\phi;P}(X_{t+1}; Y_t \mid X_t)$. However, since there is no coupling between X_{t+1} and Y_t once the confounder X_t is removed, this quantity is identically zero, regardless of π, ϕ, and P.[6]

For these reasons, many view the very notion of information flowing from a high scale to a low scale of a single system as problematic, no matter how one might try to quantify it. On the other hand, going back at least to the seminal work of Maynard Smith and others, many researchers have presumed that information *does* flow from the high (coarse-grained) scale down to the low (fine-grained) scale in biological systems (Davies, 2012; Maynard Smith, 1969, 2000; Walker et al., 2012). Indeed, Maynard Smith made strong arguments that such high- to low-scale information flow increased in each major transition in biological evolution. In addition, recent work has emphasized the critical role information flow may play in understanding the emergence of social complexity (Chapter 16) and even biological individuality (Krakauer et al., 2014) itself.

A similar phenomenon is seen in descriptions of algorithms or Turing machines. For example, consider an algorithm that determines whether its input is a prime number or not. On the one hand, the behavior of this algorithm is completely specified by its code: how it moves bits around in memory and combines them. On the other hand, the low-level bit-wise behavior of this algorithm may be described as being "determined" by its search for primality. When at some point it takes one branch of a conditional "if" statement rather than another, do we say that it took that branch because the memory was just in a certain position, or because the number 6 is not prime?

The SSC framework provides alternative ways to formalize the information flowing from a high scale down to a low scale of a single system. These are more directly grounded in the relation between the behaviors at different scales of a single system, and do not try to remove confounder effects. The overall idea starts by using SSC

---

[6] Note that the calculations of transfer entropy between scales made in Walker et al. (2012) do not adopt this simplest use of transfer entropy, in which, to use Schreiber's notation, $k = l = 1$.

to solve for the scale(s) at which to analyze the system, rather than relying on the scale(s) being prespecified in an ad hoc manner. We then define the associated information flow from the coarse scale to the fine scale by treating (ϕ and) ρ as an information channel. We refer to any such quantification based on SSC as **compression causality**, in analogy with "Granger causality."

As an example of compression causality, we might argue that the amount of information needed to construct an observable of x_t, given any particular y_t, is the entropy of X_t under the probability distribution conditioned on y_t at the previous time step (for illustrative simplicity here we consider discrete time steps). Averaging that over y_t and then over t gives us

$$-\int d\Delta t \, W(\Delta t) H_{\mathscr{P}_{\pi,\phi,\rho;P}}(\Omega_{\Delta t} \mid Y_{\Delta t-1}) \qquad (10.3)$$

where as before $W(.)$ is the weight function saying how much concern we place on future times, and as formalized below, $H_{\mathscr{P}_{\pi,\phi,\rho;P}}(\Omega_{\Delta t} \mid Y_{\Delta t-1})$ is the conditional entropy of the actual value of the observable at time Δt, given the macrostate at time $\Delta t - 1$.

A crucial thing to note about the conditional entropy in Eq. (10.3) is that the statistical coupling underlying it, relating the microstate at Δt and the macrostate at $\Delta t - 1$, arises purely through the function π, mapping the initial microstate to the initial macrostate. There is no information flow between the microstates and the macrostates at any time after that initial setting of the macrostate. Note as well the crucial role played by the choice of the observable function $\mathscr{O}$. If one leaves the dynamics P fixed, by changing the observable function, then in general one changes the compression causality for that system.

A slight variant of the measure in Eq. (10.3) arises if we wish to normalize each of the conditional entropies $H_{\mathscr{P}_{\pi,\phi,\rho;P}}(\Omega_{\Delta t} \mid Y_{\Delta t-1})$ by subtracting it from $H_{\mathscr{P}_{\pi,\phi,\rho;P}}(\Omega_{\Delta t})$. This difference tells us how much extra information about the value of ω_t is provided by y_{t-1}, beyond the prior information concerning ω_t. With this normalization our measure of information flow becomes

$$- \int d\Delta t \, W(\Delta t) I_{\mathscr{P}_{\pi,\phi,\rho,P}}(Y_{\Delta t} \, ; \Omega_{\Delta t}) \tag{10.4}$$

If $\mathcal{O}$ is the identity map, then the quantity in Eq. (10.4) reduces to the time-averaged mutual information between the macrostate and the microstate. As discussed below, this is one natural possibility for an information-theoretic accuracy cost. However, to avoid the issue that two variables can have perfect mutual information by being either perfectly correlated or perfectly anticorrelated at any given time step (see Eq. 10.4), it may make sense to replace this quantity with the mutual information of the time average or even with the conditional entropy of the time average, $H_{\overline{\mathscr{P}}_{\pi,\phi,\rho}}(\Omega \mid Y)$, a point also discussed below.

Note the essential difference between this way of formalizing information flow from a coarse scale to a fine scale and using the transfer entropy (Schreiber, 2000) from coarse to fine scales. A core concern in the definition of transfer entropy is that it reflects only how much information the random variable y_t has concerning x_t that does not arise through direct dependence of those two variables on earlier variables that affect both of them (so-called confounders). For example, if $x_{t-1} \rightarrow y_{t-1}$ is a single-valued invertible mapping, then y_t will have high mutual information with x_t, even though there is no *novel* information that flows from y_t to x_t that is not explained as "echoes" of x_t. Transfer entropy is explicitly designed to remove such effects arising from confounders.

However, in many respects, these confounders that transfer entropy is designed to ignore are *precisely* what we want to capture when considering information flow among scales. In particular, it is precisely what the measures suggested in Eqs. (10.3) and (10.4) are designed to capture.

Other Potential Applications of SSC

Several potential applications of SSC to other topics in science suggest themselves.

- **Comparing systems, from fields near and far.** SSC provides a generic method of comparing systems: if two systems (say, of the same type, e.g., two networks or two economies) have the same optimal SSC compression, this tells us something about the commonalities between the systems. More intriguingly, this statement holds even for completely different types of systems: if an economy has some good SSC that is the same as an SSC of some multicellular organism, that again highlights some commonality between the two.

- **Analyzing why natural phenomena are partitioned among scientific fields the way they are.** Loosely speaking, the set of all scientific fields can be viewed as a partition over the space of natural phenomena. In addition to considering different phenomena, any two scientific fields also use different theoretical structures to predict the future behavior of those phenomena. In other words, they use different SSCs to predict the behavior of the different sets of phenomena that lie within their purviews.

This suggests that we could use SSC to investigate how and why natural phenomena are partitioned among scientific fields the way they are. One natural hypothesis is that any phenomenon that lies in the "core" of one particular field, "far away" from the phenomena that lie in any other field, have low compression complexity (see above). That low complexity would arise using SSC that defines the theoretical structure used by practitioners of that particular field.

A related hypothesis is that some of the phenomena lying at the boundaries between any two scientific fields A and B have poor compression complexity. This would mean that any theory defined at a coarse-grained level that one might use to model those phenomena is either intractable or an inaccurate model of those phenomena. In this case, no "bridging theory" relating fields A and B would be possible. Similarly, it may be that the boundary between some particular pair of scientific fields A and B is where the optimal compression changes. In this case it may be possible to build an SSC that has good compression complexity for a set of phenomena on both sides of the boundary – a bridging theory. These questions suggest a very broad, formal analysis of what sets of "abstract natural phenomena" have boundaries of the first sort, where no good compression is possible; which have boundaries of the second sort, where boundaries are defined by changes in the optimal compression, and where it may be possible to construct bridging theories; and which have mixtures of both types of boundaries.[7]

- **Quantifying some aspects of the philosophy of scientific theories.**
  Scientific theories have many aspects of value: pragmatic, economic, aesthetic, and others. From the pragmatic viewpoint of predictability, however, one aspect that is often not considered is that of the *computation cost* of calculating a prediction within a given scientific theory. Indeed, a scientific theory that makes predictions in theory, but for which in practice even simple predictions cannot be calculated in the lifetime of the universe, is completely useless from the viewpoint of prediction. In contrast, the computational cost of prediction is a core concern of our SSC framework. Could this be used to evaluate, for example, candidate string theories? However beautiful it might be, if some candidate string theory is completely useless for predictions (even setting aside the issues of whether we'll ever be able to build a device to carry out an experiment), then perhaps we might be better off considering a variant of the theory with more feasible predictive power.

ACCURACY COSTS

In the rest of this chapter, we present fully formal definitions of accuracy cost and computation cost to clarify our arguments. We start this section by presenting some ways to use the notation introduced above to quantify accuracy cost. We do this both when an accuracy function is exogenously provided and when it is not, in which case it may make sense to use an information-theoretic accuracy cost.

We begin with an example of accuracy cost that is often appropriate when the dynamics is Markovian:

$$
\mathcal{E}(\pi, \phi, \rho; P) \equiv \int_{\Delta t > 0} d\Delta t \int dx_0\, dx\, d\omega\, dy_0\, dy\, d\omega'\, W(\Delta t)
$$
$$
\times P(x_0)\pi(y_0 \mid x_0)P_{\Delta t}(x \mid x_0)\mathcal{O}(\omega \mid x)\phi_{\Delta t}(y \mid y_0)
$$
$$
\rho(\omega' \mid y)C(\omega, \omega') \tag{10.5}
$$

---

[7] This idea arose in conversations with Robert Axtell about SSC/aggregation.

In the integrand, $C(\omega, \omega')$ is the cost if our simulation using the compressed space predicts that the observable has the value ω' when it is actually ω. As described above, $\rho(\omega' \mid y)$ is the distribution (typically a single-valued function) that converts the state y of the compressed space into such a prediction of the value of the observable. Also as discussed above, $\phi_{\Delta t}$ and $P_{\Delta t}$, here take the form of first-order Markov chains of our simulation over the compressed space and of the actual fine-grained system state, respectively. π is the distribution by which we compress the initial state of the fine-grained state, and $P(x_0)$ is the a priori probability that we will be interested in simulating the fine-grained system that starts in state x_0.

Several variants of Eq. (10.5) are possible as quantifications of accuracy cost, even in the Markovian case. For example, one might be interested in a worst-case accuracy cost over the initial states, rather than an average. More generally, if the fine-grained dynamics and/or our simulation are not first-order Markovian, then Eq. (10.5) would have to be modified accordingly. (We don't present that modification here because in general it can be very messy.)

Information-Theoretic Accuracy Costs

If an accuracy function C is not exogenously supplied, it may be appropriate to replace Eq. (10.5) with an information-theoretic definition of accuracy cost. Similarly, if in the problem at hand it's more natural to compare the *distribution* over values of predicted observables with the distribution over values of actual observables, then again a (different) information-theoretic definition of accuracy cost may be appropriate. We consider both of those variants of Eq. (10.5) in this subsection. The subtleties of information theory make this subsection necessarily somewhat technical; readers who prefer to consider only exogenously specified accuracy costs may elect to skip it.

Accuracy Cost Based on Mutual Information for Two Time Steps

We begin by focusing on the special case in which there are only two time steps, t_0 and Δt. So the sole predicted value of the observable of interest is $\omega'_{\Delta t}$, generated by applying π to x_t and then evolving it according to $\phi_{\Delta t}$. In contrast, $\omega_{\Delta t}$ is the actual value, generated by running the Markov process P over X and then applying observable distribution $\mathcal{O}$. A natural information-theoretic measure for quantifying the quality of such a prediction is the mutual information between $\omega'_{\Delta t}$ and $\omega_{\Delta t}$.

Although intuitively straightforward, the fully formal equation for this accuracy cost is a bit complicated. This is because the random variables whose mutual information we are evaluating are coupled indirectly, through an information channel that goes through the time-0 conditional distribution π. Writing it out, this accuracy cost is

$$\mathscr{E}_{\Delta t}(\pi, \phi, \rho; P) = -I_{\mathscr{P}_{\pi,\phi,\rho;P}}(\Omega'_{\Delta t} ; \Omega_{\Delta t}) \tag{10.6}$$

where the minus sign reflects the fact that large mutual information corresponds to low misfit C, and where the joint probability $\mathscr{P}_{\pi,\phi,\rho;P}(\omega'_{\Delta t}, \omega_{\Delta t})$ defining the mutual information at time Δt is given by the marginalization

$$\mathscr{P}_{\pi,\phi,\rho;P}(\omega'_{\Delta t}, \omega_{\Delta t}) \equiv \int dx_0 \, P(x_0) \mathscr{P}_{\pi,\phi,\rho;P}(\omega'_{\Delta t}, \omega_{\Delta t}, x_0) \tag{10.7}$$

where $\mathscr{P}_{\pi,\phi,\rho;P}(\omega'_{\Delta t}, \omega_{\Delta t}, x_0)$ is defined as

$$\int dy_0 dy_{\Delta t} \, \pi(y_0 \mid x_0)\phi(y_{\Delta t} \mid y_0)\rho(\omega'_{\Delta t} \mid y_{\Delta t})$$
$$P(x_{\Delta t} \mid x_0)\mathcal{O}(\omega_{\Delta t} \mid x_{\Delta t}). \tag{10.8}$$

Intuitively, the distribution $\mathscr{P}_{\pi,\phi,\rho;P}(\omega'_{\Delta t}, \omega_{\Delta t}, x_0)$ couples $\omega'_{\Delta t}$ and $\omega_{\Delta t}$ by stochastically inferring $y_{\Delta t}$ from $\omega'_{\Delta t}$, then "backing up" from $y_{\Delta t}$ to y_0 and on to x_0, and finally evolving forward stochastically from x_0 to get an $x_{\Delta t}$ and thereby $\omega_{\Delta t}$.

When there is no exogenously specified observable of interest, we may simply take the microstate itself to be the observable. In this case we would be led to quantify the accuracy cost as the mutual information between the future microstate and the future macrostate. This is essentially why mutual information has been used in other work related to SSC (Israeli and Goldenfeld, 2004, 2006; Pfante et al., 2014; Shalizi and Moore, 2003; Shalizi et al., 2001).

However, there are important differences between our mutual information and these other ones. For example, in Pfante et al. (2014) the analogous accuracy cost, defined for the values of a process at a pair of times t_0 and $\Delta t > t_0$, is the conditional mutual information $I(Y_{\Delta t}; X_{t_0} \mid Y_{t_0})$. Although there are scenarios in which both this cost and the cost $\mathscr{E}_{\Delta t}(\pi, \phi; P)$ in Eq. (10.6) are minimal,[8] there are also scenarios in which the cost $I(Y_{\Delta t}; X_{t_0} \mid Y_{t_0})$ achieves its minimal value even though the cost $\mathscr{E}_{\Delta t}(\pi, \phi; P)$ is *maximal*. For example, the latter occurs if π is pure noise, so that dynamics in y implies nothing whatsoever about dynamics of x.[9]

Accuracy Cost Based on Mutual Information for More Than Two Time Steps
The natural extension of Eq. (10.6) to multiple times is

$$\mathscr{E}(\pi, \phi, \rho; P) = \int d\Delta t \, W(\Delta t) \mathscr{E}_{\Delta t}(\pi, \phi, \rho; P)$$

$$= -\int d\Delta t \, W(\Delta t) I_{\mathscr{P}_{\pi,\phi,\rho;P}}(\Omega'_{\Delta t} ; \Omega_{\Delta t}) \qquad (10.9)$$

with $\mathscr{P}_{\pi,\phi,\rho;P}(\omega'_{\Delta t}, \omega_{\Delta t})$ defined as in Eq. (10.7) for all values of Δt. However, the following example illustrates a subtle but important problem with this formula.

Example 10.4 *Consider a discrete-time system with $X = \{0, 1\}$ with dynamics $P(x_{t+1} \mid x_t) = \delta_{x_{t+1}, x_t}$, and let $Y = X$ but with nonstationary dynamics that swaps the two values at every time step. Suppose $\pi : \{0, 1\} \ \to \ \{0, 1\}$ is the identity map. Then at the initial time t_0, the map $\rho_{even} : Y \to X$ defined by $\rho_{even}(0) = 0$ and $\rho_{even}(1) = 1$ is a*

---

[8] For example, this occurs if all of the conditional distributions π, ϕ, and $P(x_{\Delta t} \mid x_0)$ are deterministic, measure-preserving functions, so that the dynamics in y uniquely specifies dynamics in x.

[9] This distinction between these two measures reflects the fact that they are motivated by different desiderata. The cost $I(Y_{\Delta t}; X_{t_0} \mid Y_{t_0})$ is motivated by the observation that if it is zero, then there is no extra information transfer from the dynamics of X that is needed to predict the dynamics of Y, once we know the initial value y_{t_0}, and in this sense dynamics in Y is "autonomous" from dynamics in X.

perfect predictor of x_{t_0} from y_{t_0}; indeed, this same predictor works perfectly at every time that is an even number of steps from t_0. At those times t that are an odd number of steps from t_0, x_t can still be perfectly predicted from y_t, but now by a different map $\rho_{odd}: Y \to X$, which swaps the two values ($\rho_{odd}(0) = 1$ and $\rho_{odd}(1) = 0$). In such a situation, mutual information is maximal at all moments in time. However, there is no single, time-invariant map ρ that allows us to interpret y_t as a perfect prediction for the associated x_t.

One way to resolve this problem is to modify that accuracy cost to force the prediction map from Y to Ω to be time invariant. To state this formally, return to the motivation for using information theory in the first place: construct a space of codewords and an associated (prefix-free) coding function that allows us to map any value in Y to a value in Ω, taking as our accuracy cost the minimal expected length of those codewords (over all codes). To make this expected code length precise, we need to define an encoding function. So construct a Z and a (time-invariant) encoding function f such that for any y, ω, there is a $z \in Z$ such that $f(y, z) = \omega$. From one t to the next, given y_t, we have to choose a z_t so that we can recover $\omega_t = \mathcal{O}(x_t)$ by evaluating the (time-invariant) function $f(y_t, z_t)$. We then want to choose a code for Z that minimizes the length of (codewords for) zs that allow us to recover x from y, averaged over time according to W and over pairs (ω_t', ω_t) according to $\mathscr{P}_{\pi,\phi,\rho;P}(\omega_t', \omega_t)$.

So we are interested in the t-average of expectations of (lengths of codewords specifying) zs where those expectations are evaluated under $\mathscr{P}_{\pi,\phi,\rho;P}(x_t', x_t)$. This is just the expectation under the single distribution given by t-averaging the distributions $\mathscr{P}_{\pi,\phi,\rho;P}(x_t', x_t)$. Write that single t-averaged distribution as

$$\overline{\mathscr{P}}_{\pi,\phi,\rho}(\omega', \omega) \equiv \int dt\, W(t) \mathscr{P}_{\pi,\phi,\rho;P}(\omega_t', \omega_t) \tag{10.10}$$

The associated minimum of expected code lengths of zs is just $H_{\overline{\mathscr{P}}_{\pi,\phi,\rho}}(\Omega \mid \Omega')$. To normalize this we can subtract it from the entropy of the marginal, $H_{\overline{\mathscr{P}}_{\pi,\phi,\rho}}(\Omega)$. (Note that this entropy of the marginal is fixed by P, independent of π, ϕ, or ρ.) This gives us the change in the expected length of codewords for specifying values ω_t that arises due to our freedom to have those codewords be generated with a different code for each value of the prediction ω'_t. Since higher accuracy corresponds to lower accuracy cost, this motivates an information-theoretic accuracy cost given by

$$\mathscr{E}(\pi, \phi, \rho; P) = -I_{\overline{\mathscr{P}}_{\pi,\phi,\rho}}(\Omega'; \Omega) \tag{10.11}$$

This information-theoretic accuracy cost is the mutual information under the t-average of $\mathscr{P}_{\pi,\phi,\rho}(\omega'_t, \omega_t)$, rather than Eq. (10.6)'s t-average of the mutual information under the individual $\mathscr{P}_{\pi,\phi,\rho}(\omega'_t, \omega_t)$s.

Alternative Information-Theoretic Accuracy Costs
While it seems that the distribution $\mathscr{P}_{\pi,\phi,\rho;P}(\omega'_{\Delta t}, \omega_{\Delta t})$ is often the appropriate one to use to define accuracy cost, in some circumstances the associated mutual information is not the most appropriate accuracy cost. An important example of this is when the microstate Markov process $P_{\Delta t}(x \mid x_0)$ (or the observable $\mathscr{O}$, for that matter) is not deterministic, and our goal is to use x_0 to predict the future evolution of the *entire distribution* over $\omega_{\Delta t}$ given by $\mathscr{O}(\omega_{\Delta t} \mid x_{\Delta t})P(x_{\Delta t} \mid x_0)$, rather than predict the specific future values $\omega_{\Delta t}$.

In particular, in many situations it may prove useful to use Monte Carlo sampling of the distribution over $\rho(y)$ values, $\mathscr{P}_{\pi,\phi,\rho;P}(\omega'_{\Delta t}, x_0)$, as an approximation to Monte Carlo sampling of $\mathscr{O}(\omega_{\Delta t} \mid x_{\Delta t})P(x_{\Delta t} \mid x_0)$. (For example, this is often the case in the kinds of situations where we might want to use particle filters or some of the techniques of uncertainty quantification.) A natural accuracy cost for this kind of situation is

$$\mathscr{E}(\pi, \phi, \rho; P) \equiv -\int_{\Delta t > 0} d\Delta t \; W(\Delta t) \int dx_0 \; P(x_0)$$
$$\times \text{KL}[\mathscr{P}_{\pi,\phi,\rho;P}(\Omega'_{\Delta t} \mid x_0) \;\|\; \mathscr{P}_{\pi,\phi,\rho;P}(\Omega_{\Delta t} \mid x_0)]$$
$$\tag{10.12}$$

where the notation "KL[$P(A \mid b) \parallel R(A)$]" means the Kullback–Leibler divergence between the two distributions over values $a \in A$ given by $P(a \mid b)$ and $R(a)$ (Cover and Thomas, 2012; MacKay, 2003).[10]

Mutual information is the appropriate accuracy cost when one is interested in making "point predictions." As an illustration, in the context of Example 10.1 this cost would be appropriate if we wished to accurately predict the future center of mass *for one specific initial flock configuration*, with its microstate fully specified; the mutual information would give us the average accuracy of the predictions over all possible initial flock configurations. In contrast, we may be interested only in accurately predicting the *entire future distribution* of positions of the center of mass, given some distribution over initial flock configurations, without worrying about the accuracy of the prediction for any one specific initial configuration. In that case, a KL-divergence accuracy cost would be more appropriate.

In general, as with all aspects of SSC, the precise problem at hand should determine what kind of accuracy cost is used. Nonetheless, one would expect that quite often using several different accuracy costs would all provide physical insight into "what is truly driving" the dynamics across X.

COMPUTATION COSTS

The core concern of SSC is how to choose π, ϕ, and ρ in a way that minimizes computation cost without sacrificing too much accuracy cost. To quantify this goal we need to quantify the computation cost associated with any tuple $(\pi, \phi, \rho; P)$ (with the associated X, Y, and $\mathcal{O}$ being implied). This section discusses three possible quantifications, motivated by pragmatism, theoretical computer science, and thermodynamics.

---

[10] Note that there is not the same issue here involving dynamic changes to how we match each element of y with an element of x that arose in our analysis of accuracy cost based on mutual information. The reason is that both of the distributions in the Kullback–Leibler divergence are defined over the exact same space.

One of the kinds of computation cost considered in computational complexity is the running time of the computation (Hopcroft and Motwani, 2000; Moore and Mertens, 2011). This is also often a primary concern of real-world SSC, where we are often interested in expected "wall-clock" time of a simulation. As a practical issue, this measure is often accessible via profiling of the simulation ϕ as it runs on the computer. Such profiling can then be used to guide the search for a (π, ϕ, ρ) that optimizes the trade-off between computation cost and accuracy cost (see below).

Often, though, we want a more broadly applicable specification of computation cost, represented as a mathematical expression; at a minimum this is needed for any kind of mathematical analysis. One obvious way to do this is to use AIC, i. e., to quantify computation cost as the minimal size of a Turing machine that performs the desired computation (Chaitin, 2004). This has the major practical problem that how one measures the size of a Turing machine $\mathscr{T}$ (i.e., what universal Turing machine one chooses to use to emulate the running of $\mathscr{T}$) is essentially arbitrary. Furthermore, AIC is formally uncomputable, so one has to settle for results concerning asymptotic behavior. To get around this issue, people sometimes "approximate" the AIC of a string, e.g., its length after Lempel–Ziv compression. However, this in essence reduces AIC to Bayesian maximum a posteriori coding, where the prior probability distribution is implicit in the Lempel–Ziv algorithm. (There are also further problems in using either Lempel Ziv – scc, e.g., Shalizi, 2003 or AIC – see, e.g., Ladyman et al., 2013.)

There are several reasonable variants of AIC that might also be appropriate for some types of analysis of SSC. Some such variants are the many versions of Rissanen's (minimum) description length (MDL) (Barron et al., 1998; Rissanen, 1983). Another one, quite close to the measure of running time mentioned above, is logical depth (Bennett, 1988). However, logical depth is still subject to the practical difficulties associated with AIC.

A final measure of computation cost is concerned not with the space or time resources a computation requires, but rather with the amount of energy it requires, i.e., how much it "runs down the battery." This measure builds on the seminal work of Szilard, Landauer, Bennett, and many others, on the minimal thermodynamic work needed by any physical device to implement a given computation, as a function of that computation (Bennett, 2003; Landauer, 1961; Szilard, 1964). In particular, recently it has been proven that the minimal amount of work to implement a single step from t to $t + 1$ of a (potentially stochastic) computation with associated marginal distributions $\phi(y_t)$ and $\phi(y_{t+1})$ is

$$\mathscr{C}(\pi, \phi, \rho; P) = kT[H_{\pi,\phi,\rho;P}(Y_t) - H_{\pi,\phi,\rho;P}(Y_{t+1})] \qquad (10.13)$$

where $H_{\pi,\phi,\rho;P}(Y_\tau)$ is the Shannon entropy of the marginal distribution of the compressed variable Y at time τ, k is Boltzmann's constant, and T is the temperature of the environment of the computer (Wolpert, 2015). Note that in general, the smaller Y is, the smaller the maximal possible entropy of any distribution over Y. As a result, quite often the smaller Y is, the smaller the computation costs given in Eq. (10.13).

As mentioned above, the computation/measurement cost $\mathscr{C}$ can include the cost of mapping the original state to the compressed state. In this case, the computation cost $\mathscr{C}$ might include a term of the form $H_{\pi,P}(Y_0)$. As always, we emphasize that we are not advocating any one particular quantification of computation or measurement cost; as with most aspects of the SSC framework, the computation/measurement cost should be selected appropriately for the problem at hand.

THE FULL SSC OBJECTIVE FUNCTION

Formally speaking, once we have defined both an accuracy cost function and a computation cost function, we are faced with a multiobjective optimization problem of how to choose π, ϕ, and ρ in order to minimize both cost functions. There are many ways to formalize this problem. For example, as is common in multiobjective optimization, we might wish only to find the set of triples (π, ϕ, ρ) that lie on the Pareto curve of those two functions. Alternatively, we might face a constrained optimization problem. For example, we might have constraints on the maximal allowed value of the accuracy cost, with our goal being to minimize computation cost subject to such a bound. Or conversely we might have constraints on the maximum allowed computation cost (say, in terms of minutes or dollars), with our goal being to minimize accuracy cost subject to such a bound.

In this section, for simplicity we will concentrate on ways to reduce the multiobjective optimization problem into a single-objective optimization problem. To do this requires that we quantify the trade-off between computation and accuracy costs in terms of an overall SSC objective function that we want to minimize. Such an objective function maps any tuple $(\pi, \phi, \rho; P)$ (with the associated X, Y, and $\mathcal{O}$ being implied) into the reals. The associated goal of SSC is to solve for the π, ϕ, and ρ that minimize that function, for any given P, X, and $\mathcal{O}$.

The Trade-Off between Accuracy Cost and Computation Cost
Perhaps the most natural overall SSC objective function is simply a linear combination of the computation cost and accuracy cost:

$$K(\pi, \phi, \rho; P) \equiv \kappa \mathscr{C}(\pi, \phi, \rho; P) + \alpha \mathscr{E}(\pi, \phi, \rho; P). \tag{10.14}$$

When all these costs are defined information-theoretically, and $\alpha = \kappa = 1$, the quantity in Eq. (10.14) has a nice interpretation in terms of communication theory, as the minimum of the expected number of bits that must be transmitted to "complete the compression-decompression communication," i.e., the average number of bits needed to map

$$x_0 \to y_0 \to y_t \to \omega_t'. \tag{10.15}$$

There are some interesting parallels between linear objective functions of the form given in Eq. (10.14) and various "complexity measures" based on Turing machines that have been proposed in the literature and that map a (bit-string representation of) infinite time-series x_t to a real number. The cost of computing π can be viewed as (analogous to) the length of a Turing machine that takes in x_0 and produces y_0. The remainder of the computation cost can be viewed as analogous to the time it takes to run a Turing machine to construct the full sequence of values y_t starting from y_0. Finally, the accuracy cost term can be viewed as analogous to the amount of extra information that must be added to the result of running that Turing machine to generate (an approximation to) the full observable time series of interest, ω'_t.

Heuristics for Minimizing the SSC Objective Function

Because truly optimizing the SSC objective function will often be quite difficult (if not formally uncomputable), there are several heuristics one might employ. Some examples of heuristic ways to approach SSC are presented in the next section on related works. Another heuristic, which was also introduced above, is to focus on the situation where ϕ is first-order Markovian (that being a "simpler" dynamics to calculate than higher-order stochastic processes, of the type that are typically used in time-series reconstruction using delay embeddings). An obvious additional heuristic – common in real-world instances of SSC, like those discussed in the next section – is to fix Y ahead of time to some space substantially smaller than X, rather than try to optimize it. (In the case of Euclidean X, Y will also be a Euclidean space of much smaller dimension; for finite X, Y is also finite, but far smaller.) Employing this heuristic would still mean searching for the optimal π, ϕ, and ρ. However, now we would be doing that for some (small) fixed choice of Y rather than searching over Y as well, a simplification that will often decrease the size of the search space dramatically.

Another heuristic that will often make sense is to restrict the set of compression maps π that are considered, for example, to some parametrized class of maps. In particular, when Y is Euclidean, we can restrict π so that it cannot encode an arbitrary dimensional space $x \in X$ in an arbitrarily small dimensional $y \in Y$ with perfect accuracy, for example, by restricting π to a class of maps that are all continuously differentiable of a certain order, or Lipschitz. (Without such restrictions, there will often be "valid" π that depend sensitively on the infinite set of all digits of x, such as the position along a space-filling curve; as a practical matter, such π are impossible to compute.) Even if we wish to allow π to be arbitrary distributions rather than restrict them to be single-valued functions, optimizing over a parametrized family of such π (e.g., a parametrized class of Gaussian processes) might prove fruitful.

Even if we were to require zero accuracy cost, so needed only to search for the maps with minimal computation cost, we would still have to use heuristics in general. Indeed, solving for the map from x_0 to y_0 to y_t having minimal expected run time would be equivalent to solving for the optimal compilation of a given computer program down to machine code. In real computers, design of optimal compilers is still a very active area of research; calculating the cost of such an optimized compilation will not be possible in general.[11] Even calculating such costs for the abstracted version of real-world computers will likely prove intractable. However, it should be possible to put bounds on such costs, and then use those bounds as heuristics to guide the search for what compression to use. Moreover, purely pragmatically, one can run a search algorithm over the space of ϕs, finding a good (if not optimal) compilation, and evaluate its associated cost.

RELATED WORK

Causal States and Computational Mechanics
While the basic idea of coarse-graining has a long history in the physical sciences, so has the recognition of the troublingly ad hoc nature of the process (Crutchfield and McNamara, 1987). A major contribution was made by Crutchfield and Young (1989), who introduced

---

[11] Indeed, even if we allowed an infinitely expandable RAM, such a cost would be uncomputable, in general.

the notion of causal states and initiated the study of computational mechanics (Crutchfield, 1994). Causal states have been the workhorse of an influential train of work over the last 30 years; for a recent review, see Crutchfield (2012).

A **causal state** at time 0 is a maximal set of "past" semi-infinite strings $s_{\leftarrow}^0 \equiv \{\ldots, s_{-1}, s_0\}$ such that the probability distribution of the "future" semi-infinite string, $s_{\rightarrow}^0 \equiv \{s_1, s_2, \ldots\}$ conditioned on any member of that set is the same. So all past semi-infinite strings that are in the same causal set result in the same distribution over possible future semi-infinite strings. A causal state is thus an optimal compression of the original time series when our goal is perfect knowledge of the future, with dynamics given by a unifilar hidden Markov model, i.e., an ϵ-machine.

The **statistical complexity** of a process is then defined as the entropy of the stationary distribution over causal states. Statistical complexity plays a role in the objective function of Shalizi and Moore (2003) that is loosely analogous to the role played by computation cost in the SSC objective function; in those cases, the authors understood the macrostates of thermodynamics as states of a Markov process, and thus the construction is equivalent to the task of computational mechanics (Shalizi et al., 2001).

Because causal states are optimal predictors of the future, the only information they discard is that which is irrelevant to future prediction. This suggests that we identify the causal states with a coarse-graining close to, but not necessarily identical with, SSC's fine-grained space. (Indeed, the dynamics neither of the variable s_t nor of the causal states is canonically identified with the dynamics of SSC's fine-grained space.)

Causal states can be coarse-grained, now at the cost of predictive power. A number of papers (Creutzig et al., 2009; Shalizi and Crutchfield, 2002; Still et al., 2010) suggested using the information bottleneck method (Tishby et al., 2000) to reduce the complexity of an ϵ-machine representation in a principled way. Still (2014) provided a thermodynamic account of this process, and Marzen and Crutchfield

(2014) the first explicit calculations for ϵ-machines; the latter found, among other things, that it was generally more accurate to derive these coarse-grained machines from more fine-grained models, rather than from the data directly.

This new work now optimizes accuracy (measured by the average KL distance between the optimal prediction of the future distribution and that made using a coarse-graining over causal states) given a constraint on coding cost (measured by the mutual information between the new, coarse-grained causal states and the future distribution). The objective function then becomes a linear combination of coding and accuracy costs (Creutzig et al., 2009; Still et al., 2010), which can be written as a minimization of

$$\mathcal{L}(\beta) = I[\overleftarrow{X}; \mathcal{R}] - \beta I[\mathcal{R}; \overrightarrow{X}], \qquad (10.16)$$

where β provides a single parameter to govern the trade-off between the two costs. Here, we wish the internal states, $\mathcal{R}$, to be predictive of the future ($I[\mathcal{R}; \overrightarrow{X}]$ large) while minimizing the coding cost.

The memory cost, loosely analogous to SSC's computation cost, is the first term of $\mathcal{L}(\beta)$, $I[\overleftarrow{X}; \mathcal{R}]$. It measures the coding costs of a soft-clustering of the original, i.e., a probabilistic map between the original causal states and the coarse-grained space $\mathcal{R}$. One can also imagine a hard clustering (i.e., a deterministic many-to-one map) from the original space; in this case, the coding cost reduces to $H(\mathcal{R})$, the statistical complexity of the new machine.

The accuracy cost is the second term. In Eq. 10.16, we maximize mutual information between the model state and the semi-infinite future; considering all future times of equal prior importance is a common choice (Bialek et al., 2001; Still, 2014; Wiesner et al., 2012). The more general approach is to consider a weight, $W(t)$, that would make, for example, the near-term future more important to predict than times that are more distant. In contrast, as described below, in SSC's use of mutual information as an accuracy cost, we iteratively evolve the coarse-grained state further and further into the future, and at each iteration evaluate the mutual information between the coarse-grained state at that moment and the fine-grained state at that moment.

State Aggregation, Decomposition, and Projection Methods

As early as 1961 (Simon and Ando, 1961), one finds discussions of the trade-off between accuracy cost, the computation cost of a compressed model, and the cost of finding a good compressed model – precisely the three costs we consider.

This led to studies of aggregating or "lumping" the states of Markov chains in order to construct compressed (in our language) Markov chains (Auger et al., 2008; Chiappori and Ekeland, 2011; Hildenbrand, 2007; Simon and Ando, 1961; White et al., 2000). Since some systems do not admit any good aggregations, the limitations of aggregation and lumpability methods have recently been fleshed out (Kotsalis and Shamma 2011, 2012).

The same three costs are considered in control theory and reduced-order modeling (Antoulas, 2005; Antoulas and Sorensen, 2001; Bai, 2002; Beck et al., 2009; Chorin and Hald, 2013; Deng, 2013; Holmes et al., 1998; Lorenz, 1956; Moore, 1981; Mori, 1965; Schilders et al., 2008; Zwanzig, 1980), again with the same caveat that computation cost of the reduced model is typically not considered as part of the objective function being optimized and is also often considered using the simple proxy of dimension of the reduced space. While in generic cases the dimension is probably not a bad proxy for computation cost, there are many cases where the computation cost could differ considerably from what one would expect based on dimension, e.g., depending on sparsity, rank, and condition number of the reduced system, let alone other properties that might affect the computation cost of simulating the compressed system both in theory and in practice. For example, one might prefer a slightly higher-dimensional compression that is significantly sparser to a lower-dimensional but dense system. As far as we are aware, these issues have not been considered.

In fact, much of the control theory literature in this area focuses more on the cost of solving for the optimal compression (in terms of accuracy cost) than it does on the cost of actually computing trajectories in the reduced space, which is often assumed to be simple linear algebra of the appropriate dimension. Rather than aggregation, reduced-order modeling has focused mostly on variations of the following theme (see the above-referenced books and surveys for details). Compute a subspace that somehow captures key features of the dynamics, using essentially the singular value decomposition (which, in this literature, often goes by the alternative names of principal component analysis, proper orthogonal decomposition, or Karhunen–Loeve expansion), then project the dynamics onto this subspace ("Galerkin projection"). The choice of subspace is essentially governed by choosing the most significant singular directions (corresponding to the largest singular values in the SVD). There is some art in choosing how many directions to project onto, but often a sharp drop-off is observed in the singular value spectrum, and it is this cutoff that is chosen. This somewhat ad hoc, though well motivated, choice of cutoff is essentially the only place that the computation cost of the reduced system is taken into account. Balanced truncation (Balakrishnan et al., 2001; Kotsalis and Rantzer, 2010; Kotsalis et al., 2008; Lall and Beck, 2003; Moore, 1981) offers a more principled way to select the cutoff, or "truncation," point, but still does not explicitly take into account the computation cost of the compressed model.

Other methods have also been proposed, under the umbrella term of Krylov iteration, that can frequently make it easier to solve for the optimal compression (again, in terms of accuracy cost), using iterative methods rather than computing the SVD directly. But the end goal, and the way in which accuracy and computation cost of the reduced model are taken into account, are essentially the same in the pure SVD-based methods.

More recently, so-called local SVD (or POD) methods have been proposed (Peherstorfer et al., 2014), in which a model is built as above, but instead of being a model of the whole space, it is designed only to model a local portion of the state space. Many local models are then "glued" together, for example, by training a classifier (using standard techniques from artificial intelligence) to decide which of the local models should be used at any given point or trajectory.

A related set of model reduction techniques developed in the last few years (Machta et al., 2013; Transtrum and Qiu, 2014; Transtrum et al., 2015) applies information geometry (Amari and Nagaoka, 2000) to reduce the dimension of a parameter vector governing a set of probability distributions. To an extent, this approach sidesteps the use of *states* – microstates and macrostates both – and focuses entirely on stochastic models of the observables and the parameters governing them. This approach uses the Fisher information metric to determine which linear combinations of components of the parameter vector are most relevant – defined as having the largest eigenvalues, as in the other methods here – and then predicts the desired observable using only those "most relevant" parameter combinations. As with the other work discussed here, although making models conceptually and *computationally* simpler is implicit in this work based on information geometry, it is not explicitly considered; only the accuracy is explicitly considered.

The Mori–Zwanzig family of methods (Mori, 1965; Zwanzig, 1980) essentially amounts to an expansion of the dynamics in terms of time and truncation of a certain number of steps in the past. The survey of Beck et al. (2009) emphasizes the similarities between Mori–Zwanzig and the methods of reduced-order modeling discussed above. Polynomial chaos, used in uncertainty quantification, also has a similar flavor: at a high level, it involves expanding in terms of certain nice polynomials and then truncating this expansion.

The "finite state projection method" (FSP) (Munsky and Khammash, 2008, 2006) is a similar method, developed to map a microstate of a chemical reaction network evolving according to a chemical master equation to a compressed version to speed up simulation of the evolution of that network. Generalizing from these networks, the FSP is applicable whenever the set of possible microstates X is countable and the stochastic process generating x_t is a Poisson process. The idea behind the FSP is to select a large subset $X' \subset X$ and group all the states in X' into one new, absorbing state. The stochastic process rate constants connecting the elements of $X \setminus X'$ and governing the probability flow from $X \setminus X'$ into X' are not changed. The goal is to choose X' to be large and at the same time to have the total probability flow rate from $X \setminus X'$ into X' be small. While somewhat ad hoc, the FSP has been found to work well in the domain for which it was constructed.

We suspect that the methods of polynomial chaos and subspace projections will prove helpful in SSC, but it is not immediately obvious how to incorporate more general (or more nuanced) notions of computation cost directly into these methods.

Other Related Work

Another thread of work has tried to define whether a map $x_t \to y_t$ is a "valid" compression, without trying to rank such maps or find an optimal one. For example, some researchers (Derisavi, 2007; Derisavi et al., 2003; Görnerup and Jacobi, 2008, 2010; Jacobi and Goernerup, 2007) start with the set of four variables x_0, y_0, x_t, and y_t, where t is some fixed value greater than 0, and y_0 is produced by applying a proposed compression map to x_0, while y_t is produced by applying that same map to x_t. They then consider y_t to be a valid compression of x_t if the associated dynamics y_t is (first-order) Markovian.

Others (Israeli and Goldenfeld, 2004, 2006; Pfante et al., 2014) are also concerned with these four variables, x_0, y_0, x_t, and y_t, but in these cases the stochastic relationship of these four variables is used to assign a real-valued quality measure to the map $x_t \to y_t$, rather than specify whether or not it is a valid compression. This quality measure is based on the amount of extra information that is needed from x_0, in addition to the value y_0, for us to accurately predict y_t. This work does not take computation or measurement cost into account. However, it is precisely those costs, and how they trade off with accuracy cost, that lie at the heart of our SSC framework.

Similarly, in Wolpert and Macready (2000), optimal compression was implicitly defined in terms of how accurately a compressed description of a system could predict the fine-grained description. Related work in Wolpert and Macready (2007) implicitly defined optimal state compression in terms of how different probability distributions were at coarse-grained and fine-grained scales. As with the works discussed above, these works do not consider computation cost, at least not directly.

This body of previous work on state compression makes compelling points, and generally accords with intuition. However, none of it considers what a state compression of a fine-grained dynamics x_t is *for*. As a result, much of this earlier work can be vulnerable to *reductio ad absurdum*. For example, if y_t is just a constant, not evolving in time, then the dynamics y_t is perfectly Markovian, of first order. So this SSC, $x \rightarrow constant$, is a "valid" compression, according to some of this earlier work.

Similarly, say that extra information from the fine-grained value x_0 cannot provide extra help in predicting the future value y_t beyond just knowing y_0. This then implies that extra information about x_t cannot provide extra help in predicting the future value y_t beyond just knowing y_0. It might seem that in this case $x \rightarrow y$ should be deemed a "good" state compression. After all, if extra future information about future fine-grained states x_t cannot help us predict future compressed states, then dynamics in the compressed space is "autonomous," essentially independent of x_t. This is the motivation for much of the analysis in Pfante et al. (2014), for example.

However, this reasoning should be used with care. In particular, say we used this reasoning to argue along with Pfante et al. (2014) that we have a good SSC $\pi : x \rightarrow y$ if the conditional mutual information $I(y_t; x_0 \mid y_0)$ is small, i.e., if knowing x_0 does not help us predict y_t any more than knowing y_0 does. With this criterion, we would say that the compression map that sends x to a uniformly noisy value of y, which is statistically independent of x, is a "good SSC'; it results in $I(y_t; x_0 \mid y_0) = 0$. From the SSC perspective, we would argue that such a "compression" is of little practical interest, and therefore should be considered a poor compression.

There are also important features of our focus on the full compression/decompression loop that are absent from this earlier work. For instance, the earlier work considers only the compression π, with no associated "decompression" map ρ that maps Y to an observable of interest. In contrast, we consider the case where one is interested in "decompressing" future values of y, to make predictions of observable functions of x_t. In addition, this earlier work assumes that future values of y_t are obtained by iteratively applying π to x_t. Instead, we allow dynamics in y_t to evolve according to an arbitrary map ϕ from an initial value $y_0 = \pi(x_0)$. This means that rather than just assign value to a compression map π, we assign value to a triple (π, ϕ, ρ).[12]

CONCLUSIONS

This preliminary report has presented a new framework for understanding how we construct higher-level descriptions of complex systems. We have introduced the problem through a series of illustrations and defined the key quantities and their relationships. Having built an intuition for the method, we then compared this framework with a number of influential suggestions in the literature.

Our framework makes explicit two core problems for both the scientist and engineer: how accurate a theory is, and how difficult it is to work with. We have presented new theoretical results for how to quantify the answers to these questions, and how to combine them into a single objective function.

By formalizing the goals of a scientist engaged in providing a coarse-grained description of a system, SSC allows us to compare and contrast a wide variety of systems. It provides novel ways to address long-standing problems that arise both within fields and between

---

[12] A related point is that in this earlier work the Bayes net relating the four variables is $P(y_t \mid x_t)P(y_0 \mid x_0)P(x_t \mid x_0)$. In contrast, the Bayes net relating those four variables that we analyze below is $P(y_t \mid y_0)P(y_0 \mid x_0)P(x_t \mid x_0)$. This difference reflects the fact that this earlier work is ultimately concerned with issues different from those we consider.

disciplines, where the question of "how much to ignore" becomes critical.

ACKNOWLEDGMENTS

We thank Daniel Polani, Robert Axtell, Eckehard Olbrich, Nils Bertschinger, Nihat Ay, Cris Moore, and James O'Dwyer for helpful discussion. We also thank the Santa Fe Institute for support. In addition, this chapter was made possible through the support of Grant No. TWCF0079/AB47 from the Templeton World Charity Foundation. The opinions expressed in this chapter are those of the author(s) and do not necessarily reflect the view of Templeton World Charity Foundation. S.D. thanks the City University of New York's Initiative for the Theoretical Sciences for their hospitality during the course of this work. S.D. was supported in part by National Science Foundation Grant EF-1137929. J.A.G. and E.L. acknowledge the support of Santa Fe Institute Omidyar Fellowships.

REFERENCES

Amari, S., and Nagaoka, H. 2000. Methods of information geometry, volume 191 of Translations of Mathematical Monographs. *American Mathematical Society*, 13.

Antoulas, A. C. 2005. *Approximation of large-scale dynamical systems*. Vol. 6. Siam.

Antoulas, A. C., and Sorensen, D. C. 2001. Approximation of large-scale dynamical systems: an overview. *Applied Mathematics and Computer Science*, 11(5), 1093–1122.

Attanasi, A., Cavagna, A., Del Castello, L., Giardina, I., Grigera, T. S., Jelić, A., Melillo, S., Parisi, L., Pohl, O., Shen, E., et al. 2014. Information transfer and behavioural inertia in starling flocks. *Nature Physics*, 10(9), 691–696.

Auger, P., de La Parra, R. B., Poggiale, J. C., Sánchez, E., and Sanz, L. 2008. Aggregation methods in dynamical systems and applications in population and community dynamics. *Physics of Life Reviews*, 5(2), 79–105.

Ay, N., and Polani, D. 2008. Information flows in causal networks. *Advances in Complex Systems*, 11(1), 17–41.

Bai, Z. 2002. Krylov subspace techniques for reduced-order modeling of large-scale dynamical systems. *Applied Numerical Mathematics*, 43(1), 9–44.

Balakrishnan, V., Su, Q., and Koh, C. K. 2001. Efficient balance-and-truncate model reduction for large scale systems. Pages 4746–4751 of *American Control Conference, 2001 Proceedings*, vol. 6. IEEE.

Barron, A., Rissanen, J., and Yu, B. 1998. The minimum description length principle in coding and modeling. *IEEE Transactions on Information Theory*, **44**(6), 2743–2760.

Beck, C. L., Lall, S., Liang, T., and West, M. 2009. Model reduction, optimal prediction, and the Mori–Zwanzig representation of Markov chains. Pages 3282–3287 of: *Proceedings of the 48th IEEE Conference on Decision and Control, 2009 held jointly with the 2009 28th Chinese Control Conference. CDC/CCC 2009*. IEEE.

Bennett, C. H. 1988. Logical depth and physical complexity. In Rolf Herken (ed.), *The universal Turing machine: a half-century survey*, 207–235. Oxford University Press.

Bennett, C. H. 2003. Notes on Landauer's principle, reversible computation, and Maxwell's demon. *Studies in history and philosophy of science part B: studies in history and philosophy of modern physics*, **34**(3), 501–510.

Bialek, W., Nemenman, I., and Tishby, N. 2001. Predictability, complexity, and learning. *Neural Computation*, **13**(11), 2409–2463.

Bialek, W., Cavagna, A., Giardina, I., Mora, T., Silvestri, E., Viale, M., and Walczak, A. M. 2012. Statistical mechanics for natural flocks of birds. *Proceedings of the National Academy of Sciences*, **109**(13), 4786–4791.

Blagus, N., Šubelj, L., and Bajec, M. 2012. Self-similar scaling of density in complex real-world networks. *Physica A: Statistical Mechanics and Its Applications*, **391**(8), 2794–2802.

Chaitin, G. J. 2004. *Algorithmic information theory*. Vol. 1. Cambridge University Press.

Chiappori, P. A., and Ekeland, I. 2011. New developments in aggregation economics. *Annu. Rev. Econ.*, **3**(1), 631–668.

Chorin, A. J., and Hald, O. H. 2013. *Stochastic tools in mathematics and science*. Texts in Applied Mathematics. Springer.

Clark, A. 2013. Whatever next? Predictive brains, situated agents, and the future of cognitive science. *Behavioral and Brain Sciences*, **36**(3), 181–204.

Couzin, I. A. 2009. Collective cognition in animal groups. *Trends in Cognitive Sciences*, **13**(1), 36–43.

Cover, T. M., and Thomas, J. A. 2012. *Elements of information theory*. John Wiley & Sons.

Creutzig, F., Globerson, A., and Tishby, N. 2009. Past–future information bottleneck in dynamical systems. *Physical Review E*, **79**(4), 041925.

Crutchfield, J. P. 1994. The calculi of emergence: computation, dynamics and induction. *Physica D: Nonlinear Phenomena*, **75**(1), 11–54.

Crutchfield, J. P. 2012. Between order and chaos. *Nature Physics*, **8**(1), 17–24.

Crutchfield, J. P., and McNamara, B. S. 1987. Equations of motion from a data series. *Complex Systems*, **1**(417-452), 121.

Crutchfield, J. P., and Young, K. 1989. Inferring statistical complexity. *Physical Review Letters*, **63**(2), 105.

Davies, P. C. W. 2012. The epigenome and top-down causation. *Interface Focus*, **2**(1), 42–48.

DeDeo, S. 2011. Effective theories for circuits and automata. *Chaos: An Interdisciplinary Journal of Nonlinear Science*, **21**(3), 037106.

DeDeo, S. 2014. Group minds and the case of Wikipedia. *arXiv preprint arXiv:1407.2210*.

Deng, K. 2013. *Model reduction of Markov chains with applications to building systems*. Ph.D. thesis, University of Illinois at Urbana-Champaign.

Derisavi, S. 2007. A symbolic algorithm for optimal Markov chain lumping. Pages 139–154 of *Tools and algorithms for the construction and analysis of systems*. Springer.

Derisavi, S., Hermanns, H., and Sanders, W. H. 2003. Optimal state-space lumping in Markov chains. *Information Processing Letters*, **87**(6), 309–315.

Görnerup, O., and Jacobi, M. N. 2008. A method for inferring hierarchical dynamics in stochastic processes. *Advances in Complex Systems*, **11**(1), 1–16.

Görnerup, O., and Jacobi, M. N. 2010. A method for finding aggregated representations of linear dynamical systems. *Advances in Complex Systems*, **13**(2), 199–215.

Hildenbrand, W. 2007. Aggregation theory. In *The New Palgrave Dictionary of Economics*, **2**.

Holmes, P., Lumley, J. L., and Berkooz, G. 1998. *Turbulence, coherent structures, dynamical systems and symmetry*. Cambridge University Press.

Hopcroft, J. E., and Motwani, R. 2000. *Introduction to automata theory, languages and computability*. Addison-Wesley Longman.

Israeli, N., and Goldenfeld, N. 2004. Computational irreducibility and the predictability of complex physical systems. *Physical Review Letters*, **92**(7), 074105.

Israeli, N., and Goldenfeld, N. 2006. Coarse-graining of cellular automata, emergence, and the predictability of complex systems. *Physical Review E*, **73**(2), 026203.

Itzkovitz, S., Levitt, R., Kashtan, N., Milo, R., Itzkovitz, M., and Alon, U. 2005. Coarse-graining and self-dissimilarity of complex networks. *Physical Review E*, **71**(1), 016127.

Jacobi, M. N., and Goernerup, O. 2007. A dual eigenvector condition for strong lumpability of Markov chains. *arXiv preprint arXiv:0710.1986*.

Jiménez, D. A. 2005. Improved latency and accuracy for neural branch prediction. *ACM Transactions on Computer Systems (TOCS)*, **23**(2), 197–218.

Kamiński, M., Ding, M., Truccolo, W. A., and Bressler, S. L. 2001. Evaluating causal relations in neural systems: Granger causality, directed transfer function and statistical assessment of significance. *Biological Cybernetics*, **85**(2), 145–157.

Kotsalis, G., and Rantzer, A. 2010. Balanced truncation for discrete time Markov jump linear systems. *IEEE Transactions on Automatic Control*, **55**(11), 2606–2611.

Kotsalis, G., and Shamma, J. S. 2011. A counterexample to aggregation based model reduction of hidden Markov models. Pages 6558–6563 of *IEEE Conference on Decision and Control and European Control Conference*. IEEE.

Kotsalis, G., and Shamma, J. S. 2012. *A fundamental limitation to the reduction of Markov chains via aggregation*. Citeseer.

Kotsalis, G., Megretski, A., and Dahleh, M. A. 2008. Balanced truncation for a class of stochastic jump linear systems and model reduction for hidden Markov models. *IEEE Transactions on Automatic Control*, **53**(11), 2543–2557.

Krakauer, D., Bertschinger, N., Olbrich, E., Ay, N., and Flack, J. C. 2014. The information theory of individuality. *arXiv preprint arXiv:1412.2447*.

Krakauer, D. C. 2011. Darwinian demons, evolutionary complexity, and information maximization. *Chaos: An Interdisciplinary Journal of Nonlinear Science*, **21**(3), 037110.

Krakauer, D. C., Flack, J. C., DeDeo, S., Farmer, D., and Rockmore, D. 2010. Intelligent data analysis of intelligent systems. Pages 8–17 of: *Advances in intelligent data analysis IX*. Springer.

Ladyman, J., Lambert, J., and Wiesner, K. 2013. What is a complex system? *European Journal for Philosophy of Science*, **3**(1), 33–67.

Lall, S., and Beck, C. 2003. Error-bounds for balanced model-reduction of linear time-varying systems. *IEEE Transactions on Automatic Control*, **48**(6), 946–956.

Landauer, R. 1961. Irreversibility and heat generation in the computing process. *IBM Journal of Research and Development*, **5**(3), 183–191.

Lorenz, E. N. 1956. *Empirical orthogonal functions and statistical weather prediction*. Massachusetts Institute of Technology, Department of Meteorology.

Machta, B. B., Chachra, R., Transtrum, M. K., and Sethna, J. P. 2013. Parameter space compression underlies emergent theories and predictive models. *Science*, **342**(6158), 604–607.

MacKay, D. J. C. 2003. *Information theory, inference and learning algorithms*. Cambridge University Press.

Marzen, S., and Crutchfield, J. P. 2014. Circumventing the curse of dimensionality in prediction: causal rate-distortion for infinite-order Markov processes. *arXiv preprint arXiv:1412.2859*.

Maynard Smith, J. 1969. Time in the evolutionary process. *Studium generale; Zeitschrift für die Einheit der Wissenschaften im Zusammenhang ihrer Begriffsbildungen und Forschungsmethoden*, **23**(3), 266–272.

Maynard Smith, J. 2000. The concept of information in biology. *Philosophy of Science*, **67**(2), 177–194.

Moore, B. C. 1981. Principal component analysis in linear systems: controllability, observability, and model reduction. *IEEE Transactions on Automatic Control*, **26**(1), 17–32.

Moore, C., and Mertens, S. 2011. *The nature of computation*. Oxford University Press.

Mori, H. 1965. Transport, collective motion, and Brownian motion. *Progress of Theoretical Physics*, **33**(3), 423–455.

Munsky, B., and Khammash, M. 2006. The finite state projection algorithm for the solution of the chemical master equation. *Journal of Chemical Physics*, **124**(4), 044104.

Munsky, B., and Khammash, M. 2008. The finite state projection approach for the analysis of stochastic noise in gene networks. *IEEE Transactions on Automatic Control*, **53**(Special Issue), 201–214.

Peherstorfer, B., Butnaru, D., Willcox, K., and Bungartz, H. J. 2014. Localized discrete empirical interpolation method. *SIAM Journal on Scientific Computing*, **36**(1), A168–A192.

Pfante, O., Bertschinger, N., Olbrich, E., Ay, N., and Jost, J. 2014. Comparison between different methods of level identification. *Advances in Complex Systems*, **17**(2), 1450007.

Prokopenko, M., and Lizier, J. T. 2014. Transfer entropy and transient limits of computation. *Scientific Reports*, **4**, 5394.

Prokopenko, M., Lizier, J. T., and Price, D. C. 2013. On thermodynamic interpretation of transfer entropy. *Entropy*, **15**(2), 524–543.

Rissanen, J. 1983. A universal prior for integers and estimation by minimum description length. *Annals of Statistics*, 416–431.

Saunders, M. G., and Voth, G. A. 2013. Coarse-graining methods for computational biology. *Annual Review of Biophysics*, **42**, 73–93.

Schilders, W. H. A., Van der Vorst, H. A., and Rommes, J. 2008. *Model order reduction: theory, research aspects and applications*. Vol. 13. Springer.

Schreiber, T. 2000. Measuring information transfer. *Physical Review Letters*, **85**(2), 461.

Shadmehr, R., Smith, M. A., and Krakauer, J. W. 2010. Error correction, sensory prediction, and adaptation in motor control. *Annual Review of Neuroscience*, **33**, 89–108.

Shalizi, C. 2003. Complexity, entropy and the physics of gzip. Blog post. http://bactra.org/notebooks/cep-gzip.html

Shalizi, C. R., and Crutchfield, J. P. 2002. Information bottlenecks, causal states, and statistical relevance bases: how to represent relevant information in memoryless transduction. *Advances in Complex Systems*, **5**(1), 91–95.

Shalizi, C. R., and Moore, C. 2003. What is a macrostate? Subjective observations and objective dynamics. *arXiv preprint cond-mat/0303625*.

Shalizi, C. R., et al. 2001. *Causal architecture, complexity and self-organization in the time series and cellular automata*. Ph.D. thesis, University of Wisconsin–Madison.

Simon, H. A., and Ando, A. 1961. Aggregation of variables in dynamic systems. *Econometrica: Journal of the Econometric Society*, 111–138.

Still, S. 2014. Information bottleneck approach to predictive inference. *Entropy*, **16**(2), 968–989.

Still, S., Crutchfield, J. P., and Ellison, C. J. 2010. Optimal causal inference: estimating stored information and approximating causal architecture. *Chaos: An Interdisciplinary Journal of Nonlinear Science*, **20**(3), 037111.

Sutton, R. S., and Barto, A. G. 1998. *Reinforcement learning: an introduction*. MIT Press.

Szilard, L. 1964. On the decrease of entropy in a thermodynamic system by the intervention of intelligent beings. *Behavioral Science*, **9**(4), 301–310.

Tishby, N., Pereira, F. C., and Bialek, W. 2000. The information bottleneck method. *arXiv preprint physics/0004057*.

Transtrum, M. K., and Qiu, P. 2014. Model reduction by manifold boundaries. *Physical Review Letters*, **113**(9), 098701.

Transtrum, M. K., Machta, B. B., Brown, K. S., Daniels, B. C., Myers, C. R., and Sethna, J. P. 2015. Perspective: sloppiness and emergent theories in physics, biology, and beyond. *Journal of Chemical Physics*, **143**(1), 010901.

Walker, S. I., Cisneros, L., and Davies, P. C. W. 2012. Evolutionary transitions and top-down causation. *arXiv preprint arXiv:1207.4808*.

White, L. B., Mahony, R., and Brushe, G. D. 2000. Lumpable hidden Markov models-model reduction and reduced complexity filtering. *IEEE Transactions on Automatic Control*, **45**(12), 2297–2306.

Wiesner, K., Gu, M., Rieper, E., and Vedral, V. 2012. Information-theoretic lower bound on energy cost of stochastic computation. Pages 4058–4066 of *Proc. R. Soc. A*, vol. 468. The Royal Society.

Wolpert, D. H. 2015. Minimal work required for arbitrary computation. *arXiv preprint arXiv:1508.05319*

Wolpert, D. H., and Macready, W. 2000. Self-dissimilarity: an empirically observable complexity measure. In Y. Bar-Yam (ed.), *Unifying themes in complex systems*, 626–643. Perseus Books.

Wolpert, D. H., and Macready, W. 2007. Using self-dissimilarity to quantify complexity. *Complexity*, **12**(3), 77–85.

Zwanzig, R. 1980. Problems in nonlinear transport theory. Pages 198–225 of *Systems far from equilibrium*. Springer.

11 Causality, Information, and Biological Computation

An Algorithmic Software Approach to Life, Disease, and the Immune System

Hector Zenil, Angelika Schmidt, and Jesper Tegnér

Information and computation have transformed the way we look at the world beyond statistical correlations, the way we can perform experiments through simulations, and the way we can test these hypotheses. In Turing's seminal paper on the question of machine intelligence (Turing, 1950), his approach consisted in taking a computer to be a black box and evaluating it by way of what one could say about its apparent behaviour. The approach can be seen as a digital version of a *cogito ergo sum* dictum, acknowledging that one can be certain only of one's own intelligence but not of the intelligence of anyone or anything else, including machines. This may indeed amount to a kind of solipsism, but in practice it suggests tools that can be generalised. This is how in reality we approach most areas of science and technology, not just out of pragmatic considerations but for a fundamental reason. Turing, like Gödel before him (Gödel, 1958), showed that systems such as arithmetic that have a certain minimal mathematical power to express something can produce outputs of an infinitely complicated nature (Turing, 1936). Indeed, this means that while one can design software, only trivial programs can be fully understood analytically to the point where one is able to make certain predictions about them. In other words, only by running software can one verify its computational properties, such as whether or not it will crash. These fundamental results deriving from both Gödel's incompleteness theorem and Turing's halting problem point in the direction that Turing himself took in his 'imitation game', which we now identify as the *Turing test*.

Turing's halting problem implies that one cannot in general prove that a machine will ever halt or that a certain configuration will be reached on halting; therefore, one has to proceed by testing and only by testing. These fundamental theorems and results imply that testing is unavoidable; even under optimal conditions there are fundamental limits to the knowledge one can obtain simply by looking at a system, or even by examining its source code, that is, its ultimate causes. Even with a knowledge of all the causes driving a system's evolution, there are strong limits to the kind of understanding of it that we can attain.

In light of these realities, entire (relatively) new fields designated *model checking*, *systems testing*, and *software verification* seek to produce and test reliable software based on simple mathematical models. Today's airplane construction companies and other manufacturers of critical systems such as electronic voting systems use these tools all the time; they are a minimal operational requirement in these industries. Biological phenomena are likely to be at least as complicated, and our understanding of them subject to the same knowledge limitations. That is, the proof that nature and most of our models of nature can contain systems such as arithmetic and digital computers constitutes a limit to our knowledge. One must then acknowledge and deal with the fact that there are true and unavoidable limits to knowledge extraction from artificial but also from natural systems, even under ideal circumstances.

As shown in Figures 11.1 and 11.2, for an extremely simple computer program represented by an elementary cellular automaton (ECA), one needs to perform, in the best-case scenario, at least eight very precise observations at two consecutive times (hence 16) with perfect accuracy in order to hack this computer program, only to unveil its ridiculously small source code (which can be written in a few bits) and determine that the producing rule is ECA rule 30. Missing just a single observation can lead one to a rule with completely

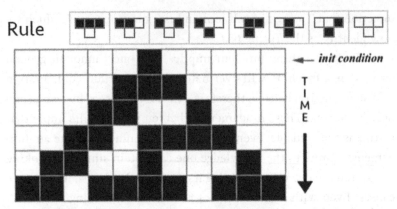

FIG. 11.1 There is a common misconception that complexity is about size, number of interacting parts or agents, and number of interactions. In this example, a minimalistic computer program starts from the simplest possible situation, yet produces random-looking (statistical) behaviour that is for all practical purposes unpredictable. This computer program, called a *cellular automaton* (CA), has a rule with code 30 (the rule icon represented in binary) according to Wolfram's enumeration (Wolfram, 2002). The first row represents the input of the program. Every black or white cell is updated according to the rule icon on top, which, according to the two nearest neighbours of the cell, proceeds to change or retain the colour of the central cell in the next row.

different behaviour, as they are highly sensitive, and by making more observations one immediately starts overfitting, leading one to a more complicated rule that may produce the same behaviour but that is ultimately different.

Figure 11.1 shows such a minimalistic example that it suggests that the same phenomenon is pervasive in physics and biology, where even the simplest conditions can generate a cascade of apparent randomness. Notice this is of even more basic nature than the phenomenon of *chaos* where the argument is that arbitrary close initial conditions diverge in behaviour over time, an additional complexity in, for example, experimental reproducibility.

Multiple sclerosis, for example, is a complex disease in which the insulating shield (the myelin sheath) of nerve cells is damaged and for which there is currently no curative treatment, yet the disease sometimes shows apparent periodic behaviour in the form of relapses.

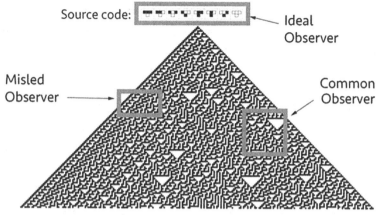

FIG. 11.2 A simple computer program such as elementary CA rule 30, with the simplest possible initial condition, can give us a sense of how complex we may expect natural phenomena to be, and it helps illuminate scientific practice and its limitations. Despite deriving from a very simple program, without knowing the *source code* (top) of the program, a partial view and a limited number of observations can be misleading or uninformative. For example, the misguided observer may think that the system is very simple and that the source code can only generate regular behaviour, but would in fact be missing most of the picture. This example shows how difficult it can be to reverse-engineer a system from its evolution in order to discover its generating code when one is privy only to partial and limited sections of the system's evolution.

Relapsing eventually becomes progressive, meaning that there is no further apparent partial recovery of the patient over time. In some cases, this might be a conflation of a misled observation coming from a compensation mechanism of the brain to take control over damaged neuromotor function rather than a strengthening of the immune system or a partial restoration of the neuron's myelin. The observation may be leading to the wrong conclusion that there are two types of multiple sclerosis when there may be, perhaps, only one behaving in different ways in different times.

An ideal observer would be able to see the source code directly, but this is virtually impossible in practice, for a number of reasons: first, because it is difficult to separate phenomena from other phenomena in nature; and second, because we never have access to first causes or, for that matter, to first causes in complete isolation (except perhaps in very artificial and controlled cases, such as are almost never encountered in nature). Moreover, true or apparent noise (e.g., measurement limitations) can have dramatic effects (see Figure 11.3B). The scientific method deals with these fundamental problems, but it does so for even more fundamental reasons than is usually believed, as these problems actually characterise even the most minimalistic and fully deterministic systems (such as these computer programs used for purposes of illustration). In other words, it is practically impossible to recover source codes in nature under noisy, imperfect, and limited conditions. And even if we manage to do so, that does not mean we can fully understand the behaviour of a system from its source code, just as nobody could imagine the complexity that rule 30 would be able to produce and nobody has been able to understand its evolution and find significant computational shortcuts.

This phenomenon is not exclusive to rule 30 but is pervasive even in the most deterministic algorithmic sciences such as mathematics and computer science, in objects such as the decimal expansion of numbers like π or the square root of 2, or in the logistic map that leads to chaos behaviour. But notice that these chaotic systems are of a slightly different nature. They are complex only when taken together with the description of their initial conditions. Here, rule 30 or the digits of π look statistically random for all intents and purposes, even though these are objects for which there is no complex initial condition or else they start from the simplest initial condition, unlike initial conditions encoded in real numbers and in descriptions of 'close initial conditions'. In nature there are no *ideal observers* like the possibly lucky one circled in Figure 11.2.

In other words, as Figure 11.2 illustrates, reverse-engineering (finding ultimate rules) is extremely difficult, and in biology there is an entire relatively new field devoted to reverse-engineering biological networks, also called network reconstruction, where rules involve genes, proteins, and metabolites, to name a few instances.

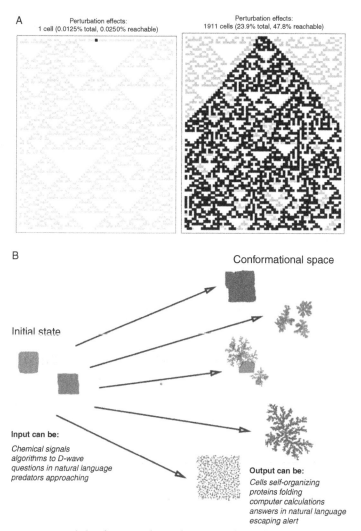

FIG. 11.3 (A) Behaviour shown by ECA rule 22 for two different small perturbations. Different perturbations have different effects on the same computer program. They either can have no impact or can propagate at maximum speed, making the system appear highly sensitive and chaotic. (B) Dealing with a black box. In this example of what can happen in and with an artificial or natural system in respect of which it is practically and fundamentally impossible to analytically determine all possible outputs for all possible inputs, we used (Terrazas et al., 2013) a Wang tile–based computing model to map a simplified parameter space of emulated physical properties determining the binding capabilities of two (assumed) types of biological molecules simulating the behaviour of porphyrin molecules to the conformation space generated when the molecules interact.

If a system such as rule 30, despite its incredible simplicity, can be relatively difficult to reverse-engineer, one can only speculate as to the complication of reverse-engineering real-world systems such as (indirect) gene–gene interactions (through gene products) from gene expression maps, that is, ascertaining which genes are (inter-) connected/coregulated with which others, down- or up-regulating their expressions and determining which proteins to produce for different cellular functions.

NATURAL COMPUTATION AND PROGRAMMABILITY

Sensitivity analysis can be useful for testing the robustness and variability of a system, and it sheds light on the relationship between input and output in a system. This is not hard to identify from a computing perspective, where inputs to computer programs produce an output, and computer programs can behave in different ways.

For example, sensitivity measures aim to quantify this uncertainty and its propagation through a system. Among common ways to quantify and study this phenomenon is, for example, the so-called Lyapunov exponent approach (Cencini et al., 2010). This approach consists of looking at the differences that arbitrarily close initial conditions produce in the output of a system, and hence is similar to the generalisation we introduce. Traditionally, if the exponent is large, the sensitivity is nonlinear, and divergence increases over time. If constant, however, the system is simple under this view. Programming systems requires them to have a nonzero Lyapunov exponent value, but in general this measure is not quite suitable, as it quantifies the separation of infinitesimally close trajectories that may be ill-defined in discrete systems, and traditionally, inputs for a program are not intended to be *infinitesimally close*, as a distance measure would then be required (even though in Zenil, 2010, one has been proposed). Furthermore, measures such as the Lyapunov exponent are meant to detect qualitative changes such as chaotic behaviour, which is not desirable for a programmable system, as is suggested by Figure 11.4. Further research on the

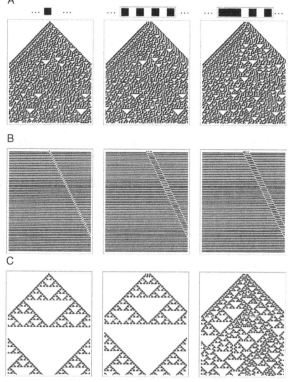

FIG. 11.4 Rule 30 (A), rule 3 (B), and rule 22 (C) ECAs showing that with slightly different initial conditions two of them lead to the same qualitative behaviour and therefore have low *programmability*, whereas rule 22 shows greater variability and sensitivity to small perturbations and can therefore be better controlled to perform tasks than can rules 3 and 30. This (in)sensitivity can be explored with an online program available at Zenil and Villarreal-Zapata (2013).

connections between programmability and these other dynamical system sensitivity measures needs to be undertaken in the future. But they only seem adequate for quantifying far greater quantitative changes than are needed for a system to be programmable while at the same time not being too sensitive to small qualitative changes (e.g., statistical, such as systems that look random even if the rate of cell changes is maximal, as in the example shown in Figures 11.1 and 11.2, top). Advances in natural and unconventional computation have made it difficult to measure how a computer or system behaves, and it is necessary to develop new tools for the purpose. In recent years there has been a heated debate concerning a company called D-wave, which claims to have built a quantum computer that it is selling as such, based on a technique called quantum annealing. But as we have seen, evaluating whether something actually is what it was designed to be is often difficult, and, interestingly, this is another case in which the said quantum computer has been evaluated as a black box (Shin et al., 2014).

In our approach to behaviourally evaluating systems for their programmability, a compression algorithm can be used as an interrogator device, where for questions one uses initial conditions of the system, while the answers are the lengths of the compressed outputs. In general, for a system to be reprogrammable, inputs with different information content should lead to outputs with different information content. If the system reacts to these stimuli in a nontrivial fashion and passes the compression/complexity test, then one should be able to declare it capable of computing and amenable to being programmed; otherwise, it is merely a system that produces similar answers (outputs) irrespective of the questions (inputs). To test this variability, one compares the information content of the answers. This means the answers can look very different but if they contain the same information they will fail the test. A system that does not react to the stimuli and produces trivial output will also clearly fail the test, and a system that produces only random output will fail too, because the information content of the answers, given their differences, will cancel each other out. This is then related to another concept in formal software engineering, the concept of *code coverage*, a measure that describes the degree to which the source code of a program is tested by a particular test suite. A program with high code coverage has been more thoroughly tested and has a lower chance of containing software bugs than does a program with low code coverage. However, code coverage, and all other software measures, traditionally requires us to know or have access to the original code. We are setting forth ideas for approaching situations when this code is not only difficult but in principle impossible to know, because there is no way to rule out alternative codes that may be concurrently performing other computations for which no tests have been designed.

These testing ideas are based on whether a system whose source code may never be known is capable of reacting to the environment – the input – as is the case in more formal implementations of this measure of *programmability*. Such a measure (Zenil, 2014, 2015) would quantify the sensitivity of a system to external stimuli and could be used to define the amenability of a system to being (efficiently) programmed. The basic idea is to replace the observer in Turing's imitation game with a lossless compression algorithm, which duplicates the relevant subjective qualities of a regular observer, in that it can only partially 'detect' regularities in the data that the algorithm determines, there being no effective (programmable in practice) universal compression algorithm, as proven by the uncomputability of Kolmogorov complexity (Chaitin, 1966; Kolmogorov, 1968). The compression algorithm looks at the evolution of a system and determines, by means of feeding the system with different initial conditions (which is analogous to questioning it), whether it reacts to external stimuli. The Kolmogorov complexity of an object is the length of the shortest program that outputs the object running on a universal computer program (i.e., a computer program that can run any other computer program, the existence of such programs having been proved by Turing (1936) himself). Then, if the evolution of a program (say, a natural phenomenon) is complex and does not react to external stimuli, all compression algorithms will fail at compressing its evolution, and no difference between different evolutions for different perturbations will be detected (e.g., by taking the differences between the compressed lengths of the objects for the two perturbations; see Figure 11.4).

For example, as is shown in Zenil (2010), certain ECA rules that are highly sensitive to initial conditions, and present phase transitions that dramatically change their qualitative behaviour when starting from different initial configurations, can be characterised by these qualitative properties. A further investigation of the relation between this transition coefficient and the computational capabilities of certain known (Turing) universal machines has been undertaken in Zenil (2010). Other calculations have been advanced in Zenil (2014, 2015).

The behavioural approach in fact generates a natural classification of objects in terms of their programmability, as sketched in Figure 11.5B. For example, while weather phenomena and Brownian motion have great variability, they are hardly controllable. On the other hand, rocks have a very low variability and hence are trivially controllable but are not therefore on the programmability diagonal and cannot count as computers. Everything on the diagonal, however, including living organisms, is programmable to some extent.

INFORMATION BIOLOGY

One of the aims of our research is to exploit these ideas in order to try to reprogram living systems so as to make them do things we would like them to do, for in the end this is the whole idea behind programming something. Figure 11.3B is an illustration of an investigation (Terrazas et al., 2013) into the possibility of mapping the conformational space of a simulation of porphyrin molecules with a view to making them self-organise in different ways. That is, it is an investigation into what it means to program a nature-like simulated system, to find the inputs for the desired output (that matched the actual behaviour of the molecule). The challenge is about mapping the parameter space to the output space, in this case the conformational space of porphyrin molecules, i.e., the space of its possible shapes under certain conditions.

In biology, the greatest challenge is the prediction of behaviour and shape ('shape' determines function in biology). Examples are protein folding or predicting whether immune cells will differentiate in one direction rather than another. In Figure 11.3B, we investigate how we could arrive at certain specific conformational configurations from an initial state by changing environmental variables, such

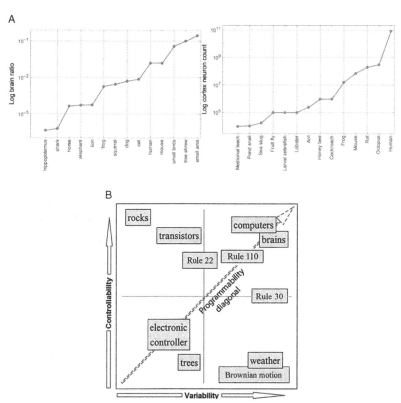

FIG. 11.5 (A) Information is key to living organisms, and therefore information theory is essential to understanding them. For example, while neither cranial capacity nor brain-to-body ratio (upper left) correspond to any biological feature, intelligence is clearly seen to be a function of information processing, when sorting species by cerebral cortex neuron count (upper right). Compiled data sources: (Wikipedia, 'Brain-to-body mass ratio' and 'List of animals by number of neurons'). (B) Programmability is a combination of variability and controllability (Zenil et al., 2013b). Behavioural measures suggest a natural classification of the programmability of natural and artificial systems, just as they suggest a classification of tumours (see Figure 11.6) by degree of programmability based on their sensitivity to their environment and external stimuli and their variability. This reclassification deletes some artificial boundaries between living and nonliving systems when it comes to behavioural and computing capacities, and between tumour types that are classified by tissue of origin rather than in terms of the error in the cell-replication process that produces them.

as temperature and other factors influencing binding properties. The proofs that certain abstract systems implemented in software can reach Turing universality constitute one example of how hardware may be seen as software. Indeed, by taking simple four-coloured tiles (called Wang tiles, after Hao Wang) and placing them on a board according to the rule that whenever two tiles touch, the sides that touch must be of the same colour, one can build an abstract machine that can simulate any other computer program. And this is a very powerful approach because one can program molecules or robots to do things like carry a payload to be released when certain specific conditions are met, release a chemical that may be used as a biological marker, fight disease, or deal with nuclear waste. While we cannot try every possible perturbation, either on computer programs or on natural systems, and therefore require some fundamental analytical approach, collections of molecular profiles from thousands of human samples, as made available by large consortia such as TCGA, TARGET, and ICG, among others, is making this process possible, because perturbations are a common mechanism of nature and natural selection.

This means that with enough data one does not need to perform perturbations but rather to find them as they occur in nature. However, while in recent decades we have come to understand that complex systems such as the immune system cannot be thoroughly analysed and fully understood as regards its components and function by using 'reductionist' approaches, and we have moved to the science of systems, such as systems and computational biology, we have nevertheless failed to make as radical a move as we ought to and apply the tools of complexity science in a holistic way. And reprogramming cells to fight complex diseases (see Table 11.1) is one of our ultimate goals.

Complex diseases (see Table 11.1) require a complex debugging system, and the immune system can be viewed to play just this role. The immune system is the highly complex and dynamic counterpart of many complex (and simple) diseases, and the most common approach to understanding it has been via evaluating its

TABLE 11.1 Properties of simple *versus* complex diseases[a]

Simple diseases	Complex diseases
• Single and isolated causes • Simple gene expression patterns • Focalised effects • High predictability	• Multiple disjoint and joint causes • Many interacting elements • External factors • Low predictability
Examples Monogenic (single gene) diseases (e.g., cystic fibrosis), chromosomal abnormalities, thalassaemia, haemophilia, Huntington disease	**Examples** Multiple sclerosis, Alzheimer disease, Parkinson disease, most cancers

[a] A good illustration of these differences can be found in Peltonen and McKusick (2001). They cannot be considered in isolation. For example, the presence of many interacting elements does not automatically imply a complex disease (system); it is the way they interact and the multiple factors involved that make for a complex and therefore unpredictable system, even if it is fully deterministic (which may or may not be the case for complex diseases).

individual components. However, as with computer programs, this kind of approach cannot always expose the way in which an immune system works under a range of possible circumstances, particularly those leading to complex diseases such as autoimmune conditions and cancer. One instance where our knowledge will come up against limitations imposed by the systems approach itself is the process by which the immune system responds to infectious agents by activating innate inflammatory reactions and dictating adaptive immune responses (the immune system is split into two branches: innate and adaptive). A similar example is how cancer systems biology has emerged to address the increasing challenge of cancer as a complex, multifactorial disease but has failed to move away from the archaic classification of cancer by tumour tissue of origin.

Interestingly, these two highly complex systems are interconnected: the immune system plays a key role in tumour development. The immune status ('immunoscore') was found to be the highly predictive for cancer patient survival (Galon et al., 2012), and evading immune destruction was added to the 'hallmarks of cancer' recently (Hanahan and Weinberg, 2011). According to the immune surveillance theory, the immune system does not only recognise and combat invading pathogens but also detects host cells that become cancerous, thus eliminating tumour cells as soon as they arise. This requires the immune system to detect 'abnormal' structures on a tumour cell that is otherwise 'self' – despite self-tolerance as an important hallmark of the immune system. In fact, the immune system can be viewed as another example of the black-box behavioural approach to systems. Immune cells 'expect' certain responses from the cells they encounter in the body, using these to identify their nature. When they do not recognise these signals as harmless, they attack the cells. In fact, sometimes this process fails and they attack the wrong cells, the healthy cells of the host body that are not foreign or abnormal; this defines what an autoimmune disease is. It is yet another example of a programmed function of the cell going wrong and requiring reprogramming by external signals (which is the way any system, including a computer, is reprogrammed by giving it a program to read, a signal).

Equipped with these ideas that suggest that we can extend concepts that properly belong only to computability and algorithmic information theory, we can devise a software-engineering approach to systems biology.

HOW NATURAL SELECTION PROGRAMS AND REPROGRAMS LIFE

At some point early in the process of replicating (copying) biological information for cell growth, which is particularly necessary for multicellular organisms, the process reaches the critical juncture of dealing with errors and redundancy that can be quantified by the so-called noisy-channel coding theorem, part of Shannon's classical information theory, determining the degree of redundancy needed to be able to deal with different degrees of noise.

Had the balance not been reached, no copying process would have conveyed the information necessary for organisms to reproduce.

The noisy-channel coding theorem (sometimes called Shannon's theorem) establishes that for any given degree of noise in a communication channel, it is still possible to communicate (digital) data nearly error-free up to a computable maximum rate by introducing redundancy into the channel. This also suggests why nature has chosen a clearly digital code for life in the form of nucleic acids, e.g., DNA and RNA. On the other hand, while the balance was positively reached, that is, the accuracy in the replication process was greater than the incidence of error, some errors did occur, and these could also be quantified using classical information theory.

While the nature of the variations (of which some can be identified as errors) can be attributed to noise, the correction is nothing but a mathematical consequence. If errors prevail, in both the primary source and replicants, then the cells and organisms have a greater chance of dying from these variations. On the other hand, new variations may confer survival or reproductive advantages in changing environmental conditions and, thus, will be positively selected during evolution. Natural selection hence effectively reprograms nature, cells, and organisms. Of course this does not explain how the whole process began, the origins of life being still an open question.

Approached as computer programs, one can explain how certain patterns, e.g., the information content in molecules, such as DNA or RNA, may have been produced in the first place (Zenil, 2011, 2013; Zenil and Delahaye, 2010; Zenil and Marshall, 2013; Zenil et al., 2012). There is also the problem of the uncaused first computer program, given that the process generating information and eventually computation required computation in the first place (e.g., the very first laws in our universe). In fact, we have explored these ideas, taking seriously the possibility that nature computes, and we have suggested a measure (Zenil, 2014) of the property of being 'computer-' or 'algorithm-like', the property of *algorithmicity*. In Zenil and Delahaye (2010) we undertook a search for statistical evidence, with interesting results.

Hacking strategies have been in place in biological systems since the beginning; viruses, for example, are unable to replicate by themselves, but they trespass the cell membrane and release their DNA content into the cell nucleus to be replicated. Cells are the basic units of life because of this property; they are the smallest unit that fully replicates by itself. No simpler unit is able to replicate in full like the cell even if RNA is suspected to be a possible precursor of replication, and possibly the first self-replicating molecule, because RNA can both store information like DNA and fold similarly to proteins and therefore serves as building blocks to build more complex structures.

Viruses are clearly (in consensus amongst biologists) nonliving structures of encapsulated DNA that evolve by undergoing a process of natural selection and are evidence of nonliving matter subject to the same process. Though they may appear so, viruses have no self-purpose or will of their own, neither as individuals or as a group. Indeed, evolution by natural selection is a mathematical inevitability independent of substrate. Despite the common flawed fashion in which the concept of evolution by natural selection may popularly be conveyed – as a biological process that carries a self-purpose meaning as an objective function for self-preservation – it is a simple statistical consequence that cannot be contested. In the case of some viruses, for example, it is the DNA in the virus that becomes more effective in the host cell nucleus, which replicates faster thereby dominating viral evolution. This basic consequence is the basis of the mechanism of evolution by natural selection. By adding limited resources, such as the number of cells that can be infected, one can extrapolate this simple mathematical fact to anything else other than viruses. Even if viruses are naturally perfect carriers of computing code, biological programming and reprogramming is in general very similar because we know everything is encoded in the cell's genetic material.

A Hacker View of Cancer and the Immune System
Equipped by these mechanisms of evolution by natural selection and the way in which self-replicating cells can be hacked, a hack

to fight diseases has been devised in the efforts to treat cancer. Traditional drug and radiation therapies have not been very successful so far but genetics is promising spectacular results. Gene therapy is a mechanism to deliver specific genes into tumours by using similar, if not exactly the same, strategies as that of viruses in their delivery of their payload. Indeed, the approach relies heavily on using viruses to deliver antitumour genes into the target cancer cells. Various hacking approaches using viruses can be devised. If the foreign code enters the nucleus, it will be replicated, but made into a buggy virus that only reaches the cytoplasm after the cell membrane, and it can interfere with protein transcription (making the enzymes in the cytoplasm to believe that the instructions come from either legitimate external signals from other cells or from the regular instructions coming from the cell nucleus). The fake instructions will then modify the cell behaviour (function) with the nucleus remaining intact and replication of its original untouched code guaranteed. Indeed, one can see how genes are effectively subroutines, computer programs that regulate the type (shape) and number of building blocks (proteins) to be produced that in turn fully determine the cell function(s). Let the instructions reach only outside the nucleus and they will interfere with messages RNA only, indicating protein production, or do so inside the nucleus and modify the actual genome of the cell DNA to replicate it (what viruses do, integrating their DNA content into the host genome and undergoing a coevolution; Ryan, 2009).

There is, therefore, a potential to develop these ideas into a more systematic information-theoretic and software-engineering view of life, cancer, and immune-related diseases based on these amazing purely computational mechanisms in biology. While it is clear that information processing is in a very fundamental sense key to essential aspects of the biology of different species (see Figure 11.5A), by applying ideas related to computability and algorithmic information theory we can take a step ahead to view processes from fresh and different angles to conceive new strategies to modify the behaviour of cells against diseases.

Cancer, for example, is like a computer programming bug that does not serve the purpose of the multicellular organism. Cancer cells are ultimately cells that grow uncontrollably and do not fulfill their contract to die or stop proliferating at the rate at which they must if the multicellular host is to remain stable and healthy. From a software-engineering perspective it is a cellular computer program that has gone wrong, resulting in an infinite loop with no halting condition, hence in effect being out of control. Normal cells have a programmed mechanism to stop their cell cycle at the point at which the population of previous cells is replaced. When the cell cycle produces more than the number of cells necessary to replace the population, the result is a mass of abnormal cells that we call a tumour.

Cancer can be seen as a purely information-theoretic problem: the information dictating the way in which a cell replicates is compromised, either because it, as it were, reneges on a contract with the multicellular organism, resulting in the cell behaving selfishly and replicating with no controls, or because noise in the environment in the form of external stimuli has broken the programming code that makes a cell's behaviour 'normal'. In other words, it is either a bug or a broken message. The question is therefore how a cell can transfer its information at replication time without generating instructions that produce cancer states when the said cell encounters noise.

Likewise, the immune system can be seen as an error-correcting code. One key aspect of the immune system is diversity, which is largely contributed by T and B lymphocytes, cell types of the adaptive immune system. By gene rearrangement of segments in their antigen receptor genes, a highly complex repertoire of different receptors expressed by individual B or T cell clones is generated, which are specific for different antigens. The B cell antigen receptor, surface bound or secreted as an antibody molecule by terminally differentiated B cells, recognises whole pathogens without any need for antigen processing, while the T cell antigen receptor is exclusively on the T cell surface and recognises processed antigens presented on so-called major histocompatibility complex (MHC) molecules.

When B cells and T cells are activated and begin to replicate, some of their progeny become long-lived memory cells, remembering each specific pathogen encountered, and can mount a strong, faster response if the pathogen is detected again. This means the memory can take the form of either passive short-term memory or active long-term memory.

The memory of the immune system stores all the information relating to all the pathogens we have encountered in our lives (Gourley et al., 2004). When B cells and T cells are activated, some will become memory cells. Throughout the lifetime of an animal these memory cells effectively form a 'database' of effective B and T lymphocytes (Vitetta et al., 1991). On interaction with a previously encountered antigen, the appropriate memory cells are selected and activated.

The major functions of the acquired immune system that involve information include:

- Pattern recognition: 'non-self' antigens in the presence of 'self', during the process of antigen presentation
- Communication: generation of responses that are tailored to maximally eliminate specific pathogens or pathogen-infected cells
- Storage (memory): development of immunological memory, in which pathogens are 'remembered' through memory cells.

Yet another illustration of how the immune system can be seen from a purely informational perspective is the fact that newborn infants have no immune memory, as they have not been exposed to microbes, and are particularly vulnerable to infection. Several layers of passive protection are provided by the mother. During pregnancy, antibodies of the IgG isotype are transported from mother to baby directly across the placenta (Sedmak et al., 1991), so human babies have high levels of antibodies even at birth, with the same range of antigen specificities as their mothers. Protective passive immunity can also be transferred artificially from one individual to another (Schlesinger et al., 1985).

This brief account reveals the extent of the role played by information in the immune system. Immune-related diseases are related to informational dysfunction, such as wrong signalling, defects in immune tolerance, or misguided pattern recognition. This can cause the immune system to fail to properly distinguish between self and non-self and attack part of the body, leading to autoimmune diseases. As described before, an important role of the immune system is also the identification and elimination of tumours. The transformed cells of tumours often express antigens that are not found in normal cells. The main response of the immune system to tumours is to destroy the abnormal cells. Tumour antigens are presented on MHC class I molecules in a similar way to viral antigens, and this allows killer T cells to recognise the tumour cell as abnormal and kill it. The immune system produces and reprograms cells to match them with pathogens. Antigens are messages. A vaccine is a way to send a fake message without the actual content.

A model created by Kogan et al. (2015) shows that the stability of a gene network stems from several major factors. These factors include 'effective' genome size, proteome turnover, and DNA repair rate but also gene network connectivity. The researchers concluded that by hacking any of these parameters, one could increase an organism lifespan, a hypothesis supported by the biological evidence (Kogan et al., 2015).

Cellular death has a strong information-theoretic component, evident in the way evolution has programmed multicellular organisms. Cancer, and indeed laboratory cell lines (the cells used in labs), are immortal (Skloot and Turpin, 2010); they continue dividing indefinitely (just as stem and germ cells too). An immortal living cell can last only as long as its error-correcting methods (such as conserving the length of its telomeres, up to the so-called *Hayflick limit* [Hayflick, 1965], and correcting DNA mutation mechanisms) preserve genome integrity and stability. In practice this is impossible, for purely chemical and physical reasons. For instance, DNA is believed to undergo more than 60,000 single mutations per day in the genomes of mammalian cells (Ames et al., 1993; Bernstein et al., 2013; Helbock et al., 1998), and a fraction of DNA will never be repaired by specific DNA enzyme-repairing molecules, hence leading to genome instability and degenerative diseases such as cancer (Corcos, 2012) and neuronal and neuromuscular diseases (Rao, 2007). Telomeres, for their part, cannot remain unchanged due to chemical degradation, so after about 40 to 60 divisions, they shorten and die (Hayflick, 1965). But all cells are potentially immortal. If they die at the end of a generation, it is because they are programmed to do so in order to serve as building blocks for multicellular organisms. The length of a cell's telomeres is also believed to play a fundamental role in the fight against cancer: the length of the telomeres may be such that they are shorter and degrade faster than the critical number of accumulated mutations, hence effectively preventing a cell from continuing to divide and replicate its errors before dying (Eisenberg, 2011; Shay and Wright, 2000). We can thus start seeing how information theory and effective programming are implicated in the most basic mechanisms of life and death. Hacking cells can be achieved simply by inhalation just as in the case of some viruses, and it is promising to be effective in a trial with cystic fibrosis caused by a gene mutation located in the chromosome 7. The hack consists in introducing the normal gene into fat globules (liposomes) that deliver the gene into the cells in the lung lining (Alton et al., 2015).

Immunity as Computation and Cancer as a Software-Engineering Problem

Cells are continually programmed and reprogrammed by the environment and by the cells themselves. Figure 11.6A illustrates how this process happens. In the example, generic signal 2 first turns a stem cell into an immune system cell (e.g., CD4-positive T helper cell), and then signal 4 finally drives it to a 'steady state', where it has completely differentiated into a cell with a specific function. This is an oversimplification of a complex system, as information and signals do not flow in a single direction. In fact, T helper cells

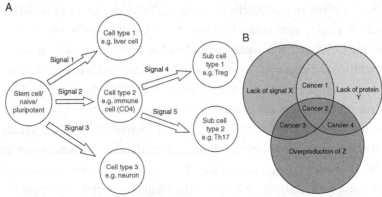

FIG. 11.6 (A) Cells are like computer programs ready to be reprogrammed, by nature or artificially, to become other cells. Cells usually start from a cell that is amenable to being programmed, as it has no specific function other than precisely to become a new type of cell and to survive until doing so by receiving the correct signals from its environment (including other cells). In this cartoon, signals 2 and 5 can produce, for example, a Th17 (T helper 17) starting from a pluripotent cell, a cell that can potentially become many different types of cell. The pseudocode to reprogram a cell involves calling computer subroutines (chemical signals) 2 and 5. End points in the tree can be seen as computer programs reaching a stable configuration, an attractor in a dynamical system that can be a fixed point, a cycle, or a strange attractor that cannot easily be reprogrammed back to previous developmental stages. (B) A computational reclassification of cancer and tumour profiling. Intersecting Venn diagrams of bug state spaces identified as leading to different diseases and tumour types (e.g., lack of signal X, lack of protein Y, overproduction of Z). This would immediately suggest the possibility of more tumour types (cancers) than source bugs (there are four intersections from three potential interacting source bugs), but the important message here is that these bugs can be debugged by type, and tackling bug 1 can actually fix many types of cancers if they are triggered only by the presence of all these bugs. This enables more general strategies than attacking each cancer in isolation, as an instance of a type (e.g., by tissue, as has been the practice to date) or using extreme approaches such as radio- or chemotherapies that are anything but localised or personalised.

in the immune system regulate other immune cells and thus send back signals to the reprogramming pipeline according to the signals they themselves receive from other cells in the body and depending on the environment (e.g., pathogens present). This also implies that one can reprogram cells with other cells and that differentiated cells are not generally impossible to reprogram, although it may be more difficult. In fact, CD4 T cell subsets have been observed to have some capabilities to be reprogrammed (Geginat et al., 2014). Signals 2 and 4 may also prompt other cells to react in different ways. Signals are simply chemicals or chemical cocktails. For example, programming a T cell towards the regulatory T cell (Treg) subset requires proteins such as the transcription factor forkhead box P3 (Foxp3) (Zhang and Zhao, 2007) and the cytokines transforming growth factor beta (TGF-β) (Josefowicz and Rudensky, 2009) and interleukin-2 (IL-2) that the body itself produces using the same or other cells, in effect producing its own signals for reprogramming itself. However, the same signals can also be artificially produced. A model based on cellular automata has shed light on the way in which the immune system can be interpreted as a computer program subject to sensitivity and robustness in the face of external changes (signals, mutations), quantifying the number of errors that keep or remove the cellular automaton from its regular basin of attraction and therefore changing or maintaining its original function interpreted as the steady state of the cellular automaton seen as a dynamical system (Cassiano and Barbosa, 2014).

As a programming strategy, one needs, for example, to induce the production of Foxp3 and TGF-β in a naive CD4 T cell to produce a Treg-like cell (under ideal conditions and in the presence of IL-2). Importantly, immune system signals also contribute to the regulation of cancer development. In fact, it is believed that TGF-β is related to cancer. Acting through its signalling pathway, it can stop the cell cycle in order to prevent proliferation, induce differentiation, and promote apoptosis (Derynck et al., 2001).

What is known as *morphological computation* is based on the universal phenomenon in biology that form is function; that is, the shape of a biological structure (epitomised by proteins) determines the structure's biological function. While all forms of computation can be traced back to classical computation and the concept of computational universality, which blurs any essential boundary between software and hardware, biology is the quintessential example. Indeed, natural and morphological/spatial computation is the realisation that in physics, but mainly in biology, hardware is also software. Everything is in some way a kind of embedded computation. For example, DNA is code and storage but other molecules are not only code and storage devices but also functional devices, such as RNA and proteins, these latter carrying different structural information with it. The code depends not only on the nucleotides but, in the DNA configuration and packaging, on histones and other structural proteins that enable or hinder fragments of code from replicating; the structure of DNA itself encodes information. RNA, in turn, is a message when encoding for proteins and an output when acting as regulator, and proteins can regulate translation and gene expression.

Using a combination of the fundamentals of information theory and the principles of natural selection one may grasp how the immune system seems to have come into existence, it being a natural error-correcting mechanism continually streamlined and reprogrammed by natural selection. This latter is again constrained as to what is possible according to the Shannon limit, determined by Shannon's afore-mentioned noisy-channel coding theorem (Cover and Thomas, 2006), for if the replication process exceeds this limit, then its accuracy will be compromised and there will be a loss of information, eventually leading to failure to recode basic functions of living organisms such as genes.

How does nature reprogram (or *unprogram*, from the point of view of their original function) cells so that they become cancerous? There are various hypotheses, one of which posits the breaking of a 'contract' that cells made when they went from being unicellular life forms, hence 'selfish', to being components of multicellular organisms such as the human body. The existence of tumour cells that lose certain functions but remain able to replicate is an indication that life is highly hierarchical and modular and robust, unlike traditional artificial computer programs that are very vulnerable to random errors introduced into their source code. In a sense, when it comes to living organisms, we seem to be dealing with a streamlined version of an object-oriented programming language.

Methods for modifying the functioning of cells have already shown promise. Recently, for example, it has been shown that the immune system can be reprogrammed to fight cancer (Lizée et al., 2013). As an example, Sharpe and Mount (2015) showed that genetically modified T cells can be used in cancer therapy. Indeed, as is pointed out, tumour cells (tumours) follow many strategies to evade the immune system, including tampering with genes that would normally regulate a function related to the sensitivity of an immune cell to certain signals produced by said tumour cells (Schreiber et al., 2011). However, tumour cells generate an immunosuppressive tumour environment that leads the immune system to neglect the tumour and therefore cause great danger to the host multicellular organism (the host body) (Sharpe and Mount, 2015).

So how might we reprogram tumour cells themselves or immune cells to recognise and naturally fight tumour cells? For a cell to replicate it needs to become redundant in the face of environmental noise, hence more simple. We can target the genes that in the replication process contribute less to the information content of the cell because of its state of redundancy. In other words, tumour cells may be identified because they are less sensitive and therefore also less programmable (just like the examples in Figure 11.4A and B).

The current common classification of cancer divides it into broad groups that are not related to the type of possible error leading to a cancer but rather to where a cancer originates and other physical properties. However, it has been found that different types of cancer under this classification can be deeply related, possessing similarities that cut across different types of tissue (Heim et al., 2014). This may be because certain cancers may have a set of common causes related to a type of bogus program. In software terms, such a classification scheme amounts to classifying all operating system errors in the same bin simply because they involve operating systems and not because, say, one has a software bug related to reference software bugs (memory access violations, null pointer dereference), an arithmetic error (e.g., division by zero, precision loss or overflow), a logic software bug (e.g., infinite loop recursion), or a syntax software bug.

Some of these software bugs may lead to incremental error accumulation. We think these types of software errors may prompt the rethinking of current cancer classifications, leading to the grouping of cancers in terms of their information/computational type, i.e., the type of error that leads to cell reprogramming, and not by their point of origin. Such an approach would make it possible to explain why so many cancers have so many things in common, while cancers that are supposed to be of the same type, e.g., liver cancers, can actually be very dissimilar. A classification strategy based on place of origin and not type of error is likely to fail because it fails at characterising the cause of the problem. It may be compared to trying to debug a program based on its application, for instance, treating all software bugs in word processors similarly, or all bugs in animation software using a common strategy, which makes no sense. It is starting to be more widely recognised that cancer should be classified or studied not by tissue type (which immediately leads to thousands of cancer types, if one takes cell subtypes into consideration), but by bug type. Not any bug will turn a cell into a cancer cell, and we just do not know how many will. This is perhaps one of the first quantification tasks and one of the first interesting research questions – to determine the bug space (see Figures 11.6B and 11.7A) leading to different diseases and tumour types. We have suggested that this can be more systematically approached by following the behavioural black box approach that we have illustrated, not because we have a penchant for seeing things as black boxes, but because, as we have explained, science has to deal with black boxes, given certain limitations that go beyond the pragmatic and are fundamental and intrinsic to even the most simple deterministic rule-based systems. Figure 11.6 shows a Venn diagram depicting the proposed informational view of a software-engineering classification of diseases such as cancer, based on bug type rather than tissue origin. The diagram is in itself a simplification. For example, in biology a lack of signal X likely represents a lack of production of a protein or its overproduction, hence bugs may themselves be produced by other bugs. But here they are considered only as direct causes, which one has to tackle layer by layer just as one would in a traditional computer program, inserting debugging breakpoints either to identify the primary cause or to design a patch – just like a software patch – to stop the propagation of the error before the first forking path leading to the undesired behaviour. Reprogramming from scratch is, of course, always more desirable than a patch, which can introduce new undesirable effects. Indeed, this can be the informatics definition of a drug with its common secondary effects. Hence we would wish to move from traditional drugs to drugs closer to the causes, such as drugs that target genes and gene products rather than drugs that target extracellular processes, which are the drugs currently dominating the market.

Some interesting initial software-engineering questions about cell biology, bugs, and cancer can be found illustrated in Figure 11.7A. While the potential cell bug space can be very large, there have to be natural selection mechanisms that prevent the bugs leading to fatal diseases, keeping them small, at least before and during reproductive age. Later in life, because there is no longer evolutionary pressure, the size of these spaces can increase freely. For example, experience with cancer tells us that about one-third of the population (believed to rise to one-half) can be considered likely to reach some cancer state during their lifetime. The immune system as well as cell-intrinsic mechanisms prevents normal cells from proliferating when they reach the bug space, yet cells and cell populations are constantly moving along these sets. What we have discussed here suggests that one cannot find infallible methods to keep cells in a completely safe bug-free zone. More complicated questions follow, questions about whether the normal functioning of the cell can dynamically lead to cancer (large and small intersections) even without the introduction of mutations, that is, whether cells may be programmed to eventually

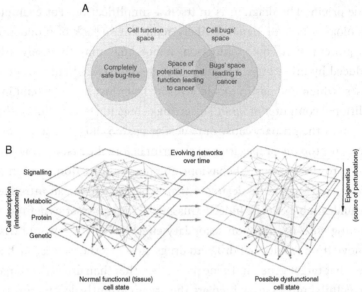

FIG. 11.7 (A) Normal function versus bug space, or abnormal cell function. The first question about cell biology, disease, and cancer from the software-engineering perspective is how large each of these sets really is relative to each other. Natural selection provides some hints; the bug space cannot be larger than the normal function space before and until reproductive age when we are subject to evolutionary pressure (natural selection simply says: if the bug space is larger than the healthy space, then we all would be dead because the population would shrink at every generation). Directions in the networks represent a gene (transcription factor) regulating other genes, e.g., upregulating, i.e., increasing its production, or downregulating, i.e., decreasing its production; proteins interacting; metabolites reacting; or signals being transported from one element to another. (B) Propagation of undesired code in a network of networks: several cell networks at different levels of description form another causal network. The full description is also called the cell *interactome*. Changes to nodes (genes, proteins, metabolites, signals) or edges (regulation, reaction, channel) in any layer lead to functional and description changes, and some of these will lead to a dysfunctional (diseased) cell, such as a tumour cell, where a normal tissue cell behaves differently from what was expected (e.g., uncontrolled replication). All these changes are reflected in the interactome and therefore in the description, and in the information content, of the cell.

reach some sort of tumour state, even if they do not do so for a long initial period of time (e.g., before reproductive age).

Diseases imply deviations of a cell towards pathological states that are encoded in the cell's descriptions. Figure 11.7B shows the intricate ways in which different descriptions of the cell interact with each other. All these causal interactions fully describing the cell constitute what is called the *interactome*. The interactome and each of its complicated interacting parts can be studied with tools from classical and *algorithmic information theory* (AIT). AIT is the subfield that characterises lossless compression and has as its core Kolmogorov complexity. It is the bridge between computation and information and it deals with an objective and absolute notion of information in an individual object, such as a cell in a network representation (Zenil et al., 2013a). We have proposed ways to study and quantify the information content of biological networks based on the related concept of algorithmic probability, and we have found that it is possible to characterise and profile these networks in different ways and with considerable degrees of accuracy (Zenil et al., 2014, 2016). This shows that the information approach may open a new pathway towards understanding key aspects of the inner workings of molecular biology that is causality-driven rather than correlation-driven.

Important sources of information are epigenetic phenomena, an additional layer of complexity reversing the traditional molecular biology dogma that describes how information is transferred from the genome all the way to the upper levels. Epigenetics shows that information can flow bottom-down from all upper layers to the lower layer (genome). This information coming back to the cell alters how genes are finally expressed even if the cell's genome (the DNA code) remains exactly the same. Epigenetics is the process in which external information is introduced to the various cell layers that can disrupt the function of it and lead to changes, including diseases. One can study the rate of information propagation by looking at the different layers and then identify the layer source of the dysfunction.

Some changes will propagate fast, inducing a different dynamic with attractors that are different from those of the normal cell function; other changes may not propagate, thanks to the qualitative robustness of biological systems (see Figure 11.3A). Small changes in lower layers are amplified in upper layers; e.g., single nucleotide polymorphisms (SNPs)[1] in the gene BRCA1 can cause a cell to become cancerous due to defects in DNA repair, leading to additional mutations in the cell that ultimately can cause its transformation to a cancer cell. This is why complex diseases may be better detected at the level of the upper layers, while 'simpler' diseases such as cancer induced by cancer-predisposing BRCA1 SNPs (a very specific case) can be detected at the lowest level. The network on which these changes propagate is poorly understood, as it requires integration of different data sources put together and studied over time, which is the direction in which we are heading by studying the interactome (the cells' full description in the form of a network of networks) over time, the ultimate goal of *network reconstruction*.

CONCLUSIONS

We have seen that uncomputability prescribes limits to what can be known about nature or models of nature, limits that are likely to apply to natural and biological systems or the models we build of them, and therefore we cannot help but develop an encompassing behavioural approach that can utilise ideas and tools from both theoretical computer science and software engineering. These ideas can then be further developed to yield new concepts, classifications, and tools with which to reprogram cells against diseases. This is a direction in which biology is moving by adopting new immunotherapies against diseases such as cancer.

An information computational approach to cancer and human diseases may be key to understanding molecular medicine from a new

---

[1] A DNA sequence variation occurring commonly within a population (e.g., 1%) in which a single nucleotide A, T, C, or G in the genome (or other shared sequence) differs between members of a biological species or paired chromosomes.

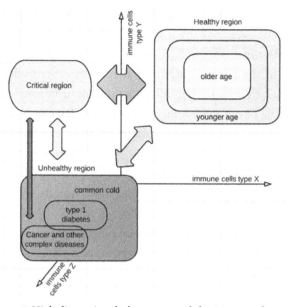

FIG. 11.8 High-dimensional phase space of the immune dynamical system: the main characteristic of the space is the existence of stable regions that represent the healthy or unhealthy state as fixed or strange attractors of the dynamical system of the immune system that is perturbed over time (both by weakening self-defenses or by introduction of foreign organisms, such as virus, bacteria, and dysfunctional cells such as tumours). Natural selection establishes that the healthy region in early ages is necessarily greater than the unhealthy region that leads to death hitting the reproductive rate (e.g., the 'cancer' region). However, after reproductive age there is no evolutionary pressure and the unhealthy region grows over time. For example, cancer is a more common disease in elderly people, which, however, is also related to the fact that usually several mutations have to be acquired over a relatively long time period in order to cause cancer. The cell state of the body crosses a critical region many times during the lifetime of an organism, and whether the organism goes in one or another direction depends on the strength and capability of the host immune system. External factors perturbing the state of an organism over time lead to changes in this high-dimensional space of immune cell types that usually respond to different threats in different ways, in both number of cells and their type, and are hence excellent candidates for disease monitoring. With some probability the organism leans towards the unhealthy region or back to the healthy one, depending on the strength of the immune system and the degree of invasion. A subset of the unhealthy region is due to complex diseases from which it is difficult to get back to the critical region towards other equilibria (e.g., healthy state) but not impossible.

perspective. The most promising approach, as in software-engineering at the design stage, may involve prevention through permanent monitoring of the immune system (see Figure 11.8) based on systematic screening of the immune status over time in order to detect trends in responses to disease, including early tumour detection. It is now clear that the best way to detect early signs of disease must inevitably involve tracking the unfolding computation of the immune system in the course of its normal operation. If the immune system produces more cells of a certain type, it is a sign of change and likely a reaction to a threat, and the trajectory of immune system behaviour is an indicator of the direction in which an individual's health will evolve. The immune system as a computational device, being our potentially best debugging system, is key to our strategy for defeating cancer and many other diseases, and thereby reprogramming our fate. Of course debugging software can break down, too, just as may happen in an artificial software debugger, and this is where autoimmune and inflammatory diseases come into play. These can likewise be detected by looking closely at the behaviour of the immune system.

ACKNOWLEDGEMENTS

We wish to thank the rest of the Unit of Computational Medicine team at Karolinska Institutet and are grateful for the support of AstraZeneca, the Strategic Area Neuroscience (StratNeuro), the Foundational Questions Institute (FQXi), the John Templeton Foundation, and the Algorithmic Nature Group, LABORES.

REFERENCES

Alton, E. W. F. W., Armstrong, D. K., Ashby, D., Bayfield, K. J., Bilton, D., Bloomfield, E. V., Boyd, A. C., Brand, J., Buchan, R., Calcedo, R., et al. 2015. Repeated nebulisation of non-viral CFTR gene therapy in patients with cystic fibrosis: a randomised, double-blind, placebo-controlled, phase 2b trial. *The Lancet Respiratory Medicine*, 3(9), 684–691.

Ames, B. N., Shigenaga, M. K., and Hagen, T. M. 1993. Oxidants, antioxidants, and the degenerative diseases of aging. *Proceedings of the National Academy of Sciences*, **90**(17), 7915–7922.

Bernstein, C., Prasad, A. R., Nfonsam, V., and Bernstein, H. 2013. DNA damage, DNA repair and cancer. In Chen, C. (ed.), *New research directions in DNA repair*. InTech.

Cassiano, K. K., and Barbosa, V. C. 2014. Error-prone cellular automata as metaphors of immunity as computation. *arXiv preprint arXiv:1401.3363*.

Cencini, M., Cecconi, F., and Vulpiani, A. 2010. *Chaos: from simple models to complex systems*. Vol. 17. World Scientific.

Chaitin, G. J. 1966. On the length of programs for computing finite binary sequences. *Journal of the ACM*, **13**(4), 547–569.

Corcos, D. 2012. Unbalanced replication as a major source of genetic instability in cancer cells. *Am. J. Blood Res.*, **2**(3), 160–169.

Cover, T. M., and Thomas, J. A. 2006. *Elements of information theory*, 2nd revised edition. Wiley-Blackwell.

Derynck, R., Akhurst, R. J., and Balmain, A. 2001. TGF-β signaling in tumor suppression and cancer progression. *Nature Genetics*, **29**(2), 117–129.

Eisenberg, D. T. A. 2011. An evolutionary review of human telomere biology: the thrifty telomere hypothesis and notes on potential adaptive paternal effects. *American Journal of Human Biology*, **23**(2), 149–167.

Galon, J., Pages, F., Marincola, F. M., Angell, H. K., Thurin, M., Lugli, A., Zlobec, I., Berger, A., Bifulco, C., Botti, G., et al. 2012. Cancer classification using the Immunoscore: a worldwide task force. *Journal of Translational Medicine*, **10**(1), 205.

Geginat, J., Paroni, M., Maglie, S., Alfen, J. S., Kastirr, I., Gruarin, P., De Simone, M., Pagani, M., and Abrignani, S. 2014. Plasticity of human CD4 T cell subsets. CD4+ T cell differentiation in infection: amendments to the Th1/Th2 axiom. *Front Immunol.*, **5**:630.

Gödel, V. K. 1958. Über eine bisher noch nicht benützte Erweiterung des finiten Standpunktes. *Dialectica*, **12**(3–4), 280–287.

Gourley, T. S., Wherry, E. J., Masopust, D., and Ahmed, R. 2004. Generation and maintenance of immunological memory. Pages 323–333 of *Seminars in Immunology*, vol. 16. Elsevier.

Hanahan, D., and Weinberg, R. A. 2011. Hallmarks of cancer: the next generation. *Cell*, **144**(5), 646–674.

Hayflick, L. 1965. The limited in vitro lifetime of human diploid cell strains. *Experimental Cell Research*, **37**(3), 614–636.

Heim, D., Budczies, J., Stenzinger, A., Treue, D., Hufnagl, P., Denkert, C., Dietel, M., and Klauschen, F. 2014. Cancer beyond organ and tissue specificity: next-generation-sequencing gene mutation data reveal complex genetic similarities across major cancers. *International Journal of Cancer*, **135**(10), 2362–2369.

Helbock, H. J., Beckman, K. B., Shigenaga, M. K., Walter, P. B., Woodall, A. A., Yeo, H. C., and Ames, B. N. 1998. DNA oxidation matters: the HPLC–electrochemical detection assay of 8-oxo-deoxyguanosine and 8-oxo-guanine. *Proceedings of the National Academy of Sciences*, **95**(1), 288–293.

Josefowicz, S. Z., and Rudensky, A. 2009. Control of regulatory T cell lineage commitment and maintenance. *Immunity*, **30**(5), 616–625.

Kogan, V., Molodtsov, I., Menshikov, L. I., Reis, R. J. S., and Fedichev, P. 2015. Stability analysis of a model gene network links aging, stress resistance, and negligible senescence. *Scientific Reports*, **5**.

Kolmogorov, A, N. 1968. Three approaches to the quantitative definition of information. *International Journal of Computer Mathematics*, **2**(1–4), 157–168.

Lizée, G., Overwijk, W, W., Radvanyi, L., Gao, J., Sharma, P., and Hwu, P. 2013. Harnessing the power of the immune system to target cancer. *Annual Review of Medicine*, **64**, 71–90.

Peltonen, L., and McKusick, V, A. 2001. Dissecting human disease in the postgenomic era. *Science*, **291**(5507), 1224–1229.

Rao, K. S. 2007. Mechanisms of disease: DNA repair defects and neurological disease. *Nature Clinical Practice Neurology*, **3**(3), 162–172.

Ryan, F. 2009. *Virolution*. Collins.

Schlesinger, J. J., Brandriss, M. W., and Walsh, E. E. 1985. Protection against 17D yellow fever encephalitis in mice by passive transfer of monoclonal antibodies to the nonstructural glycoprotein gp48 and by active immunization with gp48. *Journal of Immunology*, **135**(4), 2805–2809.

Schreiber, R. D., Old, L. J., and Smyth, M. J. 2011. Cancer immunoediting: integrating immunitys roles in cancer suppression and promotion. *Science*, **331**(6024), 1565–1570.

Sedmak, D. D., Davis, D. H., Singh, U., Van de Winkel, J. G., and Anderson, C. L. 1991. Expression of IgG Fc receptor antigens in placenta and on endothelial cells in humans: an immunohistochemical study. *American Journal of Pathology*, **138**(1), 175.

Sharpe, M., and Mount, N. 2015. Genetically modified T cells in cancer therapy: opportunities and challenges. *Disease Models and Mechanisms*, **8**(4), 337–350.

Shay, J. W., and Wright, W. E. 2000. Hayflick, his limit, and cellular ageing. *Nature Reviews Molecular Cell Biology*, **1**(1), 72–76.

Shin, S. W., Smith, G., Smolin, J. A., and Vazirani, U. 2014. How "quantum" is the D-wave machine? *arXiv:1401.7087 [quant-ph]*.

Skloot, R., and Turpin, B. 2010. *The immortal life of Henrietta Lacks*. Crown.

Terrazas, G., Zenil, H., and Krasnogor, N. 2013. Exploring programmable self-assembly in non-DNA based molecular computing. *Natural Computing*, **12**(4), 499–515.

Turing, A. 1936. On computable numbers, with an application to the Entscheidungsproblem. *Proc. of London Math. Soc.*, **42**, 230–265.

Turing, A. 1950. Computing machinery and intelligence. *Mind*, **59**, 433–460.

Vitetta, E. S., Berton, M, T., Burger, C., Kepron, M., Lee, W. T., and Yin, X. 1991. Memory B and T cells. *Annual Review of Immunology*, **9**(1), 193–217.

Wikipedia. *Brain-to-body mass ratio*. Retrieved April 5, 2015a.

Wikipedia. *List of animals by number of neurons*. Retrieved April 5, 2015b.

Wolfram, S. 2002. *A new kind of science*. Vol. 5. Wolfram Media.

Zenil, H. 2010. Compression-based investigation of the dynamical properties of cellular automata and other systems. *Complex Systems*, **19**(1), 1–28.

Zenil, H. 2011. The world is either algorithmic or mostly random. *arXiv preprint arXiv:1109.2237*.

Zenil, H. 2013. Turing patterns with Turing machines: emergence and low-level structure formation. *Natural Computing*, **12**(2), 291–303.

Zenil, H. 2014. What is nature-like computation? A behavioural approach and a notion of programmability. *Philosophy & Technology*, **27**(3), 399–421.

Zenil, H. 2015. Algorithmicity and programmability in natural computing with the Game of Life as in silico case study. *Journal of Experimental & Theoretical Artificial Intelligence*, **27**(1), 109–121.

Zenil, H., and Delahaye, J. P. 2010. On the algorithmic nature of the world. In Dodig-Crnkovic, G., and Burgin, M. (eds), *Information and computation*. World Scientific.

Zenil, H., and Marshall, J. A. R. 2013. Some aspects of computation essential to evolution and life. *ACM Ubiquity*, 1–16.

Zenil, H., and Villarreal-Zapata, E. 2013. Asymptotic behaviour and ratios of complexity in cellular automata rule spaces. *International Journal of Bifurcation and Chaos*, 23(9).

Zenil, H., Gershenson, C., Marshall, J. A. R., and Rosenblueth, D. A. 2012. Life as thermodynamic evidence of algorithmic structure in natural environments. *Entropy*, **14**(11), 2173–2191.

Zenil, H., Kiani, N. A., and Tegnér, J. 2013a. Algorithmic complexity of motifs clusters superfamilies of networks. Pages 74–76 of *Proceedings of the IEEE International Conference on Bioinformatics and Biomedicine, Shanghai, China*. IEEE.

Zenil, H., Ball, G., and Tegnér, J. 2013b. Testing biological models for non-linear sensitivity with a programmability test. Pages 1222–1223 of *Advances in Artificial Life, European Conference on Artificial Life (ECAL)*, vol. 12.

Zenil, H., Soler-Toscano, F., Dingle, K., and Louis, A. 2014. Graph automorphisms and topological characterization of complex networks by algorithmic information content. *Physica A: Statistical Mechanics and Its Applications*, **404**, 341–358.

Zenil, H. K., A., Narsis, and Tegner. 2016. Methods of information theory and algorithmic complexity for network biology. *Seminars in Cell and Developmental Biology*, **51**, 32–43.

Zhang, L., and Zhao, Y. 2007. The regulation of Foxp3 expression in regulatory CD4+ CD25+ T cells: multiple pathways on the road. *Journal of Cellular Physiology*, **211**(3), 590–597.

Part IV Complexity and Causality

12 Life's Information Hierarchy

Jessica Flack

SUMMARY

I propose that biological systems are information hierarchies organized into multiple functional space and time scales. This multiscale structure results from the collective effects of components estimating, in evolutionary or ecological time, regularities in their environments by coarse-graining or compressing time-series data and using these perceived regularities to tune strategies. As coarse-grained (slow) variables become for components better predictors than microscopic behavior (which fluctuates), and component estimates of these variables converge, new levels of organization consolidate. This process gives the appearance of downward causation – as components tune to the consolidating level, variance at the component level decreases. Because the formation of new levels results from an interaction between component capacity for regularity extraction, consensus formation, and how structured the environment is, the new levels, and the macroscopic, slow variables describing them, are characterized by intrinsic subjectivity. Hence the process producing these variables is perhaps best viewed as a locally optimized collective computation performed by system components in their search for configurations that reduce environmental uncertainty. If this view is correct, identifying important, functional macroscopic variables in biological systems will require an understanding of biological computation. I will discuss how we can move toward identifying laws in biology by studying the computation inductively. This includes strategy extraction from data, construction of stochastic circuits that map micro to macro, dimension-reduction techniques to move toward an algorithmic theory for the macroscopic output, methods for quantifying circuit collectivity, and macroscopic tuning and control.

INTRODUCTION

A significant challenge before biology is to determine whether living systems – composed of noisy, adaptive, heterogenous components with only partly aligned interests – are governed by principles or laws operating on universal quantities that can be derived from microscopic processes or reflect contingent events leading to irreducible complexity (Gell-Mann and Lloyd, 1996; Goldenfeld, 1999; Krakauer and Flack, 2010; Krakauer et al., 2011; Flack et al., 2013). We know the answer to this question for physical systems and it is useful to recall that understanding in physics was achieved only after extensive debate. This debate began with the observation that certain average quantities – temperature, pressure, entropy, volume, and energy – exist at equilibrium in fundamental relationship to each other, as expressed in the ideal gas law. This observation led to thermodynamics, an equilibrium theory treating aggregate variables. When these variables were derived from first principles using statistical mechanics – a dynamical theory treating microscopic variables – the debate about whether regularities at the macroscopic scale were fundamental was partly resolved; by providing the microscopic basis for the macroscopic variables of thermodynamics, statistical mechanics established the conditions under which the equilibrium relations are no longer valid or expected to apply.

This brief summary of the relation between thermodynamics and statistical mechanics in physics is illuminating because it raises the possibility of a potentially deep division between physical and biological systems. So far, and admittedly biology is young, biology has had limited success discovering relationships among macroscopic variables and deriving these variables from first principles rooted in physical laws or deep evolved constraints. Two areas in which there has been success include metabolic scaling (West and Brown, 2005; West et al., 1997) and neural coding (Schneidman et al., 2006; Tkavcik et al., 2013). Both allometric scaling and neural coding theories help to explain how large populations of cells achieve a coordinated maximization of a group-level property. For scaling, this property is the maximization of efficient metabolic energy use, and for neural coding the most efficient information extraction from environmental inputs. In the scaling case in particular the theory has served to validate that the macroscopic observation that mass scales with metabolic rate to the 3/4 power has a basis in mechanism – in other words the macroscopic variables can be said to be fundamental-obeying laws – rather than nominal (Krakauer and Flack, 2010), and in that sense are getting closer to temperature, pressure, entropy, etc.

Although the idea of optimization subject to simple constraints can explain a surprisingly broad range of quantifiable variation in evolved systems, from physiology to ecology, and from genetics to development (e.g., Beggs, 2008; Couzin, 2009; Frank, 2013; Shriki et al., 2013; West and Brown, 2005), the number of fundamental macroscopic variables known in biological systems remains small and limited to cases like scaling where energy plays a direct role in influencing structure. Whether there is some null expectation for how many we might expect in biological systems given the complexity of the microscopic behavior is a totally open question and perhaps not yet well posed.

THE IMPORTANCE OF INFORMATION
Beyond the obvious heterogeneity another possible reason for the so-far limited progress identifying biological laws or principles is that biological systems are collective, computational, and involve information processing (Couzin et al., 2005; Flack, 2012, 2014; Flack and Krakauer, 2011; Flack et al., 2013; Hartwell et al., 1999; Krakauer et al., 2011; Levin, 1992; Mitchell, 2011; Vetsigian et al., 2006; von Dassow et al., 2000; Walker and Davies, 2013; Yuh et al., 1998). Examples include swarming behavior in social insects (Conradt and Roper, 2005), feature detection in the peripheral visual pathway (Olshausen and Field, 1996), and voting behavior or consensus formation in primate groups (Flack and Krakauer, 2006; Brush et al., 2013). These examples hint at limitations in the scope of application of traditional physical theories. In each of these systems the basic elements,

i.e., ants, neurons, or primates, often live in relatively small populations, are adaptive with interaction rules that are subject to learning, and have functional properties at the aggregate level that feed back to influence the decision-making rules or strategies guiding behavior.

We have proposed that in systems like these functionally important macroscopic properties arise as heterogenous, adaptive components extract regularities from their environments to reduce uncertainty. This facilitates adaptation, thereby promoting survival and reproduction. Hence in biological systems functional macroscopic properties are *constructed* over evolutionary, developmental, and ecological time.

In our work, which we review briefly in Examples, we have shown that the collective effects of this regularity extraction can be captured with coarse-grained (or compressed) variables; endogenous coarse-graining represents the average perceived regularity in the environment at the microscopic level. As estimates of the regularities by components (as opposed to the scientist-observers) converge with exposure to larger data samples, the coarse-grained variables consolidate, providing a new effective background against which components tune strategies, and creating new space and time scales. This estimation and consensus assessment process makes such variables subjective, perhaps nonstationary, and inferential in character.

This recasting of the evolutionary process as an inferential one (Bergstrom and Rosvall, 2011; Krakauer, 2011) is based on the premise that organisms and other biological systems can be viewed as hypotheses about the present and future environments they or their offspring will encounter, induced from the history of past environmental states they or their ancestors have experienced (e.g., Crutchfield and Feldman, 2001; Krakauer and Zanotto, 2009). This premise, of course, holds only if the past is prologue – that is, has regularities, and the regularities can be estimated and even manipulated (as in niche construction) by biological systems or their components to produce adaptive behavior (Flack et al., 2013).

If these premises are correct, life at its core is computational, and a central question becomes: How do systems and their components estimate and control the regularity in their environments and use these estimates to tune their strategies? I suggest that the answer to this question, and the explanation for complexity, is that biological systems manipulate spatial and temporal structure to produce order – low variance – at local scales.

UNCERTAINTY REDUCTION

With these ideas in mind let's return for a moment to the question of a biological laws. Biological systems – from cells to tissues to individuals to societies – have nested organizational levels (e.g., as reviewed in Maynard Smith and Szathmary, 1998). These levels can be quantitatively described by their space and time scales, and each new level has associated with it some new or emergent function – a new feature with positive payoff consequences for the system as a whole or for its components (Flack et al., 2013). This hierarchical organization can be thought of as a nesting of functional encodings of lower-level regularities. As I argue in this chapter and elsewhere (Flack, 2012; Flack et al., 2013), these functional encodings form an *information hierarchy* (see also Walker and Davies, 2013) that results from biological systems manipulating space and time to reduce uncertainty, thereby facilitating efficient extraction of energy, promoting adaptation.

When macroscopic variables describing these levels are not directly tied to energetic constraints, as in the scaling case, but have a profoundly informational character – arising only as component estimates of regularities converge – they may not be obvious a priori from observation at the aggregate level. Discovery of the principles and possibly laws governing biological systems in these cases presumably requires we take information processing and the subjective nature of regularity extraction seriously.

To find the coarse-grainings favored by the system, we need to adopt its perspective. One way to do this is to proceed inductively, working upward from the data, starting with dynamical many-body formalisms and finding, through empirically grounded simulation and modeling, equilibrium descriptions with a few favored macroscopic degrees of freedom.

Hence I am proposing that to identify fundamental information quantities of biology, we start by identifying provisional macroscopic properties thought to have functional consequences for components, but instead of next looking for equilibrium relationships among these variables, as in physics and in the scaling case, ask instead whether

1. these provisional variables can be derived from microscopic data on strategic interactions known to be important in the system, and
2. they are tunable and 'readable' by components (hence functional) individually or collectively in evolutionary or ecological time.

If we can establish that the provisional macroscopic variables satisfy these criteria, they become good candidate fundamental *biological* variables, and the search for law-like relationships among them may be more straightforward.

EXAMPLES

To make these ideas more concrete, let's consider some examples. The first comes from my own work on conflict management in animal societies (for a review of this work, see Flack, 2012) – specifically, third-party policing in primate groups.

Policing, a form of conflict management in which an individual breaks up fights among other individuals, is the new or emergent function. The provisional macroscopic property supporting this new function is the distribution of social power, where power is operationalized as the degree of consensus in the group that an individual can win fights (see Brush et al., 2013, and references therein). When the power structure becomes effectively institutionalized (here meaning associated with a relatively slow time scale and hard to change because in order for an individual's power to change many opinions about fighting ability need change), it becomes a good predictor of the future cost of social interaction and provides information to the individuals about the kinds of conflict and conflict management behavior they can afford given how power is distributed. When the distribution is heavy tailed, policing, which is an intrinsically costly strategy, becomes affordable, at least to those individuals in the distribution's tail. These are the super-powerful monkeys who are rarely or never challenged when they break up fights (Flack et al., 2006, 2005).

A primary driver of the emergence of new functionality such as policing is the reduction of environmental uncertainty through the construction of nested dynamical processes with a range of characteristic time constants (Flack, 2012; Flack et al., 2013). In the case of the monkeys, a slowly changing status signaling network that sums up the outcomes of fights arises from the conflict interactions and encodes an even more slowly changing power structure. These nested dynamical processes arise as components extract regularities from fast, microscopic behavior by coarse-graining (or compressing) the history of events to which they have been exposed. So the monkeys coarse-grain over their fight histories with other individuals to figure out who in a pair will likely win the fight. The collective coarse-grained assessment, which changes yet more slowly, of who can win fights gives the consensus in the group about who has power and provides the basis for the power distribution (Brush et al., 2013).

Proteins offer another example from a very different level of biological organization. Proteins can have a long half-life relative to RNA transcripts and can be thought of as the summed output of translation. Cells have a long half-life relative to proteins and are a function of the summed output of arrays of spatially structured proteins. Both proteins and cells represent some average measure of the noisier activity of their constituents. Similarly, a pig-tailed macaque's estimate of its power is a kind of average measure of the collective perception in the group that the macaque is capable of winning fights, and this a better predictor of the cost the macaque will pay during fights than the outcome of any single melee, as these outcomes can fluctuate for contextual reasons. These coarse-grainings, or averages, are *slow variables* (Flack, 2012; Flack and de Waal, 2007; Flack et al., 2013; see also Feret et al., 2009, for a similar idea). Slow variables may have a spatial component as well as a temporal component, as in the protein and cell examples, or, minimally, only a temporal component, as in the monkey example.

The basic idea is that as a consequence of integrating overabundant microscopic processes, slow variables provide better predictors of the local future configuration of a system than the states of the fluctuating microscopic components. In doing so, they promote accelerated rates of microscopic adaptation. Slow variables facilitate adaptation in two ways: they allow components to fine-tune their behavior and free components to search at low cost a larger space of strategies for extracting resources from the environment (Schuster and Fontana, 1999; Rodriques and Wagner, 2009; Flack and de Waal, 2007; Flack, 2012; Flack et al., 2013). This phenomenon is illustrated by the power-in-support-of-policing example and also by work on the role of neutral networks in RNA folding. In the RNA case, many different sequences can fold into the same secondary structure. This implies that over evolutionary time, structure changes more slowly than sequence, thereby permitting sequences to explore many configurations under normalizing selection (e.g., Schuster and Fontana, 1999).

SLOW VARIABLES TO FUNDAMENTAL MACROSCOPIC PROPERTIES

As an interaction or environmental history builds up at the microscopic level, the coarse-grained representations of the microscopic behavior consolidate, becoming for the components increasingly robust predictors of the system's future state – the slow variables become fundamental macroscopic properties. We speak of a new organizational level when

1. the system's components rely to a greater extent on these coarse-grained or compressed descriptions of the system's dynamics for adaptive decision-making than on local fluctuations in the microscopic behavior *and*

2. when the course-grained estimates made by components are largely in agreement (Flack et al., 2013).

The idea is that convergence on these 'good-enough' estimates underlies nonspurious correlated behavior among the components. This, in turn, leads to an increase in local predictability and drives the construction of the information hierarchy. (Note that increased predictability can give the appearance of downward causation in the absence of careful analysis of the bottom-up mechanisms that actually produced it; see also Walker and Davies, 2013). The probability of estimate convergence should increase as the sample size grows, if the computational capacities of the components are similar, and through a feedback amplification process as new organizational levels consolidate.

CHALLENGES
Biology as Collective Computation

If, as I am arguing, life is an information hierarchy that results from biological components collectively estimating environmental or social regularities by coarse-graining or compressing time-series data, a natural (and complementary) approach is to treat the micro and macro mapping as a computation.

Describing a biological process as a computation minimally requires that we are able to specify the output, the input, and the algorithm or circuit connecting the input to the output (Flack and Krakauer, 2011; see also Mitchell, 2011; Valient, 2013). A secondary concern is how to determine when the desired output has been generated. In computer science this is called the termination criterion or halting problem.

In biology it potentially can be achieved by constructing nested dynamical processes with a range of time scales, with the slower time-scale processes (the slow variables) providing the 'background' against which the fitness of a component using a given strategy is evaluated (Flack and Krakauer, 2011). The idea is that the system makes a prediction based on its prior experience in this stable, slowly changing environment about which strategy will increase its fit (measured, for example, in terms of mutual information or fitness) to the environment and tunes its behavior to implement the strategy.

As an example, consider the distribution of social power discussed above. Because the DSP is slowly changing (compared with the interaction and status signaling rates at the individual level), it provides a stable background against which the monkeys can "predict the future cost of interaction" and hence tune their behavioral strategies. It is this ability to predict, derived from the degree of time-scale separation between the power distribution and underlying distribution of fighting abilities, that reveals when the computation is "correct." Here, in contrast to computer science, the output is continuously computed and the notion of correctness comes from the utility of the output for prediction.

A macroscopic property can be said to be an output of a computation if it can take on values that have functional consequences at the group or component level, is the result of a distributed and coordinated sequence of component interactions under the operation of a strategy set, and is a stable output of input values that converges (terminates) in biologically relevant time (Flack and Krakauer, 2011). Examples studied in biology include aspects of vision, such as edge detection (Olshausen and Field, 1996); phenotypic traits, such as the average position of cells in the developing endomesoderm of the sea urchin (e.g., Peter and Davidson, 2011); switching in biomolecular signal-transduction cascades (e.g., Smith et al., 2011); and social structures, such as the distribution of fight sizes (e.g., DeDeo et al., 2010; Flack and Krakauer, 2011) and the distribution of power in monkey societies (e.g., Brush et al., 2013).

The input to the computation is the set of elements implementing the rules or strategies. As with the output, we do not typically know a priori which of many possible inputs is relevant, and so we must make an informed guess based on the properties of the output. In the case of a well-studied phenotypic trait such as the development of a sea urchin's endomesoderm, we might start with a list of genes that have been implicated in the regulation of cell position. In the case of the distribution of fight sizes in a monkey group, we might start with a list of individuals participating in fights.

In a biological system, the input plus the strategies constitute the system's microscopic behavior. There are many approaches to reconstructing the system's microscopic behavior from raw data. The most powerful is an experiment in which upstream inputs to a target component are clamped off and the output of the target component is held constant. This allows the experimenter to measure the target component's specific contribution to the behavior of a downstream component.

When such experiments are not possible, causal relationships can be identified using time-series analysis in which clamping is approximated statistically. My collaborators and I have developed a novel computational technique, called *inductive game theory* (IGT) (DeDeo et al., 2010; Flack and Krakauer, 2011), that uses a statistical clamping principle to extract strategic decision-making rules, game structure, and (potentially) strategy cost from time-series data. (IGT is one of many approaches being developed in a growing body of literature on causal network reconstruction from time-series and correlation data.)

In all biological systems, of course, there are multiple components interacting and simultaneously coarse-graining to make predictions about the future. Hence the computation is inherently collective. A consequence of this is that it is not sufficient to simply extract from the time series the list of the strategies in play. We must also examine how different configurations of strategies affect the macroscopic output. One way these configurations can be captured is by constructing Boolean circuits describing activation rules, as illustrated by the work on echinoderm gene regulatory networks controlling embryonic cell position (the output) in the sea urchin (Peter and Davidson, 2011). In the case of our work on micro to macro mappings in animal societies, we describe the space of microscopic configurations – fight decision-making rules – using Markovian, probabilistic, social circuits (DeDeo et al., 2010; Flack and Krakauer, 2011).

Nodes in the gene regulatory circuits and social circuits described above are the input to the computation. As discussed above, the input can be genes, neurons, individuals, subgroupings of components, etc. A directed edge between two nodes in the circuit indicates that the "receiving node" has a strategy for the "sending node," and the edge weight can be interpreted as the probability that the sending node plays the strategy in response to some behavior by the receiving node in a previous time step. Hence, an edge in these circuits quantifies the strength of a causal relationship between the behaviors of a sending and receiving node.

Sometimes components have multiple strategies in their repertoires. Which strategy is being played at time t may vary with context. These metastrategies can be captured in a circuit using different types of gates specifying how a component's myriad strategies combine (DeDeo et al., 2010; Flack and Krakauer, 2011; see also Feret et al., 2009). By varying the types of gates and/or strength of allowed causal relationships, we end up with multiple alternative circuits – a family of circuits – all of which are consistent with the microscopic behavior, albeit with different degrees of precision. Each circuit in the family is essentially a model of the micro-macro relationship and so serves as a hypothesis for how strategies combine over nodes (inputs) to produce the target output. By testing the empirically parameterized circuits against each other in simulation we can determine which best recovers the actual measured macroscopic behavior of the study system and in this way discover if our provisional macroscopic variable may indeed be a candidate fundamental variable.

Circuit Logic
The circuits describing the microscopic behavior can be complicated, with many 'small' causes detailed, as illustrated by the gene regulatory circuits constructed by Eric Davidson and his colleagues.

The challenge once we have rigorous circuits is to figure out the circuit logic.

There are many ways to approach this problem. Our approach is to build what's called in physics an effective theory: a compact description of the causes of a macroscopic property. Effective theories for biological systems composed of adaptive components require an additional criterion beyond compactness. As discussed earlier in this chapter, components in these systems are tuning their behaviors based on their own effective theories – coarse-grained (or compressed) rules (see also Feret et al., 2009) that capture the regularities they perceive. If we are to build an effective theory that explains the origins of functional space and time scales, new levels of organization, and ultimately the information hierarchy, the effective theory must be consistent with *component models of average system behavior*, as these models guide component strategy choice. In other words, our effective theory should explain how the system itself is computing (see also Walker and Davies, 2013).

My collaborators and I begin the search for cognitively principled, *algorithmic* effective theories using what we know about component cognition to inform how we coarse-grain and compress the circuits (Daniels et al., 2012). This means taking into account, given the available data, the kinds of computations components can perform and the error associated with these computations at the individual and collective levels, given component memory capacity and the quality of the data sets components use to estimate regularities (Krakauer et al., 2010).

Information and Energy in Biology

As discussed above, once we have a satisfactory family of candidate effective theories for how our system is computing its macroscopic output, we need to choose from among them the one that best recovers our observable and is also the most mechanistically principled. Generally we require two criteria be met to claim a model (or circuit) is mechanistically principled. The model must capture causal relationships supported by the data – how individual strategies or some simplification of them actually combine to produce the output. The second criterion is that the model must be cognitively or computationally parsimonious. This requires knowing something about the cognitive or computational burden that each of the models assumes of the system and its components (Krakauer et al., 2010). In other words, we need to be able to measure the number of bits required to parameterize each model for a given level of performance, and for most systems we study in biology the number of allowable (given component computational capacity) bits should (probably) be relatively small.

The Informational Cost of Biological Computation

Calculating for a reasonably sized data set the number of bits required to perform the computation is generally achievable and allows comparison across the models within the study, but it is not at all clear what it means in an absolute sense to say that a particular theory or compressed representation of the system behavior requires x number of bits to encode.

For example, in our work on monkey conflict dynamics (Daniels et al., 2012), we found that about 1,000 bits of information are required to encode which individuals and subgroups are regular and predictable participants in fights, assuming a sparse coding algorithm. Our other models performed worse (required more bits). In this sense our bits measurement was a useful bar against which to compare models, but we cannot yet claim to have any idea whether a model requiring 1,000 bits is reasonable given our subjects' cognitive capacities. Establishing this requires an experimental approach.

Another open question includes how the informational cost of computation changes when the computation is collective. Work on robustness (e.g., Ay et al., 2007) and distributed computation suggests that reliability of the output may increase, but this work does not explicitly address how variance in the output at the component level affects the number of bits required to encode the collective computation. Hence the output may be more reliable, but the total informational cost could be much higher. This may not matter if there are strong, shared constraints on the component computations and/or the tuning is only at the component level.

Bits to Joules and the Energetic Cost of Computation

We understand the relationship between energy and information in the limit, as Landauer's principle tells us that there is a minimum amount of energy required to erase one bit of information (Bennett, 2003). However, even if we could in a compelling model-free way quantify the number of bits required to encode a model or make a decision, we have no idea how bits translate into watts in the adaptive, stochastic, information-processing, many-body systems of biology and society, or how this question could be approached empirically.

Yet from the perspective of evolutionary theory it seems likely that information processing is adaptive – meaning it allows biological systems to more efficiently, given constraints, extract energy and do the work required to promote survival and reproduction. Another way of putting this is that even if physical theory says information-processing is energetically costly (Parrondo et al., 2015), evolutionary theory suggests that, in the long run, or given an understanding of the full set of constraints to which a biological system is subject, information-processing *saves* watts. Unpacking this proposition may be the key to go from the information-processing mechanisms producing slow variables and emergent function to the identification of biological laws.

COLLECTIVE COMPUTATION TO STRATEGIC STATISTICAL MECHANICS FOR MANY-BODY, ADAPTIVE SYSTEMS

As understanding of the micro-macro mapping is refined through identification of cognitive effective theories that parsimoniously reduce circuit complexity and compactly encode the macroscopic output, we also refine our understanding of the natural scales of the system. This includes distinguishing strategic microscopic behavior from noise, and hence allows us to extract from (rather than imposing on) the raw data the building blocks of our system. And by investigating whether our best-performing empirically justified circuits can also account for other functionally important macroscopic properties, we can begin to establish which macroscopic properties might be biologically fundamental and whether they stand in law-like relation to one another. One can think of this approach as a *strategic statistical mechanics*, embedding complex decision rules in formalisms for calculating emergent properties and discovering law-like behavior at the aggregate level.

These ideas are very closely related to the pioneering ideas of John von Neumann (D. C. Krakauer, personal communication), who in the 1940s and 1950s began the development of a statistical mechanics of biologically inspired sensing and computing devices (von Neumann, 1987). He writes (von Neumann, 1954):

> anything that can be exhaustively and unambiguously described, anything that can be completely and unambiguously put into words, is *ipso facto* realizable by a suitable finite neural network ... we get an image of the strong limitations that our sensations, our intuitions, our logic and our language have to obey. We can put all these things in a more complete statement: The following restrictions are mutually equivalent: to be macroscopic; to be Euclidean (i.e. to adopt the parallel axiom in the way we represent space and spatial relations); to be Galileo-Newtonian in the way we represent motion, time and energy; to capture the surrounding and to act according to our sensorial-intuitive perception of reality; to use and to represent language, in both its natural and artificial variants.

Where von Neumann speaks of neural networks constrained by sensorial perception of classical observations, we consider (D. C. Krakauer, personal communication) how strategic, behavioral rules (de Waal, 1991; Flack et al., 2004), when combined into stochastic circuits expressed in the language of Markov decision processes, produce and respond to coarse-grained aggregate information through near-critical system states. In less technical language, we explore chains of probabilistic events – decisions or state transitions – that generate and respond efficiently to average features of the world. This enables these systems to tune adaptively to the needs of their social environment. My collaborators and I believe that to develop a statistical mechanics that can accommodate these kinds of adaptive, strategic systems we will need to extend existing physical theories by incorporating ideas from theoretical computer science, information theory, and evolutionary biology.

ACKNOWLEDGMENTS

These ideas were developed in collaboration with David Krakauer, who would no doubt state them some other way. Many of the ideas also developed during research projects and conversations with our postdocs and students – Bryan Daniels, Chris Ellison, Eleanor Brush, Eddie Lee, Simon DeDeo, Philip Poon, and Jess Banks – and with SFI colleagues – Eric Smith, Geoffrey West, Jim Crutchfield, and, especially, Walter Fontana. I also thank Sara Walker and Paul Davies for organizing the meeting on information that triggered this chapter. I acknowledge support from two grants to the Santa Fe Institute (SFI) from the Templeton Foundation to study complexity, a grant from the Templeton Foundation to the SFI to study the mind–brain problem, and NSF grant no. 0904863.

REFERENCES

Ay, N., Flack, J. C., and Krakauer, D. C. 2007. Robustness and complexity co-constructed in mulit-modal signaling networks. *Philosophical Transactions of the Royal Society, London, Series B*, **362**, 441–447.

Beggs, J. M. 2008. The criticality hypothesis: how local cortical networks might optimize information processing. *Philosophical Transactions. Series A, Mathematical, Physical, and Engineering Sciences*, **366**(1864), 329–343.

Bennett, C. H. 2003. Notes on Landauer''s principle, reversible computation, and Maxwell's demon. *Studies in the History and Philosophy of Modern Physics*, **34**, 501–510.

Bergstrom, C. T., and Rosvall, M. 2011. The transmission sense of information. *Biology and Philosophy*, **26**, 159–176.

Brush, E. R., Krakauer, D. C., and Flack, J. C. 2013. A family of algorithms for computing consensus about node state from network data. *PLOS Computational Biology*, **9**(7), e1003109.

Conradt, L., and Roper, T. J. 2005. Consensus decision making in animals. *Trends in Ecology & Evolution*, **20**(8), 449–456.

Couzin, I. D. 2009. Collective cognition in animal groups. *Trends in Cognitive Sciences*, **13**(1), 36–43.

Couzin, I. D., Krause, J., Franks, N. R., and Levin, S. A. 2005. Effective leadership and decision-making in animal groups on the move. *Nature*, **433**(7025), 513–516.

Crutchfield, J. P., and Feldman, D. P. 2001. Synchronizing to the environment: information-theoretic constraints on agent learning. *Advances in Complex Systems*, **4**, 251–264.

Daniels, B. C., Krakauer, D. C., and Flack, J. C. 2012. Sparse code of conflict in a primate society. *Proceedings of the National Academy of Sciences*, **109**(35).

De Waal, F. B. M. 1991. The chimpanzee's sense of social regularity and its relation to the human sense of justice. *American Behavioral Scientist*, **34**, 335–349.

DeDeo, S., Krakauer, D. C., and Flack, J. C. 2010. Inductive game theory and the dynamics of animal conflict. *PLOS Computational Biology*, **6**(5), e1000782.

Feret, J., Danos, V., Krivine, J., Harmer, R., and Fontana, W. 2009. Internal coarse-graining of molecular systems. *Proceedings of the National Academy of Sciences of the United States of America*, **106**(16), 6453–6458.

Flack, J. C. 2012. Multiple time-scales and the developmental dynamics of social systems. *Philosophical Transactions of the Royal Society B: Biological Sciences*, **367**(1597), 1802–1810.

Flack, J. C. 2014. Life's information hierarchy. *Santa Fe Institute Bulletin*, **28**, 13.

Flack, J. C., and de Waal, F. 2007. Context modulates signal meaning in primate communication. *Proceedings of the National Academy of Sciences of the United States of America*, **104**(5), 1581–1586.

Flack, J. C., and Krakauer, D. C. 2006. Encoding power in communication networks. *American Naturalist*, **168**(3).

Flack, J. C., and Krakauer, D. C. 2011. Challenges for complexity measures: a perspective from social dynamics and collective social computation. *Chaos: An Interdisciplinary Journal of Nonlinear Science*, **21**(3), 037108.

Flack, J. C., Jeannotte, L. A., and de Waal, F. B. M. 2004. Play signaling and the perception of social rules by juvenile chimpanzees (*Pan troglodytes*). *Journal of Comparative Psychology*, **118**, 149–159.

Flack, J. C., de Waal, F. B. M, and Krakauer, D. C. 2005. Social structure, robustness, and policing cost in a cognitively sophisticated species. *American Naturalist*, **165**(5).

Flack, J. C., Girvan, M., de Waal, F. B. M, and Krakauer, D. C. 2006. Policing stabilizes construction of social niches in primates. *Nature*, **439**(7075), 426–429.

Flack, J. C., Erwin, D., Elliot, T., and Krakauer, D. C. 2013. Timescales, symmetry, and uncertainty reduction in the origins of hierarchy in biological systems. *Cooperation and Its Evolution*, 45–74.

Frank, S. A. 2013. Input-output relations in biological systems: measurement, information and the Hill equation. *Biology Direct*, 1–25.

Gell-Mann, M., and Lloyd, S. 1996. Information measures, effective complexity, and total information. *Complexity*, **2**(1), 44–52.

Goldenfeld, N. 1999. Simple lessons from complexity. *Science*, **284**(5411), 87–89.

Hartwell, L. H., Hopfield, J. J., Leibler, S., and Murray, A. W. 1999. From molecular to modular cell biology. *Nature*, **402**, C47–C52.

Krakauer, D. C. 2011. Darwinian demons, evolutionary complexity, and information maximization. *Chaos*, **21**, 037110.

Krakauer, D. C., and Flack, J. C. 2010. Better living through physics. [Letter to the Editor]. *Nature*, **467**(7316), 661.

Krakauer, D. C., and Zanotto, P. 2009. Viral individuality and the limitations of life concept. Pages 513–536 of Rasmussen, M. A., et al. (eds.), *Protocells: bridging non-living and living matter*. MIT Press.

Krakauer, D. C, Flack, J. C., Dedeo, S., Farmer, D., and Rockmore, D. 2010. Intelligent data analysis of intelligent systems. Pages 8–17 of Cohen, P. R., Adams, N. M., and Berthold, M. R. (eds), *Advances in intelligent data analysis IX*. Springer.

Krakauer, D. C., Collins, J. P., Erwin, D., Flack, J. C., Fontana, W., Laubichler, M. D., Prohaska, S. J., West, G. B., and Stadler, P. F. 2011. The challenges and scope of theoretical biology. *Journal of Theoretical Biology*, **276**(1), 269–276.

Levin, S. A. 1992. The problem of pattern and scale in ecology. *Ecology*, **73**(6), 1943.

Maynard Smith, J., and Szathmary, E. 1998. *The major transitions in evolution*. Oxford University Press.

Mitchell, M. 2011. What is computation? Biological computation. *Ubiquity*, February, 1–7.

Olshausen, B. A., and Field, D. J. 1996. Emergence of simple-cell receptive field properties by learning a sparse code for natural images. *Nature*, **381**, 607.

Parrondo, J. M. R, Horowitz, J. M., and Sagawa, T. 2015. Thermodynamics of information. *Nature Physics*, **11**, 131–139.

Peter, I. S., and Davidson, E. H. 2011. A gene regulatory network controlling the embryonic specification of the endoderm. *Nature*, **474**, 635–639.

Rodriques, J. F. M., and Wagner, A. 2009. Evolutionary plasticity and innovations in complex metabolic reaction networks. *PLOS Computational Biology*, **5**, e1000613.

Schneidman, E., Berry, M. J., Segev, R., and Bialek, W. 2006. Weak pairwise correlations imply strongly correlated network states in a neural population. *Nature*, **440**(7087), 1007–1012.

Schuster, P., and Fontana, W. 1999. Chance and necessity in evolution: lessons from RNA. *Physica D: Nonlinear Phenomena*, **133**, 427–452.

Shriki, O., Alstott, J., Carver, F., Holroyd, T., Henson, R. N. A., Smith, M. L., Coppola, R., Bullmore, E., and Plenz, D. 2013. Neuronal avalanches in the resting MEG of the human brain. *Journal of Neuroscience*, **33**(16), 7079–7090.

Smith, E., Krishnamurthy, S., Fontana, W. D., and Krakauer, D. C. 2011. Nonequilibrium phase transitions in biomolecular signal transduction. *Physical Review E*, **E84**, 051917.

Tkavcik, G., Marre, O., Mora, T., Amodei, D., Berry, M. J. II, and Bialek, W. 2013. The simplest maximum entropy model for collective behavior in a neural network. *Journal of Statistical Mechanics: Theory and Experiment*, **2013**(03), P03011.

Valient, L. 2013. *Probably approximately correct*. Basic Books.

Vetsigian, K., Woese, C., and Goldenfeld, N. 2006. Collective evolution and the genetic code. *Proceedings of the National Academy of Sciences of the United States of America*, **103**(28), 10696–10701.

Von Dassow, G., Meir, E., Munro, E. M., and Odell, G. M. 2000. The segment polarity network is a robust developmental module. *Nature*, **406**(6792), 188–192.

Von Neumann, John. 1954. The general and logical theory of automata. In Taub, A. H. (ed.), *John von Neumann: collected works*. Pergamon Press.

Von Neumann, John. 1987. *Papers of John von Neumann on computing and computer theory*. MIT Press.

Walker, S. I., and Davies, P. C. W. 2013. The algorithmic origins of life. *Journal of the Royal Society, Interface*, **10**(December 2012), 20120869.

West, G. B., and Brown, J. H. 2005. The origin of allometric scaling laws in biology from genomes to ecosystems: towards a quantitative unifying theory of biological structure and organization. *Journal of Experimental Biology*, **208**(Pt 9), 1575–1592.

West, G. B., Brown, J. H., and Enquist, B. J. 1997. A general model for the origin of allometric scaling laws in biology. *Science*, **276**(5309), 122–126.

Yuh, C., Bolouri, H., and Davidson, E. H. 1998. Genomic cis-regulatory logic: experimental and computational analysis of a sea urchin gene. *Science*, **279**(5358), 1896–1902.

13 Living through Downward Causation

From Molecules to Ecosystems

Keith D. Farnsworth, George F. R. Ellis, and Luc Jaeger

Downward causation (first defined by Campbell, 1974) is both a philosophical concept and an apparent phenomenon of nature attracting great controversy. Most scientists usually assume that all observable phenomena derive from elemental fundamental physics, so that even human behaviours ultimately result from interactions of subatomic particles, via a unidirectional chain of causes and effects. On closer inspection, the act of living seems able to spontaneously generate events, breaking this chain; it is as though life possessed 'free will' by acting without a prior physical cause. In this chapter, we analyse this puzzling behaviour using information and control theory as a general framework, applying it to a range of scales of organisation in biological systems: from the molecular to the ecological. An essential element (and possibly a defining feature) of life emerges from this analysis. It is the presence of downward causation by information selection and control. Through a series of examples, we show how this phenomenon works to produce the appearance of autonomous action from information constructed and maintained by the process of living. After a brief introduction to the concept of downward causation, we set it more firmly within the concepts of biological information processing used within this volume. From this we attempt to derive a general classification of causation across scales of biological organisation. We show how selection from random processes and information embodiment in molecules, organism systems, and ecological systems combine to emerge with the properties of downward causation and the appearance of autonomy. These phenomena seem to be exclusive to life.

CAUSATION

What exactly do we mean when we say A causes B? Causal power is attributed to an agency that can influence a system to change outcomes, but does not necessarily itself bring about a physical change by direct interaction with it. In an easily grasped analogy, the Mafia boss says his rival must be permanently dealt with (the boss has causal power), but his henchman does the dirty deed. The action of the henchman is physical and dynamic, and the henchman is logically described at the same ontological level as his victim (it is not the cartel that kills the rival, not the rotten society, not the atoms in the henchman's body, but the henchman himself). A dynamic, physical cause linking agents of the same ontological level is referred to as an effective cause (alternatively, an efficient cause, following Aristotle; Falcon, 2015). So, when a snooker ball strikes another, it causes it to move, and that is an effective cause. But the laws of physics that dictate what will happen when one ball hits another are not effective causes, even though they do have causal power (as such they are called formal causes). It seems that effective cause is always accompanied by a transfer of energy: this is the only way in which a physical change can take place in the physical universe. However, for an effective cause to be realised, noneffective causes must also exercise their power. Auletta et al. (2008) introduce these distinctions as a preamble to explaining top-down causation, and they will be useful for what follows here. By definition, top-down causes are generated at a higher ontological level than that at which they are realised through an effective cause. Without the effective cause, though, nothing could happen.

When something does happen in the nonliving (abiotic) universe, we can trace its cause through a series of direct causal relationships to explain it. For example, a stone in the desert suddenly splits in two because the heat of the sun causes differential expansion, leading to internal stresses; a line of chemical bonds within the stone is the first to give way, weakened by the slow progress of entropy-maximising chemistry, rooted in those same laws of physics as caused the relative weakness of these bonds in the stone. This, however, does not explain why the stone was there in the first place, but for that we can form another chain of physical causal steps. Philosophers struggling with the question of 'free will' call each step in such an explanation one of transient causation. Following these steps back, one always returns to the basic laws of physics and (currently) to the 'Big Bang', from where our understanding runs out: this point is often referred to as the ultimate cause, but for clarity we will call it the primary cause (leaving 'ultimate' to mean the 'absolutely ultimate'). Most philosophers are interested in causes that include the animate, especially the explanation of human actions, and for this, many of them maintain (though do not necessarily agree with) a notion of agent causation, which is the point we arrive at when a living organism appears to create a cause spontaneously, breaking the chain that leads back to fundamental physics (Taylor, 1973). This notion describes the appearance of an action by an agent, the cause of which is the agent itself: it seems to act without a prior cause. Indeed this apparent behaviour is one of the mysteries common to life: it seems to have the ability to spontaneously generate events, as though it possessed free will. Here we are interested in what lies behind such weirdness. We start with a proposition (which may be tested): that the primary cause of anything is information and that the source of this information gives the apparent character of the cause as one of transient, randomly spontaneous, or agent caused.

WHAT IS SPECIAL ABOUT LIFE?

In the case of transient cause, the information at the end of the chain is to be found in the physical rules of the universe. The 'selection' of these particular laws and fundamental constant values, from among all possible, is what makes the universe what it is. As Ellis (2011) points out, the necessary simplicity of the early universe relative to its present state means that this cannot account for all causes. (Indeed, such an explanation would constitute a kind of preformationism – the term used to describe the faulty belief, once held, that living things were already fully formed, but miniature versions, in the zygote.)

In trying to account for what appear to be spontaneous actions, philosophers often come up against what some of them call 'actions without causes', where the chain of transient causes seems to halt. If there really is no cause whatsoever, then two things must be true: first, there must be more than one option (because if there were only one, then the action would be wholly determined and we would step directly on to the next link in a causal chain); second, that there must be no way (mechanism or set of circumstances) for choosing among the options (because if there were, then that 'way' would be sufficient to account for the cause). In this situation, we are left only with a random choice of the path to follow. But randomness is the phenomenon of spontaneously introducing new 'information' into the system, as is the case in quantum physics. Thus the random 'action without a conditioned cause' is itself derived from information, in this case nonstructured, random information, which we call 'information entropy'. This kind of information finds a use as the raw material for selection and therefore pattern formation, from where form, complexity, and function may be developed.

However, philosophers who believe (albeit uncomfortably) in agent causation insist that it cannot be reduced to random action (for then it is clear that the source of causation is not the agent itself, but rather an independent random process). After all, to claim that an agent possesses free will implies that it is the agent's will, and not some extraneous random process, that causes the agent's actions reliably (and not randomly).

Having established (1) the physical foundation of the universe as one source of information for primary cause and (2) randomness (information entropy) as another, we have rejected both of these as sources of primary cause in living things. We now need a third source to explain this most puzzling origin of causation. We might speculate that this one is peculiar to living systems, not found elsewhere, since the laws of physics and the entropy of the universe seem sufficient cause for abiotic phenomena (we exclude human creations, since they are logically dependent on living systems).

Cybernetic Control Systems

The easiest way to appreciate this third source is in fact through a human creation, namely, a thermostatically controlled heating or cooling system, which is a practical example of a cybernetic system. The disarmingly simple but highly significant feature of it is that the thermostat must have a set-point or reference level to which it regulates the temperature. That set-point is a piece of information and what is different about it is that it is not obviously derived from anything else: it is apparently novel information that is not random. Of course, in a heating system, this information was introduced by human intervention, and in the case of an organism's homeostatic system regulating, for example, body temperature or salt concentration, it was presumably selected for by evolution, but both of these explanations conform to the notion of new information being introduced into or by living systems. The set-point is not seen in abiotic systems, though natural dynamic equilibria are abundant, for example, the balance between opposing chemical reactions and the balance between thermal expansion and gravity maintaining a star. For these systems, the equilibrium point is a dynamic 'attractor' and the information it represents derives from the laws of physics; these are not cybernetic systems. Only in life do we see a genuine set-point that is independent, pure, novel, and functional information.

This fascinating observation has inspired a new information-based way of thinking about what life actually is. Several suggestions have been proposed as the necessary and sufficient properties of a system to be living. Perhaps the most fundamental of these is the appearance of downward causation by information, of which the set-point is clearly an example. This creates a conundrum: how can information (which is taken to be insubstantial and nonphysical) influence the physical world? One might call this 'the hard problem of life', in an echo of the 'hard problem of consciousness' (see Chapter 2). It is an important question because it seems that all cases of identified life (including artificial and hypothetical life) have in common the feature of information appearing to control aspects of the physical world. To understand this better, we now introduce the concept of information as a transcendent phenomenon.

Transcendent Complexes and Emergence

Emergence is the appearance of phenomena at some scale of system organisation that is absent from the lower (more elementary) scales within it. For example, an ecosystem goes through cycles of fluctuating productivity, diversity, and so on, but these phenomena are only the consequence of interactions among individual organisms, aggregated conceptually. Because we can make a conceptual model of an ecosystem that displays the same emergent phenomena as the real thing, we suspect that these phenomena do not strictly depend on the material underlying them. More obviously, a computer program such as a word-processor is an observable phenomenon at a scale of organisation greater than the binary bits of data from which it is composed, and it could in principle be implemented not only in any digital electronic computer, but also in hardware composed of mechanical valves and tubes of water, for example, digital plumbing. The Turing machine is a fundamental concept in computer science: it is a conceptual machine that can in principle compute anything that is computable. So when a Turing machine was created using the plastic building toy Lego, it became possible, in principle, to make the word-processor from Lego. Clearly, a word-processor is a phenomenon that is not determined by what it is made from. We shall use the term 'transcendent complex' (TC) for such entities because they transcend the particular method of implementation used. Their properties are that (1) they arise from the interaction among elements and (2) they may be realised (i.e., instantiated) in multiple ways, because they depend only on certain functions of the elements, not their whole nature. TCs are therefore multiply realisable. This concept has been described as classes of functional equivalence (Auletta et al., 2008; Jaeger and Calkins, 2012).

At least some TCs have causal power. There are many examples: the economy motivates organised production; ecological communities drive evolution; a swarm directs the movement of individuals within it. In every case, closer inspection reveals that the causal power arises from the potential for the TC to create a context for the components from which it arose. 'Context' here means the possibility of interaction among information structures that leads to additional function. This functional interaction is the key to emergent causal power.

Elements within an ensemble do not really interact with the TC of that ensemble; they interact only with each other. What makes the interactions act as something larger, controlling the behaviour of the elements, is the additional functionality provided by the particular configuration that gives rise to the TC. The configuration of the ensemble is one that instantiates functional information so that a TC emerges. The TC is the aggregate of this functional information, and it appears at the scale of organisation of the ensemble. It is the difference between a pile of jigsaw pieces and the completed puzzle. The relationship between one piece and the others in the pile is random and without function; that between one piece and the rest of the completed puzzle is special because the piece has a specific place in the configuration. The other pieces give it context: the information that it embodies 'makes sense' when placed in this context. 'Makes sense' here means that the embodied information of the piece forms part of a functional information pattern at the larger scale, once the other pieces are in the functional configuration. It is this functional pattern that is the TC. For a swarm (a flock of birds or shoal of fish) the functional configuration is the ordering of animals at specific distances and orientations (see, e.g., Aoki, 1982). In this pattern, the (translational) movement of any one of them is highly correlated with the others. It forms part of a pattern built at a higher level of organisation; that pattern is functional and its function is one of causal power – the power to direct the movements of the animals. This causal power is implemented through the effective cause of neighbouring animals, but the causal power clearly arises at the higher level of organisation because it equally applies to all of the animals in the swarm. Shortly we will show how the cell is a TC for its components, including its genetic sequences.

This is especially relevant here because information itself is found to be a TC. Physical information is instantiated as the location in space and time of physical things[1] (e.g., atoms) or their excitations relative to one another (e.g., the pattern of different magnetisation on a computer hard disk or the string of amino acids in a DNA molecule).

Clearly, we can conceptualise physical information in ways that do not depend on any particular kind of substrate for its instantiation. Physical information is independent of the components used to embody it in the physical world, but its existence still depends on there being something physical to instantiate it. Additionally, physical information creates phenomena that do not belong to the component parts instantiating it. Therefore physical information is a TC.

We can go further, since functional information is a functional relation between particular structures of physical information. That is to say: functional information is the phenomenon of function arising from one information structure providing the context for another (Farnsworth et al., 2013). Functional information is therefore a property of physical information, one that emerges from interaction among items of physical information. As a consequence, it also is a TC, this time of arrangements of physical information. Since an emergent pattern of a TC is also a TC, we can say that functional information is multiply realisable in terms of the physical location in space and time of physical things relative to one another. The result

---

[1] As a principle, this is often referred to as Landauer's dictum (Landauer, 1992), but it is then taken to be a stronger statement, meant to contrast with the information-first position, 'it from bit', attributed especially to Wheeler. When we say information is physical, we do not need to imply that 'physical' ultimately precedes information.

is that all identified kinds of information (see Chapter 4) are multiply realisable arrangements of matter in space-time.

The 'hard problem of life' then translates to the observation that multiply realisable patterns (information) apparently direct changes in the pattern of organisation in matter and energy at lower levels of organisation so as to achieve some function or purpose (Hartwell et al., 1999). Now we ask: Is this even possible?

What we have learnt about multiple realisation suggests that it is possible, if the level under (apparent) control is not the foundational substrate of existence. In other words, since each level H_n is multiply realisable in terms of a lower level H_{n-1} in the modular organisation of nature, as long as $n > 2$, there is no reason in principle why downward causation (H_n of H_{n-1}) should not be possible, since given that restriction, $H_n - 1$ is also multiply realisable. If a level is multiply realisable, then it cannot be fully deterministic, since by definition it offers more than one option as to how it is realised. That is true whether or not the system is open. Albrecht's suggestion that deterministic chaos can be traced back to quantum uncertainty (Albrecht, 2001) is interesting but not relevant here. It is relevant when we are considering the sources of raw material for information.

It is also noteworthy that many well-informed commentators (ourselves included) say that we do not yet know what H_1 is, or even if a determinable such level exists, so all known levels of natural organisation are potentially available for downward causation.

In a thought-experiment to illustrate the problem, Paul Davies puts it this way: consider the difference between flying a kite and flying a radio-controlled toy plane. The former has physical (effective) causation: it is obvious how a tug on the string directly causes physical changes in the behaviour of the kite, in the sense that all we have are physical forces and these (to a good approximation) obey Newton's laws of motion or, more generally, Hamiltonian mechanics. The toy plane is controlled via a communications channel: the only way you can influence it is via pure information (communicated via modulation of a radio signal), and there is no place for this in the

Hamiltonian description of the system. How, then, can this pure information determine the behaviour of the physical object of the plane? To understand it better, let's simplify even further and think of a remote control that just switches a lamp on and off. The remote control may send a pulse of infrared light to a tuned photodetector that, by the mechanism of receiving it, generates a small electrical current, enough to move an electromagnetic relay switch and turn the lamp on.

It now seems that we have reduced the problem to a continuous chain of physical phenomena, but more careful thought reveals that this is only the case because the whole system was specifically designed to use the information from the remote control in one very particular way. Indeed, it does not matter that the (one bit of) information was sent by light or radio – it could have been by wire – and it is true that switched signals on wires, turning other switches on and off, is the basic hardware ingredient of a digital computer. The at first hidden but essential ingredient of this control by pure information is the design of the system in which it operates. The design gives the control information a context, without which it would have no 'meaning' in the sense that it would not be functional. To be precise, functional information is that which causes a meaningful difference (see Farnsworth et al., 2013): we can now interpret that as meaning it has causal power. The pure information and the regulated system of which it is a part are inseparable. This idea can be generalised to any control system: the information causes physical effects only because it is embodied within the physical system that gives it the context necessary for it to become functional. We may even go so far as to say that design is the embodiment of functional information in the physical form of a system.

For information control to work in a cybernetic system, there must be a part of the system that performs a comparison between the current value and the set-point. This action is one of information processing (e.g., subtraction), which can be performed by a molecular switch, just as well as it can by an electronic switch in a computer. The osmoregulation of a bacterial cell is one of the most basic tasks of homeostasis in biology and makes a good example (reviewed by Wood, 2011). We see immediately that it is far from simple in practice, based not in a single molecule, but having several cascades of molecular switches in operation, each working differently to up-regulate and down-regulate the osmotic potential of the cell. To quote from Janet Wood's review:

> cytoplasmic homeostasis may require adjustments to multiple, interwoven cytoplasmic properties. Osmosensory transporters with diverse structures and bioenergetic mechanisms activate in response to osmotic stress as other proteins inactivate.

To take just one example, channels formed from proteins embedded in the cell membrane literally open and close in direct response to the internal osmotic potential, and these are crucial for relieving excess solvent. Crucially, the set-point is to be found in the shape of these molecules: they are physically constructed so as to embody it. We presume this information arose through natural selection, itself an example of downward causation (Campbell, 1974), as we will argue shortly.

The important point is that we now see how it influences the physical world. It is not a mystery at all: the set-point is embodied in a physical object (the shape of a protein), and this protein shape directly affects the flow of small molecules through the cell membrane. Pure information embodied in molecular structure has causal power within an osmoregulation system. The multiple realisability arises because it is the desired set-point that drives the system, so if this particular mechanism had not been found to work, natural selection would have substituted some other mechanism to attain the same goal, at least to a good approximation. The set-point is therefore the subject of convergent evolution (see, e.g., McGhee, 2011), as it may be represented by any of the various (and perhaps unknown) mechanisms or structures that fulfil the goal. The higher-level functionality of the set-point is the noneffective cause, as defined above, and as such is the system's biological purpose. This illustrates why multiple realisability is the key to understanding biological function in terms of linking higher and lower levels.

GENERALISING DOWNWARD CAUSATION

There is far more to downward causation than cybernetic systems and their set-points. Ellis (2011) counts such systems as one of five different mechanisms of downward causation. More generally, we can recognise causation from one level of aggregation to another in the modular organisation of nature: top-down, bottom-up, and same-level. Following Simon (1981) and Booch (2006), one can identify modular hierarchical structuring as the basis of all complexity, leading to emergent levels of structure and function based on lower level networks. Quoting Ellis (2011):

> Both bottom-up and top-down causation occur in the hierarchy of structure and causation. Bottom-up causation is the basic way physicists think: lower level action underlies higher level behaviour, for example physics underlies chemistry, biochemistry underlies cell biology and so on. As the lower level dynamics proceeds, for example diffusion of molecules through a gas, the corresponding coarse-grained higher level variables will change as a consequence of the lower level change, for example a non-uniform temperature will change to a uniform temperature. However, while lower levels generally fulfil necessary conditions for what occurs on higher levels, they only sometimes (very rarely in complex systems) provide sufficient conditions. It is the combination of bottom-up and top-down causation that enables same-level behaviour to emerge at higher levels, because the entities at the higher level set the context for the lower level actions in such a way that consistent same-level behaviour emerges at the higher level.

Thus one may give a consistent causal narrative using only concepts at that level.

In order of sophistication, the five mechanisms of top-down causation identified in Ellis (2011) are:

1. deterministic, where boundary conditions or initial data in a structured system uniquely determine outcomes
2. nonadaptive information control, where goals determine the outcomes
3. adaptive selection, where selection criteria choose outcomes from random inputs, in a given higher-level context
4. adaptive information control, where goals are set adaptively
5. adaptive selection of selection criteria, probably occurring only when intelligence is involved.

Within this type of framework, quoting Jaeger and Calkins (2012), it was proposed that:

> TDC [top-down causation] by information control and adaptive selection are at the root of converging forces that shape the evolution of living biosystems from the simplest to the most complex levels. Living systems could therefore be defined as self-reproducing systems that function via TDC by information control and adaptive selection. The functions of the cellular operating system that control cellular reproduction and DNA replication are maintained through TDC by information control leading to a converging driving force. ... [Therefore] it is anticipated that functional convergence at the molecular level might not be as rare as initially thought. ... Under the dependency of TDC by information control and adaptive selection, as long as the fundamental functions of reproduction and replication are kept, emergence of novel functions from the bottom-up is possible. ... Darwinian evolutionary processes in living systems are therefore not only ruled from the bottom up but also by fundamental emerging organizational principles that are hierarchically built up and impose necessary constraints from the top down. These principles are the key for defining organic life.

FUNCTIONAL EQUIVALENCE CLASSES AND MODULAR HIERARCHY

Coarse-graining often produces higher-level variables from lower-level variables (e.g., averages of particle behaviours). We would interpret this more specifically by saying that phenomena at a given level of organisation can produce effective higher-level variables, thus creating a higher level of organisation, this being the mechanism generating the modular hierarchy of nature. If we start with a coarse-grained (effective) view at some level, we are denied information about the details at the fine-grained level below. For this reason, many possible states at the lower level may be responsible for what we see at the higher (which is exactly the microstate/macrostate relationship we use in statistical thermodynamics). There are, therefore, multiple realisations of any higher-level phenomenon. The multiple ways of realising a single higher-level phenomenon can be collected together as a class of functional equivalence: a set of states, configurations, or realisations at the lower level, which all produce an identical phenomenon at the higher. A functional equivalence class is by definition the ensemble of entities sharing in common that they perform some defined function. But a phenomenon can be functional only in a particular context, since function is always context dependent (Cummins, 1975). This context is provided by the TC, which organises one or more of the members of one or more functional equivalence classes into an integrated whole having 'emergent properties'. The TC is an information structure composed of the interactions among its components. In practice, these make up the material body, which embodies the information that collectively constitutes the TC. It must be described in terms of functional equivalence classes because it is multiply realisable. Crucially, the TC does not integrate the lower-level components per se it integrates their effects – so the TC emerges from functional equivalence classes, not from the particular structures or states that constitute their members. Specifically, a TC is the multiply realisable information structure that gives lower-level structures the context for their actions to become functional. It is an aggregate phenomenon of functions. For it to exist, a set of components must be interacting to perform these functions; the components must collectively be members of the necessary functional equivalence classes.

For this reason there is a many-to-one relationship between lower-level phenomena and higher-level emergent phenomena. In this we see a subtle shift that generalises the notion of higher levels from mere coarse-grained aggregates of lower levels (e.g., averages) to the definition of functionally equivalent classes, containing the lower levels as members of the class. Given this concept, one can claim that the existence of multiple lower-level phenomena belonging to a functional equivalence class (e.g., those constituting the form of a bacterial cell) indicates that a higher-level phenomenon (indeed, a TC) is dictating the function of the lower-level phenomenon. One might say that the higher level provides a 'design brief' (without implying a designer) for component parts to perform a particular function (Hartwell et al., 1999) and that it does not matter what these parts are, or how they work, as long as they do the job. For more discussion, see Auletta et al. (2008) and Jaeger and Calkins (2012).

Jaeger and colleagues have described functional equivalence in the molecular machinery of cells (Jaeger and Calkins, 2012; Auletta et al., 2008) as well as in RNA molecules (Geary et al., 2011; Grabow et al., 2013; Jaeger et al., 2009). They point to functional equivalence among structurally different molecules, folding patterns of RNA-based functional molecules (e.g., RNase P), and molecular networks, including regulation 'complementation' in which regulatory systems are interchangeable. Indeed, they catalogue a host of examples found in the genetic, metabolic, and regulatory systems of bacterial cells. But functional equivalence is only one of three requirements for what they call top-down causation by information control. In this scheme, a higher level of organisation exercises its control, not via setting boundary conditions but by sending signals (information) to which lower-level components respond. The signals represent changes in the form or behaviour of lower-level elements that would benefit the functioning of the higher level according to some purpose. This is a kind of cybernetic control, which spans scales of organisation (or ontological levels, as these authors refer to them; Auletta et al., 2008; Jaeger, 2015; Jaeger and Calkins, 2012). Within this framework, it is the combination of top-down causation by information control with top-down causation by adaptive selection that enables the exploration of TCs and that according to Auletta et al. (2008) and Jaeger and Calkins (2012) might characterise life. In general, at any given level of biological organisation (other than top and bottom), there will be more than one TC sharing behaviours in common. These TCs may therefore belong to a functional equivalence class for the TC of the next level up in life's hierarchy. A nested hierarchy of functional information structures is formed this way.

QUANTIFYING DOWNWARD CAUSATION

All this may be dismissed as hand-waving philosophising if we are not able to quantitatively describe downward causation. Essentially what we need in order to make quantitative (hence falsifiable) predictions is a measure of directional influence. Effective information is quantitatively defined by mutual information, but this is symmetric, so does not capture the direction of influence: top-down cannot be separated from bottom-up causation using mutual information measures. Bayesian networks (which are implicitly causal) provided an avenue for quantifying information flows used by Ay and Polani (2008) to quantify causation. In the special case of a time-dependent process in which effect follows cause by a finite delay, this direction is relatively easy to identify, and the transfer entropy provides a solution (Schreiber, 2000). It is effectively the mutual information between one data flow, evaluated at time t, and another evaluated at $t + \tau$: it measures directed interactions between two (or more) time-varying pieces of information at earlier and later times by quantifying the effect of changes in one on the probability of change in the other (alternative, related metrics are available too, reviewed by Lungarella et al., 2007). Transfer entropy was used by Walker et al. (2012) to evaluate downward causation within a simple system for which the mean behaviour of an ensemble represents a higher level of organisation (though this coarse-graining does not imply organisation; it is an equivalence class, just as macroscopic thermodynamic variables are in relation to particle dynamics).

Their example was a system of linked logistic dynamical equations, which is a familiar model of population dynamics in ecology. Intuitively, the coarse-grained picture of an ensemble of linked logistic dynamics should follow the individual dynamics composing it, but the individual dynamics used in this study had explicitly built into them an influence from the mean field (a summary of the ensemble behaviour, often used in physics and astrophysics contexts; see Ellis, 2012). Each population $x(t)$ changed by an amount calculated from mixing its own dynamic $(1-e)f(x(t))$ with a term, in proportion to the ensemble dynamic $eM(t)$, the constant fraction e being used to control the mix. This is of course a highly contrived 'toy-model' system, but it captures important features that might describe neural assemblies or bacterial communities. The system generates complicated dynamics, for which some values of e produce downward causation in the sense that the ensemble average dynamic 'organises' the individual dynamics into (roughly) predictable patterns. The transfer entropy is calculated from the probability of $x(t + 1)$, given information about the recent history of $x(t)$ from all the members of the ensemble, and these calculations are possible only when sufficient data are available to compile statistics. That requirement sets a rather difficult task (except in the case of neural networks), but the principle of quantification is established.

BIOLOGICAL EXAMPLES

The Functional Shapes of Biopolymers
Informational biopolymers such as RNAs and proteins, which form the functional modules of living cells, are TCs with causal power. It is well established that the ability of an RNA or protein sequence to fold into a particular three-dimensional shape is crucial for its

function. The formation of a three-dimensional network of atomic interactions is effective cause for the ability of the molecule to carry a function. However, evolution does not select for the three-dimensional shape per se; rather, it selects the function embodied in the structure. Therefore, while proper folding and self-assembly of the RNA (or protein) sequence into stable three-dimensional networks is a prerequisite to function, the RNA (or protein) sequence is allowed to vary its structural information as long as the RNA (or protein) sequence can retain its biological function (e.g., holding a set-point). As such, the structural network can dramatically change in its overall three-dimensional shape or form, as it is the function carried by the RNA (or protein) sequence that is the 'biological purpose', having causal power over the structural network by constraining its evolution. The meaning of 'purpose' in this sense is defined as the context for an effective cause. The notion of function is strictly context dependent: given a particular biological context, it has a 'purpose' for the functioning of the biological system (Cummins, 1975). A function, even if it emerges from the bottom up, without an intended purpose, can become a biological function only once it is integrated (linked) within the whole biological system. As such, the beauty of biological functions is that they can become set-points with adaptability in time and space. This is borne out by the fact that several classes of biological RNAs, essential for life, can adopt significantly different three-dimensional networks of molecular interactions in related organisms, as long as the catalytic function of each class of biopolymers is retained in each organism (Auletta et al., 2008; Jaeger and Calkins, 2012; Jaeger, 2015). This is particularly striking for the mitoribosome, the complex machinery responsible for the synthesis of encoded proteins in the mitochondria of Eukaryotes (see Figure 13.1). From one eukaryotic cell to another, its overall structural morphology and RNA and protein composition can significantly vary as long as its overall function is maintained. As such, large portions of the mitoribosome structural network, which rely on RNA elements in yeast (Amunts et al., 2014), are substituted by ribosomal protein interactions in

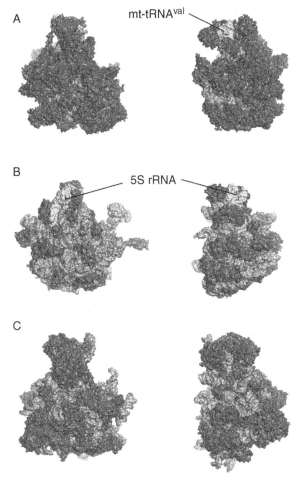

FIG. 13.1 Comparison of the overall morphologies of (A) the human mt7LSU, (B) the bacterial LSU, and (C) the yeast mt7LSU. For each ribosome, the solvent facing surface (left) and side view (right) are shown. Proteins are shown in darker grey, and rRNA in light grey. The architecture of the human mt7LSU is substantially different from both bacterial ribosomes and the yeast mitoribosome, with a much greater proportion of rRNA coated with protein (from Brown et al., 2014, figure S5). This figure demonstrates biological equivalence classes by showing how the same microbiological function can be performed by either different RNAs (in light grey) or proteins. What is selected for is the function, not the specific structure whereby it is realised. As such, the ribosome is a TC.

humans (Brown et al., 2014). Therefore, it is conservation of function and not conservation of a particular overall three-dimensional morphology that controls the sequence space within which the functional RNA structural network can evolve and change through time. This explanation is consistent with the fact that the mitoribosome as a TC emerged from the bottom up through a series of evolutionary steps (that still remain to be uncovered). Having emerged, it is the TC function, namely the translation of mitochondrial proteins, that has causal power over the whole structural complex formed of ribosomal RNA, ribosomal proteins, and other associated factors.

Downward Causation from Cells to Cellular Functions

Jaeger and Calkins (2012) and Jaeger (2015) take as axiomatic that unicellular organisms have a 'master function' giving them the high-level goal of reproduction (including replication). In other words, they claim that this master function is the real TC defining a cell such as a bacterium. If we regard the whole organism as nothing more than a coordinated network of chemicals together with their mutual interactions (reactions, binding, recognition, etc.), this teleological (goal-oriented) function can be achieved only through downward causation, since it does not exist at the level of chemical interactions. In general any network of biochemical interactions can have several potential functions because it can be an effective cause of several outcomes. According to the downward causation argument, the particular functions of the network that are observed in a living cell have been selected, by the higher level of organisation, from among all the possibilities, and of course the selection criterion is that they match best with the higher-level function. For this to work there must obviously be (1) a range of possible effective causes (the equivalence class), (2) a higher-level function from which to identify (3) a goal that can be expressed as a set-point in cybernetic terms and (4) a means of influencing the behaviour of elements in the equivalence class (5) that correlates with the difference between the present state and the goal. This 'means of influencing' is identified with 'information control'.

Jaeger and Calkins (2012) point to the substitutability of a biochemical process from one organism to another as an example.

Certainly, this example illustrates an equivalence class and shows that, of all the potential functions of the substituted element, one is expressed and it is the one that is most beneficial to the whole organism, implying selection.

How We All Became the Shape We Are: Downward Causation from Structure to Cells

Embryological development is both a marvel and the most obvious place to look for downward causation in a biological process because the differentiation of cells and their distribution in space, forming a complicated body plan, cannot easily be explained otherwise.

The embryo begins as a compact ball of identical cells: bisection of this 'blastocyst' creates identical twins. The blastocyst is a very simple structure embodying very little information, but it must become one of immense complexity by accumulating information in its form. The information held in the cells is not sufficient: they need to 'know' where they are, so as to create pattern in space (to instantiate information in form) (Gilbert and Epel, 2013). This information is taken from the only other available source: the environment of the cells. The first step arises from a difference created by the geometry of the blastocyst: those on the outside experience the external environment, whereas deeper cells experience only each other. It is, then, an example of how the geometry of a higher-level structure determines the behaviour of the cells within it: the outer cells differentiate into the trophoblast (Van de Velde et al., 2008). This initiates a series of further steps in which cells finding themselves in a particular location relative to the changing geometry of the whole respond in ways that further elaborate the geometry. It is a bootstrapping process: a disc is formed, a line is differentiated along its diameter (the 'primitive streak'), and eventually a rod (the 'notochord') is created by cells responding to gradients in signalling proteins, these gradients being a consequence of the developing geometry (Davies, 2014). As well as hormonal, field-directed morphogenesis (mechano-transduction and electro-transduction) provides integrative signals that lower-level (including genomic) systems follow (physical forces acting globally on cells result in changes of gene expression within them; see, e.g., Levin, 2012). These processes are neither solely upward nor solely downward causation; it is a combination of both whereby information is exchanged between the cellular and structural levels in a two-way dialogue.

Downward Causation from Community to Individual Organisms
The democracy used by a bee colony in choosing a new home for the hive described by Seeley (2010) provides a good example. A choice of several potential new homes, differing in quality, are explored by scout bees. Each presents a report to the hive, communicated by a waggle dance. Through a network of such communications (pure information), support for the better options builds and refines, until a quorum is achieved, and by this a decision is made. The decision is not that of a single bee, but rather an emergent phenomenon and a property of the scout bees and their communications collectively. Thus the cause of the hive choosing a particular new home exists at a higher organisational level than the effective cause, which is the physical action of relocation exercised by each individual bee. The hive as a collective controls the home location of the individuals, on the basis of information provided by scout bees – which is the crucial link to the outside world.

So where precisely is the cause of the decision to be located? Causal power should be attributed to the behavioural repertoire of the individual bees, since without that, such decision-making would be impossible, but the effective cause is the actual movement of bees en masse. Where they go is determined by which potential new home achieved a quorum of support, which in turn is the outcome of information exchange building mutual information among scout bees. Crucially, this mutual information represents the level of match between the potential home and the hive's requirements. The hive can be thought of as having a goal instantiated as a (multifactorial) set-point: pure information. Potential homes are compared with this by the scouts, who, by a series of communications of pure information, reach a threshold of mutual information, from which, according to their behavioural algorithm, the specific effective cause (moving to the best-match home) results. Again, crucially, the set-point is not a property of any one bee – it depends on the size of the colony – so its role in determining the outcome necessarily implies cybernetic control exercised by the higher level of organisation on the lower. We do not yet know how the information describing the colony size is embodied in the bees, but we can be sure that information is the source of causation at every level of abstraction in this example.

How Ecosystems Develop Downward Causation

Perhaps an even clearer example of downward causation can be found in ecology where the niche of every species (more precisely, every phenotype) is determined by its ecological community, which is of course the sum of all other species exerting an influence (plus external factors such as climate). An ecological community consists of all the organisms in it, together with all the interactions among them. While many of these interactions are of a material form (e.g., transferring food resources from prey to predator), many are not; for example, competition merely describes the influence of one organism's behaviour on the outcomes for another's, as, for example, in territorial displays. There seems to be some sort of causation by pure information on two levels here. First, there is the communication needed for one organism to influence another without a material transfer between them. Second, at a grander scale, we have the defining of a particular niche by the community, of which any species occupying it is a part.

Taking the first level briefly, it is obvious that, for example, a cat can prohibit another from sitting on the mat by staring at it in a display of dominance. This is, of course, communication; using a sign and the effect is to alter the 'mind' of the cat on the receiving end. Influence by communicated information like this is in a special category because it relies on sophisticated cognition to work. Bacteria communicate by releasing chemical signals into their environment, but this could be thought of as an extension of the chemical signalling network within the cell, to include the extracellular biotic environment. This kind of signalling may well have been an essential step in the development of multicellular organisms. The characteristic way in which such signals are implemented is through the well-known lock and key recognition of signalling molecules by their receptor molecules. We can certainly interpret this as a communication channel carrying information that in turn influences biochemical processes in the cell. At every point in the communication, we understand that the information is embodied in the form of molecules, which in turn are a material part of the living system.

The second and higher level of causation is more relevant here because it constitutes a diktat from information instantiated at a higher level of organisation, down to that at a lower level. The way it works is that species both create new niches and close them off by occupying them, so in effect an existing assembly of species sets boundary conditions for the ecological sustainability of any population of a new species. The constraint of low-level processes by boundary conditions set at a higher level is common among natural systems and inevitable among synthetic ones, a prime example being Conway's Game of Life (Gardner, 1970). Whatever happens in the game is established by the initial conditions and the automaton rules, which effectively set the boundary conditions and constraints, respectively, for what follows.

The living ecosystem is a far more interesting case since here the boundary conditions are not 'given' at the beginning; rather, they emerge as the community builds and they derive from its member populations, to which they also apply. It is like a club that changes its rules depending on who has already joined and then applies these rules to new applicants (which is true for all evolving biosystems and their evolving parts). Is this really a case of downward causation, or is it only a matter of perspective? We find it convenient to aggregate the populations into the level of a community, but perhaps there is no such thing in reality. In the present case, this is rather easy to answer, because what we call the 'community' is not simply an assembly of organisms; it specifically includes all their interactions and the organisation among them. An ecological community is specified by a very particular network of interactions, unique to it. This network exists at the community scale of organisation – indeed, it defines that scale – and it gives rise to measurable properties that are observable only at that scale. As we know, a network of this sort is functional information instantiated in the probability distribution of interactions among its nodes, ultimately among individual organisms in the case of an ecological community. It can be mathematically described by specifying a network and the connection strengths in that network. Inter alia, that description will enable us to identify network motifs and analyse the causal patterns associated with those motifs (see Alon, 2006).

We conclude that an ecological community is a clear case of instantiating information that emerged from complexity (the formation of a relatively stable network from component interactions) and therefore exists at a higher level of organisation than the component parts. Further, this information is functional in, among other things, setting the 'rules for entry' to any prospective population of organisms. In this way it adapts the members of the group to fit into the group and its purposes. It is by this means an example of downward causation.

Natural Selection as Downward Causation

Early on, Donald Campbell (1974) recognised natural selection as a case of downward causation, and we can now understand this in terms of information. Biological information manifests in autopoiesis: life is literally making itself. The essential process causing this is the filtration of randomness to create pattern, the pattern being a physical embodiment of new information. The filtration is performed by information patterns that had been created before; in their turn, they were created by earlier patterns, all the way back to the emergence of those fundamental patterns that gave form to prebiotic chemical networks.

Selective filtering works by imposing boundary conditions on a physical process; for example, the hard wall boundaries of a resonator select only those frequencies, from a random mix, that match the boundary condition of zero displacement at the walls (a classic result for acoustic and electromagnetic waves). In general, the way information is made from randomness is by filtering according to some template that selects a pattern, which matches the template. We can interpret this in terms of correlation: the potential information (some say 'information entropy') instantiated in a random distribution is selected if and only if it correlates with the information embodied by the template (the selected information is then mutual information with respect to the template). For example, a colour filter embodies information about the frequency of light that it does not absorb. By absorbing all but these frequencies, it reduces the possibility number of frequencies passing through it and therefore reduces the entropy of the transmitted light. In this way, embodied information is imposing order on the randomness generated in the universe, of which there is a plentiful supply.

In biology, natural selection follows this principle: the environment, including organisms that may feed, compete, parasitically prey on, or consume a focal species, constitutes embodied information that acts as a filter on the possibility number (the diversity) of viable designs for that focal species. The more species of different designs that form the environment for a hypothetical new entrant to the ecosystem, the more information rich is the filter acting on it. The resulting design of this new entrant is correspondingly more precisely defined. In ecological language, its niche space is more closely defined and typically narrower. So it is, perhaps, that in very species-rich ecosystems, we find the extremes of specialism where one species entirely depends on another single species (see, e.g., Thompson, 1994, pp. 9–12).

Bacteria are largely self-contained consumers of chemical resources, but in stable environments can form communities based on processing chains that allow for mutually dependent specialisation. At the molecular level, selection operates by both biotic and abiotic processes, for example, through the effect of pH on protein conformation. Within the cellular cytoplasm, this pH is closely regulated by homeostasis (which we have already classified as a case of downward causation). This regulation selects a particular pH from among all possible values, and this particular pH selects for particular protein conformations. This way, information in one form selects information in another and it is easy to interpret that as another example of downward causation since the cybernetic system of home-ostasis is a higher level of organisation than the molecular shapes over which it selects. Ellis generalises these observations (Ellis, 2011):

> Higher level environments provide niches that are either favourable or unfavourable to particular kinds of lower level entities; those variations that are better suited to the niche are preserved and the others decay away' and he refers to the process as ' top-down causation from the context to the system.

This may be interpreted as an elaboration of downward causation by boundary conditions, but a crucial difference is that now the conditions are set dynamically and by the system that, itself, is regulated by them.

Why Organisms Appear to Be Autonomous

Our view of living systems is strongly affected by seeing a nested hierarchy of organisation in nature. No organism can live in isolation, and it is also reasonable for us to see all of life as part of a continuum in the patterns that organise the whole universe (Farnsworth et al., 2013). All are affected both by their component parts, and by the environment in which they live.

But then which level is the 'real' level where the causal work gets done? One might suggest that organisms simply emerge from information processes organised at lower levels and with no real causal powers of their own. In that case, they would be no more autonomous than the patterns created in the Game of Life. Their apparent 'decisions' are not, though, entirely determined by the initial conditions as they are in Conway's game.

They are determined dynamically by conditions that are created at each level, by the constant selection for mutual information, and by information in their environment at every scale of organisation (from molecules to ecosystems). This embodied and selective information constitutes both the external and internal environment, for these are really two aspects of a continuum. The selection processes delete lower-level entities in a way that adapts the lower-level structure to higher-level conditions or purposes.

The behaviour of organisms can then be seen as another case of existing information filtering the randomness generated by thermal diffusion (and ultimately quantum fluctuations) to create and use new information at each higher level, according to the dynamics applicable at that level and the biological purposes that drive them. The organisms are constrained in both bottom-up and top-down ways, but do not lack autonomy, because they also obey their own emergent dynamics. They do not lead predetermined lives, nor are their lives entirely controlled by exogenous forces; rather, they are strongly connected with their whole environment, which acts top down to strongly influence their behaviour. Collectively, they determine the patterns of information existing around them and are in turn influenced by these. There is no identifiable beginning or master controller for this complex system. Life in this picture is all one dynamic system of information processing in which every level of organisation is both causing and being caused by other levels (as described in Farnsworth et al., 2013).

As well as this, among organisms, internal variables influenced by the same-level (effective) dynamics that motivate them also play a key role in outcomes. Many bacteria swim away from uncongenial surroundings under the influence of chemical signals that are compared with an internal set-point. Among humans, psychological dynamics or mental plans perform a far more sophisticated version of this role. The same-level dynamics emerges from the confluence of bottom-up and top-down causation: same-level choices are enabled by the randomness existing at the lower levels, described previously (Ellis, 2014). Thus every level is real and affects the outcome, as proposed by Noble (2012): they each are effective causes at their own level and also have causal power over both lower and higher levels.

Overall, the conclusion, supported by epigenetic studies and developmental biology as well as behavioural, physiological, and ecological studies, is that a core feature of the way life functions is a flow of information from higher to lower levels that enables the lower-level entities to adapt structurally and functionally to the higher-level environment in which they are imbedded. After all, without such a downwards flow of information, the lower levels would not know what to adapt to. Clearly, this must have played a key role in the evolutionary history of life on Earth, whereby all organisms are adapted to the niches in which they live. Life could not have occurred without this kind of top-down causal effect additional to the bottom-up effects whereby chemistry and structure emerge from physics. The combination of the two allows tremendous complexity – such as that of a living cell – to emerge.

ACKNOWLEDGMENTS
We would like to dedicate this chapter to Bill Stoeger (cosmologist and theologian).

REFERENCES

Albrecht, A. 2001. Quantum ripples in chaos. *Nature*, **412**(6848), 687–688.

Alon, U. 2006. *An introduction to systems biology: design principles of biological circuits.* CRC Press.

Amunts, A., Brown, A., Bai, X., Llácer, J. L., Hussain, T., Emsley, P., Long, F., Murshudov, G., Scheres, S. H. W., and Ramakrishnan, V. 2014. Structure of the yeast mitochondrial large ribosomal subunit. *Science*, **343**(6178), 1485–1489.

Aoki, I. 1982. A simulation study on the schooling mechanism in fish. *Bulletin of the Japanese Society of Scientific Fisheries (Japan)*, **48**:1081–1088.

Auletta, G., Ellis, G. F. R., and Jaeger, L. 2008. Top-down causation by information control: from a philosophical problem to a scientific research programme. *Journal of the Royal Society Interface*, **5**(27), 1159–1172.

Ay, N., and Polani, D. 2008. Information flows in causal networks. *Advances in Complex Systems*, **11**(01), 17–41.

Booch, G. 2006. *Object oriented analysis and design with application*. Pearson Education.

Brown, A., Amunts, A., Bai, X., Sugimoto, Y., Edwards, P. C., Murshudov, G., Scheres, S. H. W., and Ramakrishnan, V. 2014. Structure of the large ribosomal subunit from human mitochondria. *Science*, **346**(6210), 718–722.

Campbell, D. T. 1974. 'Downward causation in hierarchically organised biological systems. Pages 179–186 of *Studies in the Philosophy of Biology*. Springer.

Cummins, R. 1975. Functional analysis. *Journal of Philosophy*, **72**, 741–765.

Davies, J. A. 2014. *Life unfolding: how the human body creates itself*. Oxford University Press.

Ellis, G. F. R. 2011. Top-down causation and emergence: some comments on mechanisms. *Interface Focus*, **2**, 126–140.

Ellis, G. F. R. 2012. On the limits of quantum theory: contextuality and the quantum-classical cut. *Annals of Physics*, **327**(7), 1890–1932.

Ellis, G. F. R. 2014. *Necessity, purpose, and chance: the role of randomness and indeterminism in nature*. www.mth.uct.ac.za/~ellis/George_Ellis_Randomness.pdf.

Falcon, A. 2015. Aristotle on causality. In *The Stanford Encyclopedia of Philosophy* (Spring 2015 edition).

Farnsworth, K. D., Nelson, J., and Gershenson, C. 2013. Living is information processing: from molecules to global systems. *Acta Biotheoretica*, **61**(2), 203–222.

Geary, C., Chworos, A., and Jaeger, L. 2011. Promoting RNA helical stacking via A-minor junctions. *Nucleic Acids Research*, **39**(3), 1066–1080.

Gilbert, S. F., and Epel, D. 2013. *Ecological developmental biology: integrating epigenetics, medicine, and evolution*. Sinauer Associates.

Grabow, W. W., Zhuang, Z., Shea, J. E., and Jaeger, L. 2013. The GA-minor submotif as a case study of RNA modularity, prediction, and design. *Wiley Interdisciplinary Reviews: RNA*, **4**(2), 181–203.

Hartwell, L. H., Hopfield, J. J., Leibler, S., and Murray, A. W. 1999. From molecular to modular cell biology. *Nature*, **402**, C47–C52.

Jaeger, L. 2015. A (bio)chemical perspective on the origin of life and death. In Behr, J., Cunningham, C., and the John Templeton Foundation (eds.), *The role of life in death*. Cascade Books, Wipf and Stock.

Jaeger, L., and Calkins, E. R. 2012. Downward causation by information control in micro-organisms. *Interface Focus*, **2**(1), 26–41.

Jaeger, L., Verzemnieks, E. J., and Geary, C. 2009. The UA_handle: a versatile submotif in stable RNA architectures. *Nucleic Acids Research*, **37**(1), 215–230.

Landauer, R. 1992. *Information is physical*. IBM Thomas J. Watson Research Division.

Lungarella, M., Ishiguro, K., Kuniyoshi, Y., and Otsu, N. 2007. Methods for quantifying the causal structure of bivariate time series. *International Journal of Bifurcation and Chaos*, **17**(3), 903–921.

McGhee, G. R. 2011. *Convergent evolution: limited forms most beautiful*. MIT Press.

Noble, D. 2012. A theory of biological relativity: no privileged level of causation. *Interface Focus*, **2**(1), 55–64.

Schreiber, T. 2000. Measuring information transfer. *Physical Review Letters*, **85**(2), 461.

Seeley, T. D. 2010. *Honeybee democracy*. Princeton University Press.

Simon, H. A. 1981. *The sciences of the artificial*. MIT Press.

Taylor, R. 1973. *Action and purpose*. Humanities Press.

Thompson, J. N. 1994. *The coevolutionary process*. University of Chicago Press.

Van de Velde, H., Cauffman, G., Tournaye, H., Devroey, P., and Liebaers, I. 2008. The four blastomeres of a 4-cell stage human embryo are able to develop individually into blastocysts with inner cell mass and trophectoderm. *Human Reproduction*, **23**(8), 1742–1747.

Walker, S. I., Cisneros, L., and Davies, P. C. W. 2012. Evolutionary transitions and top-down causation. *arXiv preprint arXiv:1207.4808*.

Wood, J. M. 2011. Bacterial osmoregulation: a paradigm for the study of cellular homeostasis. *Annual Review of Microbiology*, **65**, 215–238.

I4   Automata and Animats

From Dynamics to Cause–Effect Structures

Larissa Albantakis and Giulio Tononi

The term "dynamical system" encompasses a vast class of objects and phenomena – any system whose state evolves deterministically with time over a state space according to a fixed rule (Nykamp, 2015). Since living systems must necessarily change their state over time, it is not surprising that many attempts have been made to model a wide range of living systems as dynamical systems (de Jong, 2002; Ermentrout and Edelstein-Keshet, 1993; Izhikevich, 2007; Kaneko, 2006; Nowak, 2006; Sumpter, 2006), although identifying an accurate time-evolution rule and estimating the relevant state variables can be hard. On the other hand, even simple state-update rules can give rise to complex spatio-temporal patterns, reminiscent of certain features associated with living systems, such as complex behavioral patterns, self-organizing behavior, and self-reproduction. This has been demonstrated extensively using a class of simple, discrete dynamical systems called "cellular automata" (CA) (Gardner, 1970; Knoester et al., 2014; von Neumann, 1966; Wolfram, 1984, 2002). CA consist of a lattice of identical cells with a finite set of states. All cells evolve in parallel according to the same local update rule, which takes the states of their neighboring cells into account (Figure 14.1).

In any dynamical system, the time-evolution rule determines the future trajectory of the system in its state space given a particular initial state. The aim of dynamical systems theory is to

The material presented in this chapter is a condensed version of an article published in a special issue of the journal *Entropy* (Albantakis and Tononi, 2015b) on "Information Theoretic Incentives for Cognitive Systems" and includes and extends data from Albantakis et al. (2014).

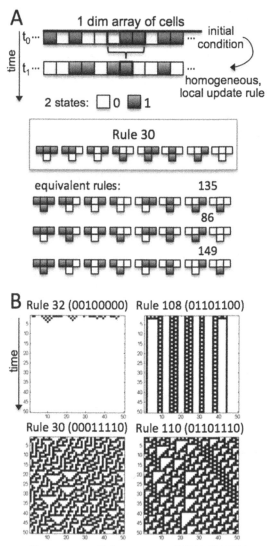

FIG. 14.1 Elementary cellular automata (ECA). (A) ECA consist of a one-dimensional chain of cells in either state 0 or 1, which are updated according to a time-evolution rule dependent on a cell's previous state and that of its two nearest neighbors. Given the $2^3 = 8$ possible past configurations, 256 different update rules exist, which are labeled by the decimal number of their rule in binary. "0" to "1" and/or left-right transformation of a rule (past and current state) lead to rules with equivalent behavior. (B) Example evolutions of four ECA ($N = 50$) with distinct long-term behavior over 50 time-steps for a random initial condition and periodic boundary conditions.

understand and characterize a system based on its long-term behavior, by classifying the geometry of its long-term trajectories. In this spirit, CA have been classified according to whether their evolution for most initial states leads to fixed points, periodic cycles, or chaotic patterns associated with strange attractors (Li and Packard, 1990; Wolfram, 1984) (Figure 14.1B). Despite the simplicity of the rules, this classification is undecidable for many CA with infinite or very large numbers of cells (Culik and Yu, 1988; Sutner, 1995). This is because determining the trajectory of future states is often computationally irreducible (Wolfram, 1985), meaning that it is impossible to predict the CA's long-term behavior in a more computationally efficient way than by actually evolving the system. In general, the relationship between the time-evolution rule, which describes the local behavior of each cell, and the *global* behavior of the entire CA remains indeterminate.

Here, we focus on the causal structure of discrete dynamical systems, treating them as directed causal graphs, that is, as systems of elements implementing local update functions. Assuming an actual physical implementation of a finite-sized CA, the time-evolution rule is equivalent to a cell's mechanism, which determines its causal interactions with neighboring cells (Figure 14.2A). While the rich dynamical repertoire (dynamical complexity) of CA has been studied extensively, their causal structure (causal complexity) has received little attention, presumably because it is assumed that all that matters causally reduces to the simple mechanism of the cells, and anything that may be interesting and complex is to be found only in the system's dynamic behavior. For the dynamical system itself, however, whether it produces interesting patterns or not might not make any causal difference.

Integrated information theory (IIT) (Oizumi et al., 2014; Tononi, 2015) offers a mathematical framework to characterize the cause–effect structure specified by all the mechanisms of a system from its own intrinsic perspective, rather than from the perspective of an extrinsic observer. In IIT a mechanism is any system element

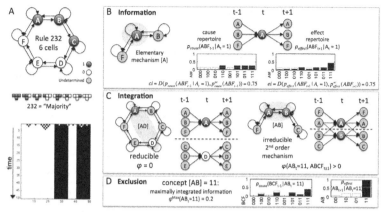

FIG. 14.2 Information, integration, and exclusion postulate at the level of mechanisms. (A) ECA 232 with $N = 6$ cells A–F and periodic boundary conditions illustrated as a network of interacting elements in the current state $A - F_t = 111000$. All cells implement rule 232, the majority rule. Edges denote connections between cells. (B) Information: element A_t in state "1" constrains the past and future states of $ABF_{t\pm1}$ compared with their unconstrained (flat) distribution (overlaid in gray). (C) Integration: elements A and D do not form a higher-order mechanism, since AD_t is reducible to its component mechanisms A_t and D_t. AB_t in state "11," however, does form a higher order mechanism, since AB_t specifies both irreducible causes and irreducible effects, as evaluated by integrated information ϕ (see text). (D) Exclusion: cause and effect repertoires are evaluated over all sets of system elements. The cause and effect repertoires that are maximally irreducible and their ϕ^{Max} value form the concept of a mechanism. In case of AB, the maximally irreducible cause repertoire is over elements BCF_{t-1}; the maximally irreducible effect repertoire over elements AB_{t+1}.

or *combination* of elements with a finite number of states and an update rule, such as a CA cell, or a logic gate, as long as it has irreducible cause–effect power within the system. This means that (1) the mechanism must constrain the past and future states of the system by being in a particular state (information) and (2) the particular way in which it does so – the mechanism's cause–effect repertoire – must be irreducible to the cause–effect repertoire of its parts (integration), as measured by its integrated information ϕ (Figure 14.2B and 14.2C). The maximally irreducible cause–effect repertoire of a mechanism in a state over the particular set of system elements that maximizes ϕ is called its "concept" (Figure 14.2D). A concept determines the mechanism's causal role within the system from the intrinsic perspective of the system itself. Elementary mechanisms specify first-order concepts; mechanisms composed of several system elements specify higher-order concepts. The set of all concepts is the cause–effect structure of a system (Figure 14.3). Integrated conceptual information Φ quantifies the irreducible cause–effect power of a system of mechanisms taken as a whole. It measures how irreducible a system's cause–effect structure is compared with the cause–effect structure of the system when it is causally partitioned (unidirectionally) across its weakest link (Figure 14.3B). Only systems with many specialized but integrated concepts can achieve high values of Φ. Φ can thus be viewed as a measure of the intrinsic causal complexity of a system.

Here, we wish to investigate the relation between the dynamical properties of a system and its intrinsic cause–effect power, both in isolated systems and in agents that evolve in and interact with an environment of rich causal structure. To that end, we (1) exhaustively characterize the cause–effect power of elementary cellular automata (ECA), one-dimensional CA with only nearest-neighbor interactions, and (2) examine the causal and dynamical properties of small, adaptive logic-gate networks ("animats") evolving in task environments with different levels of complexity. In particular, we assess the number of irreducible mechanisms (concepts) and the total amount of integrated conceptual information Φ specified by each system.

While the state evolution of isolated systems, such as the ECA, must be a product of their intrinsic mechanisms, the dynamics of behaving agents, such as the animats, is at least partially driven by the inputs they receive from their environment. We thus predict that, to have a large dynamical repertoire, isolated systems must have an integrated cause–effect structure with many concepts. Specifically, isolated systems that have few, unselective concepts with low ϕ, lack higher-order concepts, or are reducible ($\Phi = 0$) should not be able to produce interesting global dynamics. By contrast,

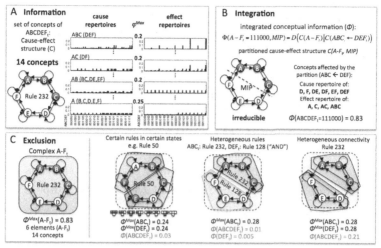

FIG. 14.3 Information, integration, and exclusion postulate at the level of systems of mechanisms. A) Information: The system $A - F_t$ in state "111000" has 14 concepts: 6 elementary concepts A_t, B_t, C_t, D_t, E_t and F_t, 4 2nd order concepts of adjacent elements, as well as AC_t and DF_t, and the 3rd order concepts ABC_t and DEF_t. B) Integration: the cause-effect structures of $A - F_t$ is irreducible to the cause-effect structure of its minimum information partition (MIP), which eliminates the effects of subset ABC onto subset DEF, as measured by the integrated conceptual information Φ, which quantifies the distance between the whole and partitioned cause-effect structure. C) Exclusion: Cause-effect structures and their Φ values are evaluated over all sets of system elements. The set of elements S_t^* with the maximally irreducible cause-effect structure forms a complex; here it is the whole system $A - F_t$. This is not always the case. For different types of rules, or more heterogeneous connections or rules, system subsets may be maximally irreducible complexes. Complexes cannot overlap. In the 3 examples on the right, DEF_t thus may form another complex beneath the main complex ABC_t; the whole system, however, is excluded from being a complex, as are subsets of elements within a complex.

nonisolated systems can exhibit complex reactive behavior driven by the environment. Analyzing the intrinsic cause–effect structure of a behaving agent can elucidate to what extent the agent itself has a complex structure and to what extent it is merely reactive. Moreover, integrated systems with a rich cause–effect structure have adaptive advantages in environments that require context-sensitivity and memory (Albantakis et al., 2014).

Importantly, while a dynamical analysis describes what is "happening" in a system from the extrinsic perspective of an observer, the analysis of its cause–effect structure reveals what a system "is" from its own intrinsic perspective, exposing its dynamical and evolutionary potential under many different scenarios.

INTRINSIC CAUSE–EFFECT POWER

In the following we illustrate the main principles and measures invoked by IIT by reference to a simple ECA with six cells and periodic boundary conditions (Figure 14.2A). For this purpose, we treat the ECA as a directed causal graph, meaning as a system of connected elements that each implement a particular function, here, for example, rule 232, the majority rule. Each cell in the ECA has a self-loop and connections to its nearest neighbors.[1] It is assumed that the transition probabilities from all $2N$ system states to all $2N$ system states over one time-step are known. In a discrete, deterministic system, such as the ECA, this corresponds to perturbing the system into all its possible states and recording all state-to-state transitions over one time-step. For further details on the mathematical and conceptual tools developed within the IIT framework, see Oizumi et al. (2014) and Tononi (2015).[2]

Central to IIT is the postulate that, in order to exist, a system in a state must have cause–effect power, since there is no point in assuming that something exists if nothing can make a difference to it or it does not make a difference to anything. To exist from its own intrinsic perspective, the system moreover must have cause–effect power on itself. To that end, the system must be comprised of mechanisms, elements that have cause–effect power on the system, alone or in combination.

[1] Note, however, that depending on the rule, some of the edges may not be causally effective. For rule 136 (10001000), for example, the state of the left neighbor is irrelevant.

[2] The PyPhi package used to calculate all IIT measures is available at https://github.com/wmayner/pyphi. For the interested reader, the IIT website (http://integratedinformationtheory.org/calculate.html) allows calculating Φ and other IIT measures for small example systems of logic gates.

MECHANISMS AND CONCEPTS

To determine whether a set of elements M_t in state m_t forms a mechanism with cause–effect power on the system, we first need to evaluate whether (1) the past state of the system makes a difference to the set M_t and (2) the state of the set $M_t = m_t$ makes a difference to the future state of the system. Condition (1) is satisfied if M_t, by being in its current state m_t, constrains the probability distribution of possible past states of a set of system element Z_{t-1}, specified by the cause repertoire $p_{\text{cause}}(z_{t-1}|m_t)$.[3] The cause repertoire of element A in state "1" of the ECA shown in Figure 14.2A over its inputs ABF_{t-1}, for example, reflects A's update rule: only states in which two or more of A's inputs are "1" are possible past states of $A_t = 1$. The *cause information* (*ci*) of $A_t = 1$ quantifies the distance[4] between its cause repertoire and the unconstrained (uc) cause repertoire in which each past state of ABF_{t-1} is assumed equally likely (see Figure 14.2B). Condition (2) is satisfied if M_t, by being in its current state m_t, constrains the probability distribution of possible future states of a set of system elements Z_{t+1}, specified by the effect repertoire $p_{\text{effect}}(z_{t+1}|m_t)$. Again, the *effect information* (*ei*) of, for example, $A_t = 1$ quantifies the distance between its effect repertoire and the unconstrained effect repertoire, which considers all input states to all system elements to be equally likely. In order to have cause–effect power, the *cause–effect information* (*cei*) of a set of elements, the minimum of its *ci* and *ei*, must be positive.

Even a set of elements with positive *cei* does not have cause–effect power of its own, and is thus not a mechanism, if it is reducible to its submechanisms. Consider, for example, elements A and D in the ECA system of Figure 14.2. A_t and D_t together do not constrain the past or future states of the system more than A_t and D_t taken

---

[3] By contrast to correlational measures, which use observed state distributions, in IIT cause–effect power is quantified taking all possible system perturbations into account with equal probability.

[4] Distances in IIT are measured by the so-called Earth mover's distance (*emd*). The *emd* quantifies the cost of transforming one probability distribution into another, using the Hamming distance between states as the underlying metric.

separately. Partitioning[5] between them does not make a difference. The elements A and B together, however, do form a higher-order mechanism AB_t, since no matter how the mechanism's connections are partitioned, AB_t's cause–effect repertoire is changed. *Integrated cause information* ϕ_{cause} quantifies the distance[6] between a cause repertoire and the cause repertoire under the minimum information partition (MIP) of the candidate mechanism M_t. The MIP is the partition that makes the least difference. Likewise ϕ_{effect} quantifies *integrated effect information*. An irreducible set of elements with positive integrated information ϕ, the minimum of ϕ_{cause} and ϕ_{effect}, is a mechanism that has cause–effect power on the system. Note that ϕ_{cause} and ϕ_{effect} can be measured for cause–effect repertoires over all possible sets of system elements $Z_t \pm 1$. The cause–effect repertoire of M_t that is maximally irreducible over the sets of elements $Z^*_{(t\pm1)}$, with $max(\phi_{\text{cause}})$ and $max(\phi_{\text{effect}})$, and its integrated information ϕ^{Max} form the "concept" of mechanism M_t. It can be thought of as the causal role of the mechanism in its current state from the intrinsic perspective of the system itself. The concept of AB_t in state "11" in the ECA system, for example, is specified by the cause repertoire over the set of elements BCF_{t-1} and the effect repertoire over the set of elements AB_{t+1}, with $\phi_{\text{Max}} = 0.2$. For comparison, the cause–effect repertoire over all inputs and outputs of AB_t is less irreducible with $\phi = 0.17$ and thus not further considered.

Cause–Effect Structures

Our objective is to determine whether and how much a system exists (has cause–effect power) from its own intrinsic perspective. As stated above, to have intrinsic cause–effect power, the system must have elements with cause–effect power on the system. In other words, the

---

[5] "Partitioning" here means making connections causally ineffective, and can be thought of as "injecting independent noise" into the connections.

[6] The distance between two cause–effect structures is evaluated using an extended version of the *emd*. It quantifies the cost of transforming one cause–effect structure into another, taking into account how much the cause–effect repertoires and ϕ^{Max} values of all concepts change through the partition; see Oizumi et al. (2014).

system must have concepts. The set of concepts of all mechanisms within a system shapes its *cause–effect structure*. The 6-node, rule 232 ECA $A - F_t$ in state "111000," for example, has 14 concepts, 6 elementary and 8 higher-order concepts (Figure 14.3A).

As with a mechanism, a system exists as a whole from the intrinsic perspective only if it is irreducible to its parts. Specifically, every subset of the system must have cause–effect power on the rest of the system, otherwise there may be system elements that never affect the system or are never affected by the system. This is tested by unidirectionally partitioning the connections between every subset of the system and its complement. Integrated conceptual information Φ quantifies the distance between the cause–effect structure of the intact system $S_t = s_t$ and the cause–effect structure of the partitioned system. Again, we search for the MIP, the system partition that makes the least difference to the cause–effect structure. In the example system $A - F_t$ the MIP renders the connections from subset ABC to DEF causally ineffective. Although this partition is the one that least affects the cause–effect structure of $A - F_t$, it still alters the cause–effect repertoires of many concepts (Figure 14.3B). Only the concepts $B_t = 1$, $E_t = 0$, $AB_t = 11$, and $BC_t = 11$ remain intact. $A - F_t$'s cause–effect structure is thus irreducible with $\Phi = 0.83$.

Even if a system is irreducible, there may be subsets (or supersets, if the system was embedded in a larger set of elements) that also specify irreducible cause–effect structures. To avoid causal overdetermination (counting the cause–effect power of the same mechanism multiple times) and thereby the multiplication of "entities" beyond necessity (Occam's razor), only one of all overlapping irreducible cause–effect structures is postulated to exist – the one that is maximally irreducible with the highest Φ value (Φ^{Max}). A set of elements that has Φ^{Max} and is thus maximally irreducible is called a complex. In our example, the whole set $A - F_t$ is a complex, since it has the highest Φ value compared with all of its subsets. This is not always the case: given different ECA rules, systems with heterogeneous rules, or systems with slightly different connectivity, smaller sets may have the most irreducible cause–effect structure (Figure 14.3C).[*] Once the major complex with Φ^{Max} is identified, nonoverlapping sets of elements can form additional, minor complexes, as shown for the system with heterogeneous rules (Figure 14.3C, middle). The whole system $A - F_t$, however, is excluded from being a complex in this example, since it overlaps with the major complex ABC_t.

In sum, IIT postulates that a set of elements exists for itself to the extent that is has maximally irreducible cause–effect power, evaluated by Φ^{Max}. In order to have high cause–effect power, it is necessary to have many internal mechanisms, which form a rich cause–effect structure with many elementary and higher-order concepts.

In the following, we assess the cause–effect structure of certain isolated and adaptive discrete dynamical systems and determine how they relate to their dynamic complexity. The results shed light on the distinction between the intrinsic perspective of the system itself and the extrinsic perspective of an observer, and highlight key aspects of the IIT formalism, such as the notions of causal selectivity, composition, and irreducibility and their significance in the analysis of discrete dynamical systems.

BEHAVIOR AND CAUSE–EFFECT POWER OF ELEMENTARY CELLULAR AUTOMATA

There are 256 possible time-evolution rules of ECA, which can be grouped into 88 different equivalency classes. Each class contains maximally four rules, which show identical behavior under "0" to "1" transformations, left-right transformations, or both transformations applied at once (Figure 14.1A). Equivalent rules have identical Φ values and equivalent cause–effect structures for complementary states. In Figure 14.4A and B, the average Φ^{Max} values of all 88 ECA rule classes are plotted for systems with five and six cells. Even for simple systems with a low number of cells N, evaluating the cause–effect structure and its Φ value for a given state is computationally

---

[*] When subsets of elements of a larger system are considered, Φ is calculated for the set, treating the other elements as fixed background conditions.

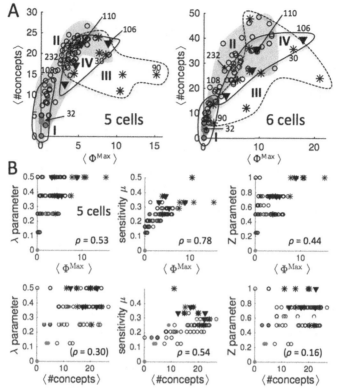

FIG. 14.4 Relation of $\langle \Phi^{Max} \rangle$ and $\langle \#concepts \rangle$ to Wolfram ECA classes I–IV and the rule-based parameters λ, μ, and Z. (A) $\langle \#concepts \rangle$ plotted against $\langle \Phi^{Max} \rangle$ for all 88 ECA equivalence classes for ECA implementations with $N = 5$ and $N = 6$ cells. Each equivalency class is labeled by its lowest-numbered member rule. The symbol of each rule indicates its Wolfram class: rules with (I) uniform fixed points are shown by gray circles, (II) nonuniform fixed points and periodic behavior by open circles, (III) random, chaotic behavior by asterisks, and (IV) complex behavior by black triangles. Rules shown in Figure 14.1B, the example rule 232, and rule 90, which strongly depends on the parity of the number of cells, are highlighted. (B) Correlation of $\langle \Phi^{Max} \rangle$ and $\langle \#concepts \rangle$ with Langton's λ parameter, sensitivity μ, and Wuensche's Z parameter. ρ is Spearman's rank correlation coefficient, with $p < 0.001$ for all correlations, except for values in parentheses, which were not significantly correlated when corrected for multiple comparisons. Symbols indicate Wolfram classes as in (A).

costly. For this reason, average values were calculated from only $N+1$ states out of the $2N$ possible states with different numbers of cells in states "0" and "1."[7] Each equivalency class is labeled according to its Wolfram classification I–IV for random initial states (Wolfram, 2002) indicating whether almost all initial conditions lead to (I) the same uniform stable state (gray circle), (II) a nonuniform stable state or periodic behavior (open circle), (III) pseudorandom or chaotic behavior (asterisks), or (IV) complex behavior with a mixture of randomness and order (black triangles). Figure 14.1B shows the time evolution of four example rules from classes I–IV, which are also highlighted in Figure 14.4A. For ECA with five cells, simple rules from class I all have $\Phi=0$ or low $\langle \Phi^{Max} \rangle$ and only a limited number of concepts. Rules from class II have low to intermediate values of $\langle \Phi^{Max} \rangle$, but can have a high number of concepts. Class III rules show the highest values of $\langle \Phi^{Max} \rangle$ with an intermediate number of concepts, and the complex class IV rules have high numbers of concepts with intermediate to high $\langle \Phi^{Max} \rangle$ values. The number of concepts and $\langle \Phi^{Max} \rangle$ of ECA tends to grow with the number of cells N (see below). Nevertheless, the distribution of rules on the $\langle \Phi^{Max} \rangle / \langle \#concepts \rangle$ plane for ECA with six cells stays the same for classes I, II and IV and most rules of class III. Exceptions are class III rules 18, 90, 105, and 150, which are strongly dependent on N being even or odd, with lower $\langle \Phi^{Max} \rangle$ values and $\langle \#concepts \rangle$ for even N.

Overall, the cause–effect structures of ECA systems with five and six cells suggest that a minimum number of concepts and $\langle \Phi^{Max} \rangle$ may be necessary for rules to have the capacity for intricate class III and IV patterns. On the other hand, the cause–effect structures of certain class I and II rules are too simple to produce complex global dynamics, since they are either reducible ($\Phi = 0$) or lack composition of elementary mechanisms into higher-order concepts. Examples for reducible rules are class I rule 0 (00000000), or class II

---

[7] Since all elements in an ECA specify the same rule and symmetric states have redundant cause–effect structures, this sample of states is representative of a large number of the $2N$ possible states.

rule 204 (11001100), the identity rule. In systems implementing these rules, individual cells do not interact with each other and therefore can always be partitioned without loss of cause–effect power ($\Phi = 0$). Such systems cannot exist as a whole from the intrinsic perspective of the system itself. Several linear class I and II rules generally specify only first-order concepts (e.g., rule 32 (00100000)), which leads to low Φ values in all states. Nevertheless, certain rules that behaviorally lie in class II have similar $\langle \#concepts \rangle$ and $\langle \Phi^{Max} \rangle$ levels as more complex or random rules. This suggests an evolutionary potential of these class II rules for complex behavior, meaning that only small changes are necessary to transform them into class III or IV rules. Indeed, the class II rule with the highest Φ value, rule 154, has been classified as (locally) chaotic by other authors (Li and Packard, 1990; Schüle and Stoop, 2012).

Taken together, rules that specify cause–effect structures with higher $\langle \Phi^{Max} \rangle$ tend to have more higher-order concepts and more selective concepts with high $\langle \Phi^{Max} \rangle$ at all orders. Note that considering only causes and effects of individual cells (elementary, first-order mechanisms) or of the ECA system as a whole (highest order) would not expose these differences in complexity across ECA. This highlights the importance of assessing the causal composition of a system across all orders.

Behaviorally, class II CA can be further subdivided into rules with stable states and rules with periodic behavior (Li and Packard, 1990). In terms of $\langle \#concepts \rangle$ and $\langle \Phi^{Max} \rangle$, however, there is no inherent causal difference between simple periodic rules and rules with nonuniform stable states. For example, the majority rule 232 (11101000), which evolves to a nonuniform stable state, is causally equivalent to rule 23 (00010111), which is periodic with cycle length 2. Rule 23 is the negation and reversion of rule 232. Analyzing the cause–effect structures of ECA here reveals additional equivalences between rules: all rules that under negation or reversion transform into the same rule are causally equivalent to their transformation (e.g., in class III 105 to 150, or in class I

51 to 204). These additional symmetries between rules have recently been proposed in a numerical study by Krawczyk (2015), which equates ECA rules if they show equivalent compressed state-to-state transition networksfor finite ECA across different numbers of cells N. Since compressing the transition network is based on grouping states with equivalent causes and effects, the approach is related to IIT measures of cause–effect information but lacks the notion of integration. Intrinsic causal equivalencies between rules that converge to fixed points and rules that show simple periodic behavior challenge the usefulness of a distinction based on these dynamical aspects. At least for the system itself such differences in dynamic behavior do not make a difference.

Integrated conceptual information Φ is moreover related to several rule-based quantities that have been proposed as indicators of complex behavior during the past three decades and ultimately rely on basic notions of causal selectivity (see below). In Figure 14.4B we show the correlation of $\langle\Phi^{\text{Max}}\rangle$ and $\langle\#concepts\rangle$ to three prominent measures (de Oliveira et al., 2000): Langton's λ parameter (Langton, 1990), the sensitivity measure μ (Binder, 1993), and the Z-parameter (Wuensche and Lesser, 1992). For ECA, the λ parameter simply corresponds to the fraction of "0" or "1" outputs in the ECA rule. Rule 232 (11101000), for example, has $\lambda = \frac{1}{2}$. Class III and IV rules typically have high λ, and the rules with the highest $\langle\Phi^{\text{Max}}\rangle$ values also have the maximum value of $\lambda = \frac{1}{2}$ (Figure 14.4B, first panel). The parameter μ measures the sensitivity of a rule's output bit to a one-bit change in the state of the neighborhood, counting across all possible neighborhood states (Binder, 1993). Nonlinear rules that depend on the state of every cell in the neighborhood, such as the parity rule 150 (10010110), have the highest values of μ. Finally, the Z parameter assesses the probability with which a partially known past neighborhood can be completed with certainty to the left or right side (Wuensche and Lesser, 1992). Sensitive rules with high μ also have high Z. However, Z can also be high for some simple rules, such as the identity rule 204 (11001100), which in addition has the highest $\lambda = \frac{1}{2}$.

All three rule-based quantities are related to each other and the IIT measres to some extent through the notion of causal selectivity. A mechanism is maximally selective if it is deterministic and non-degenerate, which means that its current state perfectly constrains the past and future state of the system (Hoel et al., 2013). Causal selectivity decreases with indeterminism (causal divergence) and/or degeneracy (causal convergence). While CA are fully deterministic systems, many rules show degeneracy, which means they are deterministically convergent, mapping several past states into the same present state. Even in fully deterministic systems, individual mechanisms comprised of a subset of system elements typically cannot constrain the future state of the system completely, if there are degenerate mechanisms in the system; conditioning on $M_t = m_t$ in this case may still leave the remaining inputs to the elements in Z_{t+1} undetermined ("noised," i.e., equally likely to be "0" or "1") (Figure 14.2B). Low values of λ, μ, and Z all indicate high degeneracy in the system. This means that, on average, the system's mechanisms and current states do not constrain the past and future states of the system much. Unselective cause–effect repertoires lead to concepts with low Φ^{Max}, fewer higher-order concepts in the system, and thus overall less integrated conceptual information Φ.

Of the three rule-based measures plotted in Figure 14.4B, μ shows the strongest correlation with $\langle \Phi^{Max} \rangle$ and $\langle \#concepts \rangle$. This is because, similar to the IIT measures, μ assesses the causal power of each cell in the rule neighborhood, by testing whether perturbing it makes a difference to the output. Unlike the λ and Z parameter, μ thus assigns low (but still not zero) values to rules with selective but trivially reducible causes and effects such as the identity rule 204 or its negation rule 51, which depend only on a cell's own past value but not that of its neighbors. However, compared with the IIT measures, the sensitivity parameter μ lacks the notion of causal composition, according to which higher-order mechanisms can have irreducible cause–effect power. Generally, while λ, μ, and Z are largely based on empirical considerations, measures of information integration are derived from one underlying principle: intrinsic, irreducible cause–effect power.

Apart from rule-based measures and the classification of a rule's long-term behavior, a CA's dynamical complexity can also be evaluated based on the morphological diversity (Adamatzky and Martinez, 2010) and Kolmogorov complexity (Zenil and Villarreal-Zapata, 2013) of its transient patterns. Morphological diversity measures the number of distinct 3×3 patterns in an ECA's evolution for a particular initial state. This is related to the ECA's cause–effect information for cell triplets, albeit inferred from the observed distribution rather than from the uniform distribution of all possible states. Again, $\langle \Phi^{\text{Max}} \rangle$ and $\langle \#concepts \rangle$ correlate in a necessary but not sufficient manner with morphological complexity, which can be low while the IIT measures are high ($\rho = 0.60/0.35$ Spearman's correlation coefficient, $\rho < 0.001/0.05$ for $\langle \Phi^{\text{Max}} \rangle / \langle \#concepts \rangle$ for $N = 5$ ECA systems) (see Figure 14.5). Finally, there is also a significant correlation between the IIT measures and the Kolmogorov complexity of an ECA rule, approximated by its Block Shannon entropy ($\rho = 0.65/0.40, p < 0.001/0.005$) or compressibility ($\rho = 0.61/0.37, p < 0.001/0.05$) averaged over several different initial conditions (Zenil and Villarreal-Zapata, 2013).

To date, there is no universally agreed-on classification of (elementary) CA based on their dynamical behavior that uniquely assigns each rule to one class (but see Schüle and Stoop, 2012). Part of the problem is that depending on the initial conditions, a rule can show patterns of very different complexity. The left panel in Figure 14.5A, for example, displays the trivial evolution of four different rules for the typically applied initial condition of a single cell in state "1." Random initial conditions, however, reveal that only rules 0 and 40 typically tend to uniform stable states (Wolfram class I), while rules 104 and 232 belong to Wolfram class II. Moreover, since CA are deterministic, finite CA will eventually all arrive at a steady state or periodic behavior (for binary ECA after at most 2^N states). Small systems with few cells thus cannot unfold the full dynamical

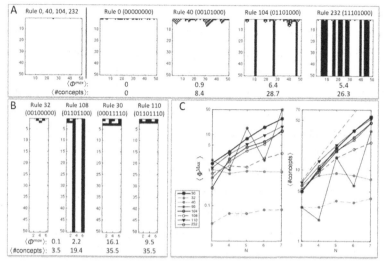

FIG. 14.5 "Being" versus "happening." (A) Different rules show the same behavior under certain initial conditions. Rules 0, 40, 104, and 232 all immediately converge to the state "all 0" for the standard initial condition of a single cell in state "1." Perturbing the system into random initial conditions, however, reveals that rules 0 and 40 belong to Class I, while rules 104 and 232 belong to Class II. (B) Rules from the four different Wolfram classes all converge to a steady state for small ECA with $N = 6$ cells (cf. Figure 14.1B). In both A and B, the IIT measures reflect the classification ("being"), i.e., the potential dynamical complexity, rather than the actual behavior ("happening"). IIT measures are obtained from the respective $N = 6$ ECA of each rule. (C) $\langle \Phi^{Max} \rangle$ and $\langle \#concepts \rangle$ across $N = 3\text{--}7$ cells. Note the logarithmic scale of the y-axis. Already small systems reveal typical features of an ECA's cause–effect structure. An ECA's increase of $\langle \Phi^{Max} \rangle$ and $\langle \#concepts \rangle$ is further indicative of the rule's cause–effect structure, independent of N (see text).

potential of their rule, although the local interactions of each cell are the same as in a larger system (Figure 14.5B). In order to predict the actual complexity of a CA's dynamical evolution accurately, the initial state and the size of the system must be known in addition to the rule. This is why purely rule-based measures can, overall, provide only necessary but not sufficient conditions for complex behavior.

The problems encountered in classifying rules based on their dynamical complexity stem from a discrepancy between "being" and "happening" (Tononi, 2015). Patterns describe what is happening in the CA system following a particular initial state. A classification, however, is about what the system is. As the examples in Figure 14.5 show, classifying a rule for arbitrary initial conditions reveals its *potential* dynamical complexity, rather than the states actually observed in a particular dynamic evolution. Since it requires perturbing the system into many different initial states, this approach is somewhat related to the causal approach of IIT. More generally, the objective of IIT is to reveal and characterize how, and how much, a system of mechanisms in its current state exists ("is") from its own intrinsic perspective, rather than from the perspective of an external observer observing its temporal evolution. Intrinsic cause–effect power relies on causal composition and requires irreducibility (Oizumi et al., 2014; Tononi, 2015). According to IIT, the particular way the system exists is given by its cause–effect structure, and to what extent it exists as a system is given by its irreducibility, quantified as its integrated conceptual information Φ^{Max}.

While IIT measures do depend on the size of the system and its state, the average values obtained for ECA with a small number of cells and for a subset of states already reveal the general causal characteristics underlying different rules (Figure 14.5). Rules with the capacity for complex or random behavior show relatively high values of $\langle \Phi^{Max} \rangle$ already for small systems with three to seven nodes. The way $\langle \Phi^{Max} \rangle$ and $\langle \#concepts \rangle$ increase with increasing system size is also characteristic of a rule's cause–effect structure. Class I rule 32, for example, has only first-order concepts; no combination of system elements can have irreducible cause–effect power ($\Phi = 0$). Its maximum number of concepts thus increases linearly, while its $\langle \Phi^{Max} \rangle$ value stays low, since no matter where the system is partitioned, only one concept is lost. By contrast, rules with higher-order concepts at each set size, such as rule 30 or 110, show an almost exponential increase of $\langle \#concepts \rangle$ and $\langle \Phi^{Max} \rangle$. The maximum number of potential concepts in any system is determined by the powerset of $2^N - 1$ candidate mechanisms (combinations of elements) in the system (indicated by the top-most dashed line in Figure 14.5C, right panel). While the ⟨#*concepts*⟩ and ⟨Φ^{Max}⟩ will continue to grow for rules like rule 30 and 110, the number of impossible concepts[8] also increases, because ECA are limited to nearest-neighbor interactions. Finally, ⟨#*concepts*⟩ and ⟨Φ^{Max}⟩ of rule 90, for example, increase with N but vary with the parity of N. This is because rule 90 (01011010) depends nonlinearly on the past state of both of its neighbors (XOR mechanism). Therefore, only combinations of elements can have cause–effect power. For even N, only second-order concepts exist, while for odd N the periodic boundary conditions allow the system to form many additional higher-order concepts about itself.

In many cases, the temporal evolution of CA is computationally irreducible (Wolfram, 1985), which makes it impossible to predict their dynamical behavior. Similarly, calculating the cause–effect structure of a system becomes computationally intractable already for a small number of elements. On the other hand, IIT measures can in principle be extended to higher-dimensional CA with more than two states and larger neighborhoods. Certain general properties of the cause–effect structures of an ECA system are determined by its ECA rule and will hold for any number of cells. As Figure 14.5C shows, these general features can thus already be inferred from very small CA systems.

In summary, the cause–effect structure of a system and its Φ describe what a system "is" from its intrinsic perspective. An ECA's cause–effect structure can reveal the potential for dynamical complexity of the rule it is implementing by making its specific causal features explicit. While we found strong relations between the IIT measures of ECAs and their Wolfram classes, having many concepts and high ⟨Φ^{Max}⟩ are not sufficient conditions for a system to actually show complex behavior under every initial condition. Nevertheless, employing IIT measures of causal complexity can significantly reduce the search space for complex rules, since they appear to be necessary for class III and IV behavior. Finally, class II rules with high ⟨Φ^{Max}⟩

---

[8] Combinations of cells, without shared inputs and outputs, such as AD in Figure 14.2C, are reducible by default with $\Phi = 0$.

and many concepts typically share certain rule-based features with more complex class III and IV rules, which can be interpreted as a high potential for complex behavior under small adaptive changes in an evolutionary context.

BEHAVIOR AND CAUSE–EFFECT POWER OF ADAPTING ANIMATS

CA are typically considered as isolated systems. In this section, we examine the cause–effect structures of small, adaptive logic-gate systems ("animats"), which are conceptually similar to discrete, deterministic CA. By contrast to typical CA, however, the animats are equipped with two sensor and two motor elements, which allow them to interact with their environment (Figure 14.6A). Moreover, the connections between an animat's elements, as well as the update rules of its four hidden elements and two motors, are evolved through mutation and selection within a particular task environment over several 10,000 generations (Figure 14.6B).[9] The task environments the animats were exposed to are variants of "active categorical perception" (ACP) tasks, where moving blocks of different sizes have to be distinguished (Albantakis et al., 2014; Beer, 2003; Marstaller et al., 2013). Solving the ACP tasks successfully requires combining different sensorial inputs and past experience. Adaptation is measured as an increase in fitness: the percentage of correctly classified blocks ("catch" or "avoid").

As demonstrated in the previous section using ECA, isolated ECA must have a sufficiently complex, integrated cause–effect structure to exhibit complex global dynamics. On the other hand, animats are not isolated but connected to the environment, hence they can receive arbitrarily complex inputs. In principle, this allows animats to exhibit complex dynamics even if their internal structure is causally trivial and/or reducible, as would be the case, for example, for an animat whose hidden elements merely copy the input to each sensor.

[9] An animated example of animat evolution and task simulation can be found at http://integratedinformationtheory.org/animats.html.

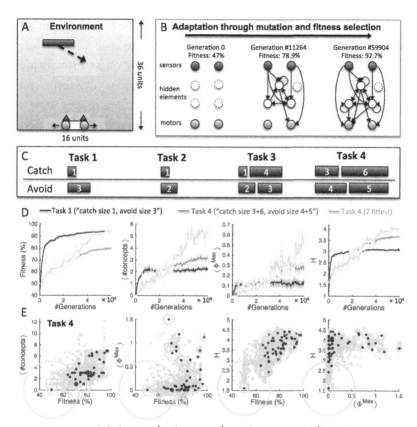

FIG. 14.6 Animats adapting to task environments with varying require-ments for internal memory. (A) Schematic of animat in example environ-ment. On each trial, the animat has to recognize the size of the downward moving block and either catch or avoid it. Blocks continuously move downward and either to the right or to the left, at a speed of one unit per time-step (periodic boundary conditions). The animat has two sensors with a space of one unit between them and thus a total width of three units. Its two motors can move it one unit to the left or right, respectively. (B) Animat evolution. Each animat is initialized at generation 0 without connections between elements. Through mutation and fitness selection, the animats develop complex network structures with mechanisms that enable them to solve their task. Animats were allowed to evolve for 60,000 generations. (C) Illustration of the four Task environments from Albantakis et al. (2014) with increasing difficulty and requirements for internal memory from left to right. (D) Fitness, (#concepts), (Φ^{Max}), and state entropy H for Tasks 1 and 4 from Albantakis et al. (2014). The final fitness achieved by the animats after 60,000 generations corresponds to the task difficulty. The light gray trace shows data from a subset of Task 4 trials with the same high average fitness as Task 1. Animats that evolved

Thus, thanks to external inputs, the dynamic behavior of an animat could become entirely dissociated from its intrinsic cause–effect structure. However, as will be shown below, as soon as animats are forced to evolve under adaptive pressure, their system's dynamics and intrinsic cause–effect structures become linked in interesting ways.

An animat's behavior is deterministically guided by the perceptual stimuli it receives from the environment. An animat sensor turns on if a block is located vertically above it; otherwise it is off. The hidden and motor elements are binary Markov variables, whose value is specified by a deterministic input–output logic. However, an animat's reaction to a specific sensor configuration is context dependent in the sense that it also depends on the current state of the animat's hidden elements, which can be considered as memories of previous sensor and hidden element configurations. Albantakis et al. (2014) evaluated the cause–effect structure and integrated conceptual information Φ of animats evolved to four different task environments that differed primarily in their requirements for internal memory, as illustrated in Figure 14.6C. For each task environment, we simulated 50 independent evolutions with 100 animats at each generation. The probability of an animat to be selected into the next generation was proportional to an exponential measure of the animat's fitness (roulette wheel selection) (Albantakis et al., 2014; Marstaller et al., 2013).

At the end of each evolution, the line of descent (LOD) of one animat from the final generation was traced back through all

---

Caption for FIG. 14.6 (*cont.*) to the more difficult Task 4 developed significantly more concepts, higher $\langle \Phi^{Max} \rangle$ values, and more state entropy H than those animats that evolved to Task 1. Shaded areas around curves denote *SEM* across 50 independent evolutions (LODs). (E) Scatter plots of all evaluated generations of animats from all 50 LODs of Task 4 illustrating the relation of the $\langle \#concepts \rangle$, $\langle \Phi^{Max} \rangle$, and H to fitness, and H to $\langle \Phi^{Max} \rangle$. The circle size is proportional to the number of animats with the same pair of values. Black dots denote the final generation of animats from all 50 independent evolutions. Panels A, C, and D were adapted with permission from Albantakis et al. (2014).

generations, and the cause–effect structures of its ancestors were evaluated every 512 generations. In the more difficult task environments that required more internal memory to be solved (Task 3 and particularly Task 4), the animats developed overall more concepts and higher $\langle \Phi^{\text{Max}} \rangle$ than in the simpler task environments (Tasks 1 and 2),[10] as shown in Figure 14.6D, comparing Tasks 1 and 4. This is even more evident when the tasks are compared at the same level of fitness (light gray line in Figure 14.6D).

As an indicator for the dynamical repertoire (dynamical complexity) of the animats in their respective environments, we measured the state entropy $H = \sum_s p_s \log p_s$ of the animats' hidden and motor elements for the different task environments (see right panel of Figure 14.6D for Tasks 1 and 4). The animats' state entropy increases with adaptation across generations and with task difficulty across the different task environments, similar to the IIT measures. The maximum possible entropy for six binary elements is $H = 6$, if all system states have equal probability to occur. Note that the animats are initialized without connections between elements, and elements without inputs cannot change their state.

During adaptation, the number of connected elements increases, particularly in the more difficult tasks that require more memory. More internal elements mean a greater capacity for memory, entropy, and a higher number of concepts and integrated conceptual information. In this way, fitness, dynamical complexity, and causal complexity are tied together, particularly if the requirement for internal memory is high, even though, in an arbitrary, nonisolated system, the state entropy H could be dissociated from the system's cause–effect structure. This relation is illustrated in Figure 14.6E,

---

[10] For a particular animat generation in one LOD, the IIT measures were evaluated across all network states experienced by the animat during the 128 test trials, weighted by their probability of occurrence. While Φ^{Max} is the integrated conceptual information of the set of elements that forms the major complex (MC), the number of concepts shown in Figure 14.6 was evaluated across all of the animat's elements including sensors and motors, not only the MC. Values for the number of MC concepts behave similarly and can be found in Albantakis et al. (2014).

where $\langle\#concepts\rangle$, $\langle\Phi^{Max}\rangle$, and the state entropy H are plotted against fitness for every animat of all 50 LODs of Task 4. All three measures are positively correlated with fitness ($p = 0.80/0.79/0.54$ Spearman's rank correlation coefficient for $H/\langle\#concepts\rangle/\langle\Phi^{Max}\rangle$ with $p < 0.001$).[11]

In contrast to the state entropy H, the entropy of the sensor states H_{Sen} is mostly task dependent: during adaptation, H_{Sen} increases only slightly for Tasks 3 and 4 and decreases slightly for Tasks 1 and 2 (see Albantakis et al., 2014, figure S4). The entropy of the motor states H_{Mot} represents the behavioral repertoire (behavioral complexity) of the animats and is included in H. H_{Mot} increases during adaptation, but terminates at similar values ($\sim$1.6) for all tasks. This reflects the fact that the behavioral requirements ("catch" and "avoid") are similar in all task environments (see Albantakis et al., 2014, figure S4).

More elements allow for a higher capacity for state entropy H and also higher $\langle\Phi^{Max}\rangle$. Nevertheless, H is also directly related to $\langle\Phi^{Max}\rangle$, since the highest level of entropy for a fixed number of elements is achieved if, for each element, the probability to be in state "0" or "1" is balanced. As we saw above for ECA, balanced rules that output "0" or "1" with equal probability are more likely to achieve high values of $\langle\Phi^{Max}\rangle$ (Figure 14.4B, λ parameter). This is because mechanisms with balanced cause–effect repertoires have on average higher ϕ values and lead to more higher-order concepts, and thus cause–effect structures with higher $\langle\Phi^{Max}\rangle$. Likewise, as shown for Task 4 in Figure 14.6E, right panel ($\rho = 0.66$, $p < 0.001$), animats with high $\langle\Phi^{Max}\rangle$ also have high entropy H.

---

[11] Note that animats from the same LOD are related. The black dots in Figure 14.6E highlight the final generation of each LOD, which are independent of each other. Taking only the final generation into account, H and $\langle\#concepts\rangle$ still correlate significantly with fitness. However, the correlation for $\langle\Phi^{Max}\rangle$ is not significant after correcting for multiple comparisons ($p = 0.63/0.56$ for $H/\langle\#concepts\rangle$ with $p < 0.001$), since having more $\langle\Phi^{Max}\rangle$ even at lower fitness levels has no cost for the animats.

In the last section, we noted that for isolated ECA systems, having a certain level of $\langle \Phi^{Max} \rangle$ and $\langle \#concepts \rangle$ is necessary in order to have the potential for complex dynamics, and thus high state entropy. The animats, however, receive sensory inputs that can drive their internal dynamics. Consequently, also animats with modular, mainly feedforward structures ($\Phi = 0$) can have high state entropy H while they are behaving in their world. Keeping the sensory inputs constant, animats converge to steady states or periodic dynamics of small cycle length within at most seven time-steps. The average length of these transients, measured for the final generation of animats of all 50 LODs in Tasks 1–4, tends to correlate with the average $\langle \Phi^{Max} \rangle$ calculated from all states experienced during the 128 test trials, especially for the simpler tasks 1 and 2 ($\rho = 0.45/0.46/0.43/0.39$ Spearman's rank correlation coefficient for Tasks 1–4 with $p = 0.04/0.03/0.067/0.19$ after correcting for multiple comparisons). Interestingly, there is no correlation between the transient length and the animats' fitness. This is because, in general, high fitness requires a rich behavioral repertoire only while interacting with the world, but not in isolation.

In addition to the state entropy, Albantakis et al. (2014) also assessed how the sensory-motor mutual information (I_{SMMI}) (Ay et al., 2008) and predictive information (I_{Pred}) (Bialek et al., 2001) of the animats evolved during adaptation.[12] I_{SMMI} measures the differentiation of the observed input–output behavior of the animats' sensors and motors. I_{Pred}, the mutual information between observed past and future system states, measures the differentiation of the observed internal states of the animats' hidden and motor elements. Both high I_{SMMI} and high I_{Pred} should be advantageous during adaptation to a complex environment, since they reflect the animats' behavioral and dynamical repertoire, in particular how deterministically one state leads to another. I_{Pred} in the animats is indeed closely tied to the state entropy: it increases during adaptation with increasing fitness and a higher number of internal elements. I_{SMMI}, however, may actually

---

[12] I_{SMMI} and I_{Pred} are calculated as defined in Albantakis et al. (2014) and Edlund et al. (2011).

decrease during adaptation in the animats, since an increase in internal memory may reduce the correlation between sensors and motors, which are restricted to two each (see Albantakis et al., 2014, figure S4). Both I_{SMMI} and I_{Pred} are correlational measures, which depend on the observed distributions of system states. By contrast, analyzing the cause–effect structure of a system requires system perturbations that reveal the causal properties of the system's mechanisms under all possible initial states. The cause–effect structure thus takes the entire set of possible circumstances the animat might be exposed to into account and not just those observed in a given setting. As for CA, an animat's cause–effect structures, evaluated by its $\langle \#concepts \rangle$ and $\langle \Phi^{Max} \rangle$, quantify its intrinsic causal complexity and its dynamical potential.

Under external constraints on the number of available internal elements, having many concepts and high integrated conceptual information Φ proved advantageous for animats in more complex environments (Figure 14.6D and Albantakis et al., 2014). While the simpler Tasks 1 and 2 could be solved (100% fitness) by animats with either integrated ($\Phi > 0$) or modular ($\Phi = 0$) network architectures, only animats with integrated networks reached high levels of fitness in the more difficult Task 3 and particularly Task 4, which required more internal computations and memory (Albantakis et al., 2014). This is because integrated systems can implement more functions (concepts) for the same number of elements, since they can make use of higher-order concepts, irreducible mechanisms specified by combinations of elements.

When particular concepts are selected for during adaptation, higher-order concepts become available at no extra cost in terms of elements or wiring. This degeneracy in concepts may prove beneficial in responding to novel events and challenges in changing environments. Degeneracy here refers to different structures that perform the same function in a certain context (Edelman, 1987; Tononi et al., 1999). Contrary to redundant structures, degenerate structures can diverge in function under different contexts. Animats with integrated networks with many degenerate concepts may already be equipped to master novel situations. In principle, this allows them to adapt faster to unpredicted changes in the environment than animats with modular structures, which first have to expand and rearrange their mechanisms and connectivity (Albantakis and Tononi, 2015a).

In the context of changing environments, large behavioral and dynamical repertoires are advantageous not only at the level of individual organisms, but also at the population level. Albantakis et al. (2014) found that the variety of network connectomes, mechanisms, and distinct behaviors was much higher among animats that solved Tasks 1 and 2 perfectly with integrated network structures ($\langle \Phi^{\text{Max}} \rangle > 0$, high degeneracy) than among animats with the same perfect fitness, but $\langle \Phi^{\text{Max}} \rangle = 0$ (low degeneracy). In Task 1, for example, integrated solutions were encountered in 6 out of 50 LODs; modular solutions in 7 out of 50 LODs. Nevertheless, in analyzing all animats with perfect fitness across all generations and LODs, animats with $\langle \Phi^{\text{Max}} \rangle > 0$ showed 332 different behavioral strategies, while animats with $\langle \Phi^{\text{Max}} \rangle = 0$ produced only 44 different behavioral strategies. The reason is that integrated networks are more flexible and allow for neutral mutations that do not lead to a decrease in fitness. By contrast, modular networks showed very little variability once a solution was encountered. Having more potential solutions should give a probabilistic selective advantage to integrated networks and should lead to more heterogeneous populations, which provide an additional advantage in the face of environmental change.

Taken together, in causally rich environments that foster memory and sensitivity to context, integrated systems should have an adaptive advantage over modular systems. This is because under naturalistic constraints on time, energy, and substrates, integrated systems can pack more mechanisms for a given number of elements, exhibit higher degeneracy in function and architecture, and demonstrate greater sensitivity to context and adaptability. These prominent features of integrated systems also link intrinsic cause–effect power to behavioral and dynamical complexity at the level of individuals and populations.

Conclusion

One hallmark of living systems is that they typically show a wide range of interesting behaviors, far away from thermodynamic equilibrium (e.g., Schrödinger, 1944). How living systems change their states in response to their environment can be seen as a form of natural computation (Still et al., 2012). Among the oldest model systems for the study of natural computation are small, discrete, dynamical systems, CA (Kari, 2005; von Neumann, 1966). CA have revealed that complex dynamical patterns can emerge from simple, local interactions of small homogeneous building blocks. It is thus not surprising that the extensive body of research dedicated to CA has focused mainly on the systems' dynamical properties, investigating what is "happening" during their temporal evolution. While the scientists studying CA may observe intriguing patterns computed by the system, it has been pointed out that these patterns have no relevance for the CA itself in the absence of some kind of "global self-referential mechanism" (Pavlic et al., 2014). IIT provides a framework for establishing precisely to what extent a system "makes a difference" to itself, from its own intrinsic perspective. The cause–effect structure of a system and its integrated conceptual information characterize what a system "is" – how much and in which way it exists for itself, independent of an external observer – rather than what a system happens to be "doing." Consequently, even inactive systems, or systems in a steady state that do not appear to be "doing" anything from the extrinsic perspective, can nevertheless specify rich cause–effect structures (Oizumi et al., 2014) from their own intrinsic perspective. The cause–effect structure of a system thus forms the causal foundation for the system's dynamic behavior, which may or may not manifest itself under the observed circumstances. For purposes of dynamical analysis, evaluating cause–effect structures may help to identify systems that are candidates for complex dynamic behavior.

Acknowledgments

We thank Will Mayner for help with the Python IIT software (PyPhi) and William Marshall for helpful discussions. This work has been supported by the Templeton World Charities Foundation (grant #TWCF 0067/AB41).

REFERENCES

Adamatzky, A., and Martinez, G. J. 2010. On generative morphological diversity of elementary cellular automata. *Kybernetes*, **39**(1), 72–82.

Albantakis, L., and Tononi, G. 2015a. Fitness and neural complexity of animats exposed to environmental change. *BMC Neurosci.*, **16**:P262.

Albantakis, L., and Tononi, G. 2015b. The intrinsic cause–effect power of discrete dynamical systems from elementary cellular automata to adapting animats. *Entropy*, **17**(8), 5472–5502.

Albantakis, L., Hintze, A., Koch, C., Adami, C., and Tononi, G. 2014. Evolution of integrated causal structures in animats exposed to environments of increasing complexity. *PLOS Comput Biol.*, **10**(12), e1003966.

Ay, N., Bertschinger, N., Der, R., Güttler, F., and Olbrich, E. 2008. Predictive information and explorative behavior of autonomous robots. *European Physical Journal B*, **63**(3), 329–339.

Beer, R. D. 2003. The dynamics of active categorical perception in an evolved model agent. *Adaptive Behavior*, **11**(4), 209–243.

Bialek, W., Nemenman, I., and Tishby, N. 2001. Predictability, complexity, and learning. *Neural Computation*, **13**(11), 2409–2463.

Binder, P. M. 1993. A phase diagram for elementary cellular automata. *Complex Systems*, **7**(3), 241.

Culik, K. II, and Yu, S. 1988. Undecidability of CA classification schemes. *Complex Systems*, **2**(2), 177–190.

De Jong, H. 2002. Modeling and simulation of genetic regulatory systems: a literature review. *Journal of Computational Biology*, **9**(1), 67–103.

De Oliveira, G. M. B., de Oliveira, P. P. B., and Omar, N. 2000. Guidelines for dynamics-based parameterization of one-dimensional cellular automata rule spaces. *Complexity*, **6**(2), 63–71.

Edelman, G. M. 1987. *Neural Darwinism: the theory of neuronal group selection.* Basic Books.

Edlund, J. A., Chaumont, N., Hintze, A., Koch, C., Tononi, G., and Adami, C. 2011. Integrated information increases with fitness in the evolution of animats. *PLOS Comput Biol.*, **7**(10), e1002236.

Ermentrout, G. B., and Edelstein-Keshet, L. 1993. Cellular automata approaches to biological modeling. *Journal of Theoretical Biology*, **160**(1), 97–133.

Gardner, M. 1970. Mathematical games: the fantastic combinations of John Conway's new solitaire game "life." *Scientific American*, **223**(4), 120–123.

Hoel, E. P., Albantakis, L., and Tononi, G. 2013. Quantifying causal emergence shows that macro can beat micro. *Proceedings of the National Academy of Sciences*, **110**(49), 19790–19795.

Izhikevich, E. M. 2007. *Dynamical systems in neuroscience*. MIT Press.

Kaneko, K. 2006. *Life: an introduction to complex systems biology*. Springer.

Kari, J. 2005. Theory of cellular automata: a survey. *Theoretical Computer Science*, **334**(1), 3–33.

Knoester, D. B., Goldsby, H. J., and Adami, C. 2014. Leveraging evolutionary search to discover self-adaptive and self-organizing cellular automata. *arXiv preprint arXiv:1405.4322*.

Krawczyk, M. J. 2015. New aspects of symmetry of elementary cellular automata. *Chaos, Solitons & Fractals*, **78**, 86–94.

Langton, C. G. 1990. Computation at the edge of chaos: phase transitions and emergent computation. *Physica D: Nonlinear Phenomena*, **42**(1), 12–37.

Li, W., and Packard, N. 1990. The structure of the elementary cellular automata rule space. *Complex Systems*, **4**(3), 281–297.

Marstaller, L., Hintze, A., and Adami, C. 2013. The evolution of representation in simple cognitive networks. *Neural Computation*, **25**(8), 2079–2107.

Nowak, M. 2006. *Evolutionary dynamics*. Harvard University Press.

Nykamp, D. 2015. *Dynamical system definition*. http://mathinsight.org/definition/dynamical_system, accessed on 15 May 2015.

Oizumi, M., Albantakis, L., and Tononi, G. 2014. From the phenomenology to the mechanisms of consciousness: integrated information theory 3.0. *PLOS Computational Biology*, **10**(5), e1003588.

Pavlic, T. P., Adams, A. M., Davies, P. C. W., and Walker, S. I. 2014. Self-referencing cellular automata: a model of the evolution of information control in biological systems. *arXiv, arXiv:1405.4070*.

Schrödinger, E. 1944. *What is life?* Cambridge University Press.

Schüle, M., and Stoop, R. 2012. A full computation-relevant topological dynamics classification of elementary cellular automata. *Chaos: An Interdisciplinary Journal of Nonlinear Science*, **22**(4), 043143.

Still, S., Sivak, D. A., Bell, A. J., and Crooks, G. E. 2012. Thermodynamics of prediction. *Physical Review Letters*, **109**(12), 120604.

Sumpter, D. J. T. 2006. The principles of collective animal behaviour. *Philosophical Transactions of the Royal Society of London B: Biological Sciences*, **361**(1465), 5–22.

Sutner, K. 1995. On the computational complexity of finite cellular automata. *Journal of Computer and System Sciences*, **50**(1), 87–97.

Tononi, G. 2015. Integrated information theory. *Scholarpedia*, **10**(1), 4164.

Tononi, G., Sporns, O., and Edelman, G. M. 1999. Measures of degeneracy and redundancy in biological networks. *Proceedings of the National Academy of Sciences*, **96**(6), 3257–3262.

Von Neumann, J. 1966. *Theory of self-reproducing automata*. University of Illinois Press.

Wolfram, S. 1984. Universality and complexity in cellular automata. *Physica D: Nonlinear Phenomena*, **10**(1), 1–35.

Wolfram, S. 1985. Undecidability and intractability in theoretical physics. *Physical Review Letters*, **54**(8), 735.

Wolfram, S. 2002. *A new kind of science*. Vol. 5. Wolfram Media.

Wuensche, A., and Lesser, M. 1992. *The global dynamics of cellular automata: an atlas of basin of attraction fields of one-dimensional cellular automata*. Andrew Wuensche.

Zenil, H., and Villarreal-Zapata, E. 2013. Asymptotic behavior and ratios of complexity in cellular automata. *International Journal of Bifurcation and Chaos*, **23**(09), 1350159.

15 Biological Information, Causality, and Specificity

An Intimate Relationship

Karola Stotz and Paul E. Griffiths

The lack of a rigorous account of biological information as a proximal causal factor in biological systems is a striking gap in the scientific worldview. In this chapter we outline a proposal to fill that gap by grounding the idea of biological information in a contemporary philosophical account of causation. Biological information is a certain kind of causal relationship between components of living systems. Many accounts of information in the philosophy of biology have set out to vindicate the common assumption that nucleic acids are distinctively informational molecules. Here we take a more unprejudiced approach, developing an account of biological information and then seeing how widely it applies.

In the first section, 'Information in Biology', we begin with the most prominent informational idea in modern biology – the coding relation between nucleic acid and protein. A deeper look at the background to Francis Crick's Central Dogma, and a comparison with the distinction in developmental biology between permissive and instructive interactions, reveals that 'information' is a way to talk about specificity. The idea of specificity has a long history in biology, and a closely related idea is a key part of a widely supported contemporary account of causation in philosophy that grounds causal relationships in ideas about manipulability and control. In the second section, 'Causal Specificity: An Information-Theoretic Approach', we describe the idea of 'causal specificity' and an information-theoretic measure of the degree of specificity of a cause for its effect. Biological specificity, we suggest, is simply causal specificity in biological systems. Since we have already argued that 'information' is a way to talk about biological specificity, we conclude that causal relationships are 'informational' simply when they are highly specific. The third section, 'Arbitrariness, Information, and Regulation', defends this identification against the claim that only causal relationships in which the relation between cause and effect is 'arbitrary' should count as informational. Arbitrariness has an important role, however, in understanding the regulation of gene expression via gene regulatory networks. Having defended our identification of information with specificity, we show in the final section, 'Distributed Specificity', that information is more widely distributed in biological systems than is often supposed. Coding sequences of DNA are only one source of biological specificity, and hence only one locus of biological information.

INFORMATION IN BIOLOGY

One of the best-known uses of 'information' in biology occurs in Crick's 1958 statement of the 'central dogma of molecular biology':

> *The Sequence Hypothesis* ... In its simplest form it assumes that the specificity of a piece of nucleic acid is expressed solely by the sequence of its bases, and that this sequence is a (simple) code for the amino acid sequence of a particular protein ...

> *The Central Dogma* This states that once 'information' has passed into protein *it cannot get out again*. In more detail, the transfer of information from nucleic acid to protein may be possible, but transfer from protein to protein, or from protein to nucleic acid is impossible. Information means here the *precise* determination of sequence, either of bases in the nucleic acid or of amino-acid residues in the protein.
>
> *(Crick, 1958, pp. 152–153, emphasis in original)*

Here Crick simply identifies the specificity of a gene for its product with the information coded in the sequence of the gene. By doing so, he linked the idea of information very closely to one of the fundamental organising concepts of biology. Biological specificity is nothing less than the 'orderly patterns of metabolic and developmental reactions giving rise to the unique characteristics of the individual and of its species' (Kleinsmith, 2014). From the second half of the nineteenth to the first half of the twentieth century specificity was 'the thematic thread running through all the life sciences' (Kay, 2000, p. 41), starting with botany, bacteriology, immunology, and serology. By mid-century quantum mechanics had provided the necessary insight to explain the observed structural complementarity between molecules in terms of the quantum-physical forces that underlie ability of enzyme and substrate to form a certain number of weak hydrogen bonds. This development of quantum chemistry, majorly driven by Linus Pauling and Max Delbrck in the 1940s, transformed the stereochemical concept of specificity based on the abstract and intuitive side-chain receptor theory (developed by Paul Ehrlich), and their lock-and-key interaction with a ligand (an image suggested by Emil Fischer, both at the turn of the century), into stereochemical specificity based on weak intermolecular forces (Pauling and Delbrück, 1940).

Crick introduces a new, more abstract conception of high selectivity or absolute specificity in terms of how one molecule can precisely specify the linear structure of another. For him it is the *colinearity* between DNA, RNA, and amino acid chains that embodies its specificity. The information that specifies the product is no longer carried by a three-dimensional structure but instead by the linear, one-dimensional order of elements in each sequence. Amongst other consequences, this means that specificity becomes independent of the medium in which this order is expressed (i.e., DNA, RNA, or amino acid chain) and of the kind of reaction by which the specificity is transmitted (i.e., transcription or translation). The same information/specificity flows continuously through these three media and two processes.

According to Crick the process of protein synthesis contains 'the flow of energy, the flow of matter, and the flow of information'.

While he notes the importance of the 'exact chemical steps', he clearly separated this transfer of material substances from what he regarded as 'the essence of the problem', namely the problem of how to join the amino acids in the right order. The flow of 'hereditary information', defined as 'the specification of the amino acid sequence of the protein', solved for him this critical problem of 'sequentialization'.

In his later paper, 'Central Dogma of Molecular Biology', Crick clarified these earlier arguments:

> The two central concepts which had been produced ... were those of sequential information and of defined alphabets. Neither of these steps was trivial ... This temporarily reduced the central problem from a three dimensional one to a one dimensional one ... The principal problem could then be stated as the formulation of the general rules for information transfer from one polymer with a defined alphabet to another.
>
> *(Crick, 1970, p. 561)*

The philosopher Gregory Morgan[1] corresponded with Crick late in his career about his original inspiration to use the term 'information'. Crick's replies of March 20 and April 3, 1998 show the consistency of his view over 40 years. He states that his use of 'information' was influenced by the idea of Morse code rather than Shannon's information theory, which he sees as more concerned with the reduction of noise during transmission. Like Shannon, however, he was not using the idea of information to express the 'meaning' or 'aboutness' of genes. Rather, information was 'merely a convenient shorthand for the underlying causal effect', namely the 'precise determination of sequence'. Information for him solely meant 'detailed residue-by-residue determination'.

---

[1] Personal communication. We are extremely grateful to Morgan for making this correspondence available to us.

The concept of information in terms of the precise determination of sequence primarily offered Crick a way to reduce the transfer of specificity from a three-dimensional to a one-dimensional problem by abstracting away from the biochemical and material connotations of specificity. The conception of biological information defended in this chapter takes this abstraction of the idea of specificity a stage further but is very much in the spirit of Crick's original proposal.

Another biological field in which the concepts of information and specificity have been entwined is developmental biology, although here the idea of information is less tightly associated with DNA. We refer here particularly to the problem of tissue differentiation. Interaction between neighboring cells or tissues in development can lead to further differentiation in one, the responder, as a result of its interaction with the other, the inducer. Developmental biologists commonly distinguish between 'instructive' (or active, explicit, directive) induction, on the one hand, and 'permissive' (or passive, implicit), on the other.

The notion of the specificity of interaction is closely associated with the terms 'instructive' and 'permissive' interaction. When the action system is largely responsible for the specificity of the interaction through the transfer of a specific message, to which the reaction system responds by entering into a particular pathway of differentiation, we speak of an instructive action. When, on the other hand, the specificity of a reaction is largely due to the state of the competence of the reaction system, so that even rather unspecific messages can serve as signals to open up new developmental pathways, we speak of a permissive action (Nieuwkoop et al., 1985, p. 9).

Papers on this subject cite as the two original sources of the distinction between instructive and permissive interactions either Holtzer (1968) or Saxén (1977). All seem to agree that instructive interactions provide instructions or messages simply because these interactions have a high degree of specificity. But the informational language also enters this context regularly:

> Embryonic induction is generally described as an instructive
> event. The problem itself is often posed in terms implying the
> transmission of *informational* molecules [either proteins or
> nucleic acids] from one cell to another cell.
>
> *(Holtzer 1968, p. 152, emphasis added)*

Gilbert's treatment of the vital question regarding the source of specificity illustrates nicely how the instructive/permissive distinction is explained in terms of both specificity and information: 'Instructive partners provide specificity to the reaction, whereas permissive partners ... do not provide specificity ... [They are therefore not] on the same informational level' (Gilbert, 2003).

We conclude from these examples that there are at least some contexts in which the language of information is a way to talk about the relatively high degree of specificity seen in some causal processes in biology. This matters to us, since in the next section we will present an information-theoretic analysis of specificity. If the argument of this last section is correct, then what follows is also an information-theoretic analysis of biological information.

CAUSAL SPECIFICITY: AN INFORMATION-THEORETIC APPROACH

James Woodward (2010) and ourselves (Griffiths and Stotz, 2013; Stotz, 2006) have argued that the idea of causal specificity is closely related to the idea of biological specificity. Causal specificity is an idea from the contemporary philosophy of causation. The philosophy of causation has many concerns, some of them entirely in the domain of metaphysics. The interventionist (or sometimes 'manipulability') account of causation, however, is primarily aimed at explaining why science cares about causation, and using that explanation to think more clearly about causation in scientific practice. Because of its applicability to actual cases of scientific reasoning it has been widely applied to problems in the contemporary philosophy of the life and social sciences. This account of causation focuses on the idea that 'causal relationships are relationships that are potentially exploitable for purposes of manipulation and control' (Woodward, 2010, p. 314). Causation is conceived as a relation between variables in an organized system that can by represented by a directed graph. A variable X is a cause of variable Y when a suitably isolated manipulation of X would change Y. This theory of causation, in it simplest form, can be used to pick out which variables are causes rather than merely correlates. However, a great many things get identified as causes. So, for example, a gene might be a cause for a phenotype, because a mutation (a 'manipulation') would change the phenotype. But equally, a change in the environment (another 'manipulation') will be picked out as a cause if it changes that phenotype.

A comprehensive theory of causation doesn't just distinguish cause from noncause, but can also differentiate between causes in various ways – to identify ones that 'are likely to be more useful for many purposes associated with manipulation and control than less stable relationships' (Woodward, 2010, p. 315). A number of different ways to distinguish types of causes have been suggested, and two of these – stability and specificity –are particularly relevant to understanding biological information. Stability refers to whether an intervention continues to hold across a range of background conditions, and we will not pursue it here. Specificity refers to the fine-grained control that an intervention might have, controlling a gradient of change, rather than a simple on-off switch, for example (Griffiths and Stotz, 2013; Stotz, 2006; Walker and Davies, 2013; Waters, 2007; Woodward, 2010).

The intuitive idea is that interventions on a highly specific causal variable C can be used to produce any one of a large number of values of an effect variable E, providing what Woodward terms 'fine-grained influence' over the effect variable (Woodward, 2010, p. 302). The ideal limit of fine-grained influence, Woodward explains, would be a bijective mapping between the values of the cause and effect variables: every value of E is produced by one and only one value of C and vice versa. The idea of a bijective mapping does not admit of degrees, but in earlier work with collaborators we have developed an information-theoretic framework with which to measure the specificity of causal relationships within the interventionist account (Griffiths et al., 2015). Our work formalises the simple idea that the more specific the relationship between a cause variable and an effect variable, the more information we will have about the effect after we perform an intervention on the cause. This led us to propose a simple measure of specificity:

Spec: the specificity of a causal variable is obtained by measuring how much mutual information interventions on that variable carry about the effect variable.

The mutual information of two variables is simply the redundant information present in both variables. Where $H(X)$ is the Shannon entropy of X, and $H(X|Y)$ the conditional entropy of X on Y, the mutual information of X with another variable Y, or $I(X; Y)$, is given by:

$$I(X; Y) = H(X) - H(X|Y) \tag{15.1}$$

Mutual information is symmetrical: $I(X; Y) = I(Y; X)$. So variables can have mutual information without being related in the manner required by the interventionist criterion of causation. However, our measure of specificity measures the mutual information between interventions on C and the variable E. This is not a symmetrical measure because the fact that interventions on C change E does not imply that interventions on E will change C: $I(\hat{C}; E) \neq I(\hat{E}; C)$, where $\hat{C}$ is read 'do C' and means that the value of C results from an intervention on C (Pearl et al., 2009).

This measure adds precision to several aspects of the interventionist account of causation. Any two variables that satisfy the interventionist criterion of causation will show some degree of mutual information between interventions and effects. This criterion is sometimes called 'minimal invariance' – there are at least two values of C such that a manipulation of C from one value to the other changes the value of E. If the relationship $C \to E$ is minimally invariant, that is, invariant under at least one intervention on C, then C has some specificity for E, that is, $I(\hat{C}; E) > 0$. Moreover, our measure of specificity is a measure of what Woodward calls the 'range of invariance' of a causal relationship – the range of values of C and E across which the one can be used to intervene on the other. Relationships with a large range of invariance have high specificity according to our measure (Griffiths et al., 2015).[2]

In light of the examples in the section, 'Information in Biology', we propose that causal relationships in biological systems can be regarded as informational when they are highly causally specific. Biological specificity, whether stereochemical or informational, seems to us to be simply the application of the idea of causal specificity to biological systems. The remarkable specificity of reactions in living systems that biology has sought to explain since the late nineteenth century can equally be described as the fact that living systems exercise 'fine-grained control' over many variables within those systems. Organisms exercise fine-grained control over which substances provoke an immune response through varying the stereochemistry of recognition sites on antibodies for antigens. They catalyse very specific reactions through varying the stereochemistry of enzymes for their substrates, or of receptors and their ligands. Organisms reproduce with a high degree of fidelity through the informational specificity of nucleic acids for proteins and functional RNAs. Genes are regulated in a highly specific manner across time and tissue through the regulated recruitment of trans-acting factors and the combinatorial control of gene expression and posttranscriptional processing by these factors and the cis-acting sites to which they bind. These are all important aspects of why living systems appear to be 'informed' systems, and what is distinctive about all these processes is that they are highly causally specific.

---

[2] Here we give a simple, absolute measure of specificity. Normalised relatives of our measure are available, as we discuss in this chapter.

ARBITRARINESS, INFORMATION, AND REGULATION

In this section we consider another property that has been said to essentially characterise informational relationships in biology. This is 'arbitrariness', the idea that the relationship between symbols and the things they symbolise represent only one permutation of many possible relationships between them. This is a familiar property of human languages – 'cat' could equally well be used to mean 'cow' and vice versa. Like Crick, we have so far eschewed ideas of meaning and representation, so with respect to our proposal, arbitrariness would mean that the systematic mapping between values of C and E is only one of many possible systematic mappings.

Sahotra Sarkar imposes just such a condition on the informational relationships in biology. Sarkar, known for his critical stance towards the use of informational language in biology, argued that '[e]ither informational talk should be abandoned altogether or an attempt must be made to provide a formal explication of "information" that shows that it can be used consistently in this context and, moreover, is useful' (Sarkar, 2004, p. 261). He makes a serious attempt to provide the required formal explication, a definition of information that both performs a significant explanatory or predictive role and applies to information as it is customarily used. He proposes two adequacy conditions for a biological or genetic account of information:

> Whatever the appropriate explication of information for genetics is, it has to come to terms with specificity and the existence of this coding relationship ... Along with specificity, this arbitrariness is what makes an informational account of genetics useful.
>
> (Sarkar, 2004, pp. 261 and 266)

Sarkar's analysis of specificity is similar to Woodward's and we would urge that he adopt our information-theoretic extension of that analysis. His second condition, arbitrariness, relies on his interpretation of the Central Dogma, according to which it introduces two different types of specificity, namely, 'that of each DNA sequence for its complementary strand, as modulated through base pairing; and that of the relationship between DNA and protein. The latter was modulated by genetic information' (Sarkar, 1996b, p. 858). Sarkar needs to distinguish these two because the relationship between DNA and RNA is not arbitrary – it is dictated by the laws of chemistry. Only the relationship between RNA and protein is arbitrary, because it depends on the available t-RNAs. Many different t-RNAs are available, and substituting these would lead to different genetic codes.

In our view, however, Crick clearly states that 'genetic information' applies to the specification 'either of bases in the nucleic acid or in amino acid residues in the protein' (Crick, 1958, p. 153). DNA provides informational specificity for RNA as much as RNA provides specificity for amino acid chains. Ulrich Stegmann agrees that the difference between the two is 'irrelevant to the question of whether they carry information: they all do' (Stegmann, 2014, p. 460). There is just one type of informational specificity, and what distinguishes it from conformational specificity is its independence from the medium in which it is expressed or the mechanism by which it is transferred. Hence if arbitrariness should be regarded as an important condition for informational language in biology, it should be for the reason of this medium independence in general, rather than the coding relationship between RNA and amino acids in particular. The coding relationship between RNA and amino acid is not the reason that led to Crick's use of the idea of information in formulating the central dogma.

Like ourselves, Sarkar aims to explicate the notion of information in such a way as to make it a useful tool for biology. But adding the second condition of arbitrariness, at least when applied just to the coding relationship, to his definition of information seems to us to come with some substantial costs. It may exclude the concept of information from what seems to us one of its most useful roles, namely as a way to compare different sources of biological specificity, as we do in the last section. This is because many of these alternative sources of specificity, like the DNA-RNA relationship, are not arbitrary.

This is not to say that arbitrary relationships play no vital role in biology. It is interesting that the notion of arbitrariness has been introduced in another area of biology that regularly deploys informational language, namely, the regulation of gene expression through gene regulatory networks.

The pioneers of research into gene regulation, François Jacob and Jacques Monod, derived a notion of arbitrariness from their operon model (Jacob and Monod, 1961). The biosynthesis of the enzyme galactosidase is indirectly controlled by its substrate, β-galactosides. This indirect control is made possible by the intervening repressor of the gene, an allosteric protein, which is rendered inactive by its effector, the substrate of the enzyme expressed by the gene. The repressor thereby indirectly transduces the controlling signal.

> There is no chemically necessary relationship between the fact
> that β-galactosidase hydrolyses β-galactosides, and the fact that
> its biosynthesis is induced by the same compounds.
> Physiologically useful or 'rational', this relationship is chemically
> arbitrary – 'gratuitous', one may say. This fundamental concept of
> gratuity – i.e., the independence, chemically speaking, between
> the function itself and the nature of the chemical signal
> controlling it – applies to allosteric proteins.
>
> *(Monod, 1971, p. 78)*

Most controlling environmental stimuli have only an indirect controlling effect on gene expression, which is mediated or transduced by the processes of transcription, splicing, or editing factors. The latter relay the environmental information to the genome. So the role of allosteric proteins in signal transduction due to their chemical arbitrariness that Monod has identified could be assigned to many signalling molecules in biological signal transduction systems, just as is the case for many human-designed signalling systems. It is this arbitrariness that renders the system flexible and evolutionarily evolvable.

> The result – and this is the essential point – is that ... everything is possible. An allosteric protein should be seen as a specialized product of molecular 'engineering' enabling an interaction, positive or negative, to take place between compounds without chemical affinity, and thereby eventually subordinating any reaction to the intervention of compounds that are chemically foreign and indifferent to this reaction. The way hence in which allosteric interactions work permits a complete freedom in the choice of control. And these controls, subject to no chemical requirements, will be the more responsive to physiological requirements, by virtue of which they will be selected according to the increased coherence and efficiency they confer on the cell or organism. In short, the very gratuitousness of the systems, giving molecular evolution a practically limitless field for exploration and experiment, enabled it to elaborate the huge network of cybernetic interconnections which makes each organism an autonomous functional unit, whose performances appear to transcend, if not to escape, the laws of chemistry.
>
> *(Monod, 1971, pp. 78–79)*

The mutual information between the specificity of the environmental signal for the regulatory factor, on the one hand, and the specificity of the regulatory factors for a certain gene via its regulatory sequence, on the other hand, are chemically arbitrary and subject to the convention of an intervening allosteric biomolecule.

The central feature of such a relationship between any two pathways is that it is subject to heritable variation. This means that an environmental stimulus may lead in future to a quite different, adaptive response by the system, if mediated by a novel signalling protein that has evolved independent specificities to both the environmental stimulus (its effector) and the appropriate regulatory sequence (its substrate). We can understand the regulation of gene expression as an internal signalling game where sender and receiver are not two organisms but parts within one plastic organism (Calcott, 2014). The organism encounters two environments, and a different behaviour is optimal in each environment. The sender is a sense organ, or transducer, reacting to the environment by sending a signal inside the organism. The receiver is an effector converting the signal into some behaviour that changes how the organism as a whole interacts with that environment. Signalling occurs inside the organism, and the evolution of a signalling system allows it to optimally map the different environments to the appropriate behaviour. Signalling arose because the modular structure – the separation of transducer and effector – created a coordination problem. For the organism to respond adaptively, it needed to coordinate these parts, and a signalling system provided the solution. Signalling, from this internal perspective, is a way of building adaptive, plastic organisms.

What such a signalling system allows is the decoupling of informational dynamics from the dictates of local chemistry. According to Walker and Davies, one of the hallmarks of biological versus nonbiological systems is the separation between their informational and mechanical aspects (Walker and Davies, 2013, p. 4). This reminds us of Crick's insistence on the importance of the medium independence of informational specificity. But more important, it stresses the relationship between arbitrariness and informational control.

So arbitrariness is, indeed, an important feature of information processing in living systems. It is at last one of the fundamental keys to evolvability. But this, we would argue, is not a good reason to add arbitrariness to the definition of biological information. Arbitrary relationships are prevalent in biological signalling networks because of their biological utility, not because of the definition of information!

DISTRIBUTED SPECIFICITY
Griffiths and Stotz (2013) have termed the encoding of specificity 'Crick information'. If a cause makes a specific difference to the linear sequence of a biomolecule, it contains Crick information for

that molecule. This definition embodies the essential idea of Crick's sequence hypothesis, without in principle limiting the location of information to nucleic acid sequences, as Crick does. Our definition of Crick information can clearly be applied to other causal factors that affect gene expression. However, it is a specifically biological conception of information, rather than a general one such as Shannon's mutual information, or our measure of causal specificity, because by definition it applies only to causes that specify the order of elements in a biomolecule.

Crick's Central Dogma was based on a very simple picture of how the specificity of biomolecules is encoded in living cells. We now know that in eukaryotes, coding regions are surrounded by a large number of noncoding sequences that regulate gene expression. The discrepancy between the number of coding sequences and the number of gene products leads to the insight that the informational specificity in coding regions of DNA must be amplified by other biomolecules in order to specify the whole range of products. 'Precise determination' implies a one-to-one relationship, and if we focus on coding sequences alone, we find a one-to-many relationship between sequence and product. Different mechanisms of gene regulation co-specify the final linear product of the gene in question, first by activating the gene so it can get transcribed, second by selecting a chosen subset of the entire coding sequence (e.g., alternative splicing), and third by creating new sequence information through the insertion, deletion, or exchange of single nucleotide letters of the RNA (e.g., RNA editing). Thus, specificity, and hence Crick information, is distributed between a range of factors other than the original coding sequence: DNA sequences with regulatory functions; diverse gene products such as transcription, splicing, and editing factors (usually proteins); and noncoding RNAs (Stotz, 2006).

Absolute specificity turns out to be not inherent in any single biomolecule in these molecular networks, but induced by regulated recruitment and combinatorial control. And it is here that we will find that the networks cannot be reduced to DNA sequences plus gene products, because many of the latter need to be recruited, activated, or transported to render them functional. The recruitment, activation, or transportation of transcription, splicing, and editing factors allow the environment to have specific effects on gene expression (being 'instructive' rather than merely 'permissive' in the terms introduced in the first section). Some gene products serve to relay environmental (Crick) information to the genome. While in embryology and morphogenesis it is often acknowledged that environmental signals play a role in the organisation of global activities, they are rarely seen to carry information for the precise determination of the nucleic acid or amino acid chains in gene products. But this is precisely what occurs. Not just morphogenesis at higher levels of organisation, but even the determination of the primary sequence of gene products is a creative process of (molecular) epigenesis that cannot be reduced to the information encoded in the genome alone (Griffiths and Stotz, 2013; Stotz, 2006).

Interestingly, concurrent with Crick's Central Dogma, the ciliate biologist David L. Nanney acknowledged that the 'library of specificities' found in coding sequences needed to be under the control of an epigenetic control system. In other words, in addition to requiring both an analogue and a digital conception of specificity, the study of biological development requires two sources of information. In an immediate response to Crick's new picture of sequential information coded in DNA, Nanney pointed out:

> This view of the nature of the genetic material ... permits, moreover, a clearer conceptual distinction than has previously been possible between two types of cellular control systems. On the one hand, the maintenance of a 'library of specificities', both expressed and unexpressed, is accomplished by a template replicating mechanism. On the other hand, auxiliary mechanisms with different principles of operation are involved in determining which specificities are to be expressed in any particular cell. 'To simplify the discussion of these two types of

systems, they will be referred to as "genetic systems" and "epigenetic systems".'

(Nanney, 1958, p. 712)

In a similar vein, Crick's biographer Robert Olby remarks of the Central Dogma:

> Clearly, in concentrating on this aspect of informational transfer he was setting aside two questions about the control of gene expression – when in the life of a cell the gene is expressed and where in the organism. But these are also *questions of an informational nature*, although not falling within Crick's definition.
>
> (Olby, 2009, p. 251, emphasis added)

As it has turned out, many epigenetic mechanisms are strongly associated with DNA. Developmental biologist Scott Gilbert argues that the specificity of a reaction 'has to come from somewhere, and that is often a property of the genome' (2003, 349). But since all cells start with exactly the same genetic library of specificities, that can't be the whole story of differentiation. Nanney describes this as a developmental paradox: 'How do cells with identical genetic composition acquire adaptive differences capable of being maintained in clonal heredity?' (Nanney, 1989). Gilbert indeed acknowledges that the action of a gene itself 'depends upon its context. There are times where the environment gets to provide the specificity of developmental interaction' (2003, 350). So we conclude that while genes are seen as a key source of specificity, in biology causes are not regarded as informative merely because they are genetic, but whenever they are highly specific.

Many years later Nanney looked back on this period in the late 1950s as one in which the powerful image of the double helix caused a 'near disruption of an incipient merging of cybernetics with regulatory biology'. It 'may have hindered the exploration of the systemic components of living systems, which are not just creatures reified from the "blueprints", but essential complementary components of life that reciprocally regulate the nucleic system' (Nanney, 1989). In recent years, however, our image of how biological systems exercise fine-grained control over their internal processes has developed to the point where his description of the two complementary control systems seems quite conservative.

It is now clear that the epigenetic control system, if we still want to call it that, not only regulates when and where the specificities encoded in the library are to be expressed.[3] It also substantially augments the information of the literal coding sequence. A strange aspect of the management of genetic information is that the epigenetic control system – which Paul Davies likens to 'an emergent self-organizing phenomenon' (see Davies, 2012, p. 42) – does not just provide a supervising function on the expression of the specificities encoded in the DNA, in the sense of when, where, and how much will be expressed. Since the information encoded in the DNA does not entail a complete set of instructions for which biomolecules shall be synthesised, the epigenetic control system amplifies the information of the literal code (Davidson, 2002). Genes are not only switched on and off, even though this already 'leads to exponentially more information being stored in the system (since a set of N genes can have $2N$ distinct states)' (see Davies, 2012, p. 43). Eukaryotes have epigenetic mechanisms that allow them to produce many products from a single coding region, ranging from just two up to thousands of isoforms of the resulting protein.

Most epigenetic mechanisms are now fairly well understood at the molecular level. Most of them include chemical modifications of the DNA or the tails of the histone protein around which the DNA is wrapped. The posttranscriptional processing mechanisms,

---

[3] Woodward suggests that specificity includes both the 'systematic dependencies between a range of different possible states of the cause and different possible states of the effect, as well as dependencies of *the time and place of occurrence* of E on the time and place of C' (Woodward, 2010, pp. 304–305, emphasis added). So even in Nanney's original vision, the epigenetic system is an additional source of specificity.

mainly alternative slicing and RNA editing that create this large range of gene products, are also fairly well understood. But if epigenetic mechanisms are simply a set of physical modifications of DNA, isn't the organism still an expression of its genome, even if the genome is a little more complex than initially supposed? This will not do because the molecular mechanisms and epigenetic marks are just the final stages of regulatory processes that start far from the genome. For instance the up- or down-regulation of the glucocorticoid receptor gene in the hypothalamus of a rat pup is proximally caused by the increased or decreased methylation state of the receptor's promoter region. This in turn is influenced by the increased or decreased expression and activation of the transcription factor NGF1-A. Increased expression of NGF1-A is due to an increased serotonin tone in the hippocampus. But this in turn is being caused by the mother rat's licking and grooming of her pup, which in turn reflects the more or less stressed state of the mother due to the environment in which she finds herself. The mother's maternal care behavior comprised part of the environmental context of the rat pup. The increased serotonin tone represents a change of the overall state of the whole system, with a range of downstream effects, one of which is a change in the expression of the glucocorticoid receptor. This in turn produces a range of bottom-up effects on the system in terms of a changed behavioural repertoire. This is just one example of how the environment or the system as the whole is ultimately affecting the expression of genes (Meaney, 2001; Weaver et al., 2007). Therefore we can say that a substantial amount of information needed to construct an organism is derived from elsewhere, such as the organism's environment. This information augments or amplifies the information inherited via the genome.

INFORMATION AND 'DOWNWARD CAUSATION'

We have argued that additional specificity, or information, is derived from the environmental context, but it may also be generated de novo by physical processes of self-organisation. Self-organisation is the spontaneous formation of well-organised structures, patterns, or behaviors. In biology it means the self-maintaining organisation of constraints that harness flows of matter and energy and allow the 'constrained release of energy into relatively few degrees of freedom' (Kauffman, 1969, p. 1094). Biological systems, in Kauffman's term, '[act] on their own behalf' when they constrain exergonic processes in a specific way to produce work, which can be used to generate endergonic processes, which in turn generate those constraints canalising exergonic processes.[4] It has often been suggested that such processes are an additional source of order in biological systems.

Walker and Davies have recently characterised life by 'context-dependent causal influences, and in particular, that top-down (or downward) causation – where higher-levels influence and constrain the dynamics of lower-levels in organizational hierarchies – may be a major contributor to the hierarchal structure of living systems' (Walker and Davies, 2013, p. 1).

Downward causation shouldn't be understood as the direct dynamic interaction of the whole with some of their parts. It has long been acknowledged in the physical sciences that in dynamic – efficient – causation, only the interaction between parts at the same ontological level has causal effectiveness. The way that the overall biological system is still able to exert real causal effects is by way of informational control via feedback mechanisms that influence the dynamic interaction between the parts (Auletta et al., 2008). Philosophers Carl Craver and William Bechtel (2006) have advocated this view more generally, in an attempt to rid the idea of downward causation of any mysterious overtones. They suggest that interlevel relationships, such as the interactions between parts and the whole, should not be understood as causal relationships at all, even though these relationships exert real influences on the system at different levels. Both top-down and bottom-up causation

---

[4] An endergenic reaction absorbs and stores energy from the surroundings. During exergenic reactions, stored energy is released to drive various functions.

describe mechanistically mediated effects. Mechanistically mediated effects are hybrids of constitutive and causal relations in a mechanism, where the constitutive relations are interlevel, and the causal relations are exclusively intralevel.

(Craver and Bechtel, 2007)

A system as a whole – a higher-level entity – is engaged in a process that would not happen without some aspects of the organisation of that system, and which therefore needs to be understood at the higher level. But this system is composed of parts, and as the system as a whole changes, so do the parts, obviously. The relation between the process going on at the systems level and a change in one part is not because of an additional causal relation between system as a whole and that part (over and above the interaction of the part with other parts) but because of the relation of constitution between the system and its parts.

It is in this sense that we understand and endorse Walker and Davies' claim that 'algorithmic information gains direct, context-dependent, causal efficacy over matter' (Walker and Davies, 2013, p. 2). That does not just mean that the digital information within the genetic code just by itself gains such control over matter. After all, as Nanney had realised some 65 years ago, the expression of the repository of information within DNA is in need of epigenetic control. 'The algorithm itself is therefore highly delocalised, distributed inextricably throughout the very physical system whose dynamics it encodes' (Walker and Davies, 2013, p. 5). The causal efficacy is achieved through some 'unique informational management properties ... Focusing strictly on digital storage therefore neglects this critical aspect of how biological information is processed' (Walker and Davies, 2013, pp. 2–3).

Conclusion

Sarkar has argued that the conventional account of biological information as coded instructions in the sequence of DNA nucleotides lacks explanatory power. He calls for, first, the development of a 'systematic account of specificity', and, second, an 'elaboration of a new informational account' with wider applicability than nucleic acid alone (Sarkar, 1996a, p. 222). If the latter course were to be adopted, he suggested, it would be 'highly unintuitive not to regard [epigenetic specifications] as "transfers of information" if "information" is to have any plausible biological significance' (Sarkar, 1996a, p. 220). Our proposal in this chapter represents a synthesis between Sarkar's two ways forward, namely, a systematic account of specificity and a new approach to biological information (Griffiths et al., 2015).

Biological specificity is simply causal specificity in biological systems. Causal specificity is a degree property of causal relationships – the more specific a relationship, the more apt it is for the exercise of fine-grained control over the effect. In the second section we gave a brief summary of how this property can be measured using tools from information theory. Informational language in biology represents a way to talk about specificity. No doubt informational language is used for many other purposes in biology as well, but the cases we have presented in which it relates to specificity are central to molecular and developmental biology. As a result we feel justified in calling our information-theoretic analysis of specificity an analysis of biological information.

What is distinctive about living systems, we would argue, is that they are structured so that many of their internal processes have an outstanding degree of causal specificity when compared with most nonliving systems. This underlies the phenomenon that first attracted the label of 'specificity' in biology – the ability of organisms to develop in a very precise way and to respond in a very selective and precise way to their circumstances. The idea that living systems differ from nonliving systems by being 'informed' – under the control of information – makes a great deal of sense in terms of our analysis of biological information as causal specificity. However, there is a great distance between a broad, philosophical interpretation like this and an actual scientific theory of the informational nature of living systems. In the final two sections we have reviewed some of the ideas that we think may form part of such a theory.

Acknowledgements

This publication was made possible through the support of a grant from the Templeton World Charity Foundation, 'Causal Foundations of Biological Information', TWCF0063/AB37. The opinions expressed in this chapter are those of the authors and do not necessarily reflect the views of the Templeton World Charity Foundation.

REFERENCES

Auletta, G., Ellis, G. F. R., and Jaeger, L. 2008. Top-down causation by information control: from a philosophical problem to a scientific research programme. *Journal of the Royal Society Interface*, **5**(27), 1159–1172.

Calcott, B. 2014. The creation and reuse of information in gene regulatory networks. *Philosophy of Science*, **81**(5), 879–890.

Craver, C. F., and Bechtel, W. 2007. Top-down causation without top-down causes. *Biology & Philosophy*, **22**(4), 547–563.

Crick, F. H. C. 1958. On protein synthesis. *Symposia of the Society for Experimental Biology*, **12**:138–163.

Crick, F. H. C. 1970. Central dogma of molecular biology. *Nature*, **227**(5258), 561–563.

Davidson, N. O. 2002. The challenge of target sequence specificity in C U RNA editing. *Journal of Clinical Investigation*, **109**(3), 291–294.

Davies, P. C. W. 2012. The epigenome and top-down causation. *Interface Focus*, **2**(1), 42–48.

Gilbert, S. F. 2003. Evo-devo, devo-evo, and devgen-popgen. *Biology and Philosophy*, **18**(2), 347–352.

Griffiths, P. E., and Stotz, K. 2013. *Genetics and philosophy: an introduction.* Cambridge University Press.

Griffiths, P. E., Pocheville, A., Calcott, B., Stotz, K., Kim, H., and Knight, R. 2015. Measuring causal specificity. *Philosophy of Science*, **82**(4):529–555.

Holtzer, H. 1968. Induction of chondrogenesis: a concept in quest of mechanisms. In In R. Gleischmajer and R. E. Billingham (eds.), *Epithelial-mesenchymal interactions*, 152–164. Williams & Wilkins.

Jacob, F., and Monod, J. 1961. Genetic regulatory mechanisms in the synthesis of proteins. *Journal of Molecular Biology*, **3**(3), 318–356.

Kauffman, S. A. 1969. Metabolic stability and epigenesis in randomly constructed genetic nets. *Journal of Theoretical Biology*, **22**(3), 437–467.

Kay, L. E. 2000. *Who wrote the book of life?: a history of the genetic code.* Stanford University Press.

Kleinsmith, L. J. 2014. *Biological specificity.* Retrieved 29 January 2015 from www.accessscience.com/content/biological-specificity/082900.

Meaney, M. J. 2001. Maternal care, gene expression, and the transmission of individual differences in stress reactivity across generations. *Annual Review of Neuroscience*, **24**(1), 1161–1192.

Monod, J. 1971. *Chance and necessity: an essay on the natural philosophy of modern biology.* Translated from the French by Austryn Wainhouse. Knopf.

Nanney, D. L. 1958. Epigenetic control systems. *Proceedings of the National Academy of Sciences*, **44**(7), 712–717.

Nanney, D. L. 1989. Metaphor and mechanism: 'epigenetic control systems' reconsidered. Paper presented at the Symposium on the Epigenetics of Cell Transformation and Tumor Development, American Association for Cancer Research, Eightieth Annual Meeting, San Francisco, California, May 26, 1989, San Francisco, CA. www.life.illinois.edu/nanney/epigenetic/sanfrancisco.html.

Nieuwkoop, P. D., Johnen, A. G., and Albers, B. 1985. *The epigenetic nature of early chordate development: inductive interaction and competence.* Cambridge University Press.

Olby, R. C. 2009. *Francis Crick: hunter of life's secret.* Cold Spring Harbor Laboratory Press.

Pauling, L., and Delbrück, M. 1940. The nature of intermolecular forces operative in biological processes. *Science*, **92**, 77–79.

Pearl, J., et al. 2009. Causal inference in statistics: an overview. *Statistics Surveys*, **3**, 96–146.

Sarkar, S. 1996a. Biological information: a skeptical look at some central dogmas of molecular biology. *Boston Studies in the Philosophy of Science*, **183**, 187–232.

Sarkar, S. 1996b. Decoding "coding": information and DNA. *BioScience*, **46**(11), 857–864.

Sarkar, S. 2004. Genes encode information for phenotypic traits. In Hitchcock, C. (ed.), *Contemporary debates in the philosophy of science.* Blackwell.

Saxén, L. 1977. Directive versus permissive induction: a working hypothesis. *Society of General Physiologists Series*, **32**, 1.

Stegmann, U. 2014. Causal control and genetic causation. *Noûs*, **48**(3), 450–465.

Stotz, K. 2006. Molecular epigenesis: distributed specificity as a break in the central dogma. *History and Philosophy of the Life Sciences*, **28**(4): 533–548.

Walker, S. I., and Davies, P. C. W. 2013. The algorithmic origins of life. *Journal of the Royal Society Interface*, **10**(79), 20120869.

Waters, C. K. 2007. Causes that make a difference. *Journal of Philosophy*, **104**(11), 551–579.

Weaver, I. C. G., D'Alessio, A. C., Brown, S. E., Hellstrom, I. C., Dymov, S., Sharma, S., Szyf, M., and Meaney, M. J. 2007. The transcription factor nerve growth factor-inducible protein A mediates epigenetic programming: altering epigenetic marks by immediate-early genes. *Journal of Neuroscience*, **27**(7), 1756–1768.

Woodward, J. 2010. Causation in biology: stability, specificity, and the choice of levels of explanation. *Biology & Philosophy*, **25**(3), 287–318.

Part V From Matter to Mind

16 Major Transitions in Political Order

Simon DeDeo

> [T]hey then threw me upon the bed, and one of them (I think it
> was Mary Smith) kneeled on my breast, and with one hand held
> my throat; Mary Junque felt for my money; by my struggling
> about, they did not get it at that time; then they called another
> woman in ... when she came in, they said cut him! cut him! –
> Evidence of Benjamin Leethorp in the trial of Mary Junque and
> Mary Smith for grand larceny, Old Bailey Criminal Court,
> London, England; April 4, 1779 (Hitchcock et al., 2012)

Unless we are historians, the eighteenth-century world of Junque,
Smith, and Leethorp is almost impossible to imagine. In stealing
from Leethorp, the two women put themselves at risk not only of
imprisonment but of indentured servitude in the colonies and even
death. Leethorp, for his part, began his evidence by explaining to
the jury how he was seeking a brothel different from the one in
which he was throttled, stripped, and robbed. Junque and Smith were
without benefit of legal counsel, and Smith's witnesses, unaware of
the trial date, did not appear. The court condemned them to branding
and a year's imprisonment in less than 500 words. The indictment,
formally for a nonviolent offense, was one of hundreds of its kind that
decade marked by assault, knives, and (sometimes) freely-flowing
blood.

In the risks they ran and the things they were ashamed of, the
three are alien to us; in its casual violence, so was the society that
enclosed them. Yet this world gradually, continually, evolved into
one far less tolerant of violence and yet far more protective of an
individual's rights – into the world, in other words, of most readers
of this volume. How witnesses, victims, and defendants spoke about
both facts and norms in the law courts of London shifted, decade by

decade, over the course of one hundred and fifty years (Klingenstein et al., 2014). This shift in speech paralleled a similar decline in how people behaved toward each other, as the state came, increasingly, to enforce its monopoly on violence (Elias, 1982).

These changes took place in the decentralized common-law courts, among hundreds of thousands of victims and defendants. Acts of Parliament, sensational crimes, the invention of the criminal defense lawyer – these changed the courts, but in the moments of their introduction showed little effect on changes in the speech and practices themselves. We are predisposed to see the introduction of a law as identical to the recognition and enforcement of the moral sentiments it invokes. Yet it is, in the final analysis, individuals who constitute a social world. Laws and formal practices may be created by a small group that can unilaterally enforce its will, but they often lag behind the conditions they ratify; when laws do appear, they have unpredictable effects on the minds of the people they concern (Bowles, 2008; Gneezy and Rustichini, 2000).

Evidence from the quantitative behavioral and social sciences accumulates daily for the existence of a complex relationship between individual minds and the persistent social worlds they create. Over decades of development, new styles of prose nucleate on the periphery of the generation that came before, as writers perceive the patterns of the past and struggle with their influence (Hughes et al., 2012). French revolutionaries borrow words such as *contract*, *rights*, and *the people* from Enlightenment philosophers to both signal and make possible their shifting political alliances (Baker, 1990); these same words appear, hundreds of years later, as signals in the House and Senate of twenty-first-century America (Correia et al., 2015). Pre-Hispanic Mexico and twenty-first-century Europe have similar patterns in the distribution of city sizes, outputs, and infrastructure, showing how widely varying cultures find similar solutions to the management of social contact over more than three millennia (Ortman et al., 2014).

Such phenomena are often called political, but *Homo sapiens* is not the only political animal. Increasing evidence from the behavioral sciences shows that social animals such as pig-tailed macaques and monk parakeets interact not only with each other but with the creations of their society as a whole. As we approach our own branch on the evolutionary tree, we find a sequence of transitions in the nature of the relationship between the individual and the group: individuals come to know coarse-grained facts about their social worlds; they gain the ability to reason normatively, from a collective ought; they gather their norms into self-reinforcing bundles. New research provides a quantitative window onto the distinct and traceable imprints each of these transitions leaves on the logic of society.

In their book *Major Transitions in Evolution*, John Maynard Smith and Eörs Szathmáry (1997) argued that leaps in complexity over evolutionary time were driven by innovations in how information is stored and transmitted. Our *social feedback hypothesis* extends their argument to account for the major transitions in political order. We argue that these later transitions are driven by innovations in how information is *processed*.[1]

Our attention to information processing focuses in particular on the lossy compression of large numbers of individual-level facts to produce coarse-grained representations of the world. Understanding what coarse-graining is, and how it works, is essential. We begin there.

COARSE-GRAINING THE MATERIAL WORLD

To build a scientific account of the origins and major transitions in political order, we begin in a perhaps unexpected place, with a question at the heart of twentieth-century physics: What is the charge of the electron? This apparently simple problem of measurement is far

---

[1] We do not, however, describe the evolutionary pressures that might drive the creation of these novel abilities; most notably, the collective action problem (Blanton and Fargher, 2007; Carballo et al., 2014), whose study has formed the basis of fruitful contact between the anthropological and political sciences.

more subtle than it appears, and its resolution was a major advance with unexpected implications.

With the classical theory of electromagnetism – the one taught in high school – it is simple to devise any number of experiments that can measure the electron charge, which appears constant no matter how it is studied. But the extensions of electromagnetism to the quantum domain are far less tractable; depending on the calculations one does, the apparent charge varies and can even, when the mathematics are worked out, diverge.

In response to this unacceptable state of affairs, physicists considered the idea that the charge of the electron might vary depending on the scale – literally, the physical size – on which the experiment is done. Rather than construct an explicit, mechanistic account of the electron's substance, they developed a theory that described the dynamics of a smoothed-out version of the electromagnetic fields it creates. If you average those fields on centimeter scales, you find one theory, and one effective charge; if you average on scales of meters or nanometers, you find others. As you retain information about smaller and smaller distances, the implied properties of the electron shift rather than stabilize.[2] Such shifts allow us to predict the fields we see without us ever having to know the fundamental mechanism itself.

Electrodynamics is just one example of how physicists built a theory not on a detailed account of underlying mechanisms but on the rules obeyed by averaging their effects. To do this averaging in the case of electromagnetism, physicists were naturally drawn to the idea of a spatial average. When mechanisms are local – when a point X can influence a point Y only via intermediate points between X and Y – one can retain a great deal of predictive power by averaging together points that are physically nearby. Because of how influence propagates, it makes little sense to average together two distant

---

[2] The process by which these properties changed was, for historical reasons, given the name "renormalization"; see Kadanoff (2000) for a simple introduction and Fisher (2004) for extended discussion.

points; conversely, we can build a reliable, if only partial, theory from considering the interactions between neighborhoods.

A simple example of this spatial coarse-graining is provided by cellular automata. These discrete, spatially-organized systems are governed by a local mechanism that evolves deterministically. The state of any point in the system is determined by its neighbours at the time-step before. If we "squint" – i.e., if we blur the system, averaging nearby points and reducing the resolution – the objects of the new, coarse-grained system will obey a different set of laws. Among other things, this lossy compression means that some of the information necessary to predict the fine-grained evolution will be lost; in general, this loss of predictivity will affect the coarse-grained level as well, making a system that is fundamentally deterministic appear to follow probabilistic laws.

An example is shown in Figure 16.1; we begin with the exact solution, down at the mechanism scale (panel A). If we coarse-grain in space (panel B) or both space and time (panel C), we have fewer blocks to keep track of while still preserving some of the gross features of the system (such as the transition, in this figure, to diagonal order around the midway point). In dropping fine-grained complexity, however, our new logic becomes probabilistic, not deterministic. We have gained simplicity at the cost of predictive accuracy. The so-called critical points of this phenomenon, for the case of cellular automata, have been investigated in detail by Edlund and Jacobi (2010).

Spatial coarse-graining is not the only way to simplify a system, and in many cases may not be appropriate. When we move from the physical to the biological or social sciences we find systems that are fundamentally long-range in nature or have mechanisms that tie together distant locations. Averaging nearby points might simplify the system, but destroy any possibility of finding a reasonable model to relate these coarse-grained states. In a cell, for example, a fragment of RNA should not be averaged with nearby molecules to describe a cell in terms of the local density of its cytoplasm; better descriptions might summarize the counts of different RNA sequences within

A   B   C

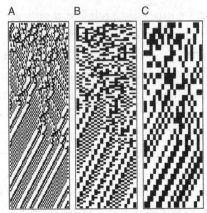

FIG. 16.1 Coarse-graining a one-dimensional cellular automata. (A) One hundred iterations of the $r = 2$, $W = 360A96F9$ rule, with random initial conditions; (B) the same run, coarse-grained by Kadanoff decimation along the spatial axis. While perceptual and memory costs are reduced by a factor of three, the coarse-grained system becomes harder to predict and deterministic rules become probabilistic in nature. (C) The same run, coarse-grained in both time and space. The system is now simplified by a factor of nine; we preserve approximate relationships, and rough, probabilistic logics of evolution.

the cell, even when they occur at large spatial separations. A lossy compression is to be evaluated not only on how much it simplifies a system but on the extent to which that resulting system obeys reasonably reliable (and hopefully simple) laws.

This two-fold criterion – simplification and prediction – extends to the coarse-graining of systems that are nonspatial but still topological in nature, meaning that there is some notion of what is "nearby," closer or further. A classic example is provided by the science of complex networks. Within this field, an entire industry is devoted toward the problem of community detection and network clustering, which tries to group nodes on the basis of the larger network topology. Nodes that are connected to each other are considered "nearby" in some important sense, and the community detection problem amounts to developing innovative ways to summarize these relationships and to group nodes in larger clusters, or communities (Fortunato, 2010). However, this is only part of the problem: when deciding between different community detection algorithms for use on a dynamically evolving system, we should also ask about the extent to which the new coarse-graining obeys reliable dynamical laws (see Wolpert et al., 2014, and references therein). Going beyond the network case, it is worth nothing that not all processes of interest take place on a topological space. Many systems lack spatiality altogether – we may, for example, even wish to coarse-grain a computer program (DeDeo, 2011).

The laws that obtain for a coarse-grained system are known as *effective theories*. The choice of the lossy compression is dictated by the goal of producing objects that lead to good effective theories that allow for description, prediction, and explanation. When we cluster a high-dimensional dataset, we usually hope to find simplified descriptions of its patterns that provide predictive leverage. If a scientist uses k-means, say, her goal is to find a simpler description of the world. Rather than a list of dozens of coordinates, she might find "three clusters with means μ_i, variance σ_i." The process is a success if, for example, a point's membership in a cluster reveals useful or unexpected features of its origin or future development.

When clustering is "hard" – i.e., when any particular fine-grained description falls under a single coarse-grained category – it can be represented as a tree, or hierarchy. An evolutionary phylogeny provides a simple example, where distinct species can be grouped on the basis of their common ancestors. Whatever the algorithm – k-means, phylogenetic reconstruction, graph clustering, multidimensional scaling, latent Dirichlet allocation – the new description is simpler and more compact. It is a form of lossy compression that discards much of the original information and, among other things, makes it impossible to reconstruct the original in all its glory.

When we go beyond the physical sciences, we should not be surprised to discover that we sometimes wish to coarse-grain by destroying *long*-range order. When we describe texts in a bag-of-words model, for example, we count words but throw away all information about word proximity; the arc of a narrative is lost as the words that

(A) What a piece of work is a man! how noble in reason! how infinite in faculty! in form and moving how express and admirable! in action how like an angel! in apprehension how like a god! the beauty of the world! the paragon of animals!

(B) how like: 2, how infinite: 1, and admirable: 1, piece of: 1, noble in: 1, the paragon: 1, . . .

(C) how: 5: in: 5: of: 3: a: 3: the: 3: and: 2: like: 2, moving: 1, noble: 1, is: 1, reason: 1, . . .

FIG. 16.2 Coarse-graining a text. Rather than keep track of full word order (A), we can count occurrences, summarizing the text by its vocabulary (C). Less aggressively, we can summarize the abundance of word pairs (B); 2-grams retain more of the structure of the original text while discarding long-range syntactic order.

appear at the beginning and the end are mixed together in a single probability distribution. Tracking n-grams – pairs, triplets, and n-word units – can be considered forms of coarse-graining less destructive than simple bag-of-words, preserving more of the original structure, while dropping longer-range correlations (see Figure 16.2). Coarse-graining a text through bag-of-words is often, for example, a good first start toward finding out which were most likely written by the same author, or in the same time period, or as the raw material for accounts of cultural dynamics (Rule and Cointet, 2015).

An ideal coarse-graining not only summarizes the full system at any point in time but provides descriptions with a useful – if probabilistic – logic connecting them together. Much remains to be done in understanding the relationship between how we coarse-grain and why, the ways in which a particular desire (summary, prediction, explanation, understanding) in a particular field (social, biological, physical) suggests a particular algorithm.

Rate distortion theory is one of the simplest mathematical accounts, where the loss is quantified in terms of a single utility function that can be understood in terms of an organism's action policy (Marzen and DeDeo, 2015); we might, however, want to go beyond this canonical paradigm to consider coarse-grainings that are predictive, comprehensible, or easy to compute with (Wolpert et al., 2014).

This is the domain of machine learning, broadly conceived, and these questions remain at the forefront of the field.

This section has considered the problem of how to coarse-grain, or lossily compress, in an optimal fashion (given constraints such as memory, processing power, and risk tolerance). But how do individuals – intelligent agents such as humans or the non-human animals – actually coarse-grain their world? How do their brains work when they try, and when they do try, what do they end up doing? Optimal models may provide upper bounds to the correct answers, but this is at heart a problem for cognitive science, neuroscience, and psychology. It is also, as we shall see in the next section, the crucial step needed for us to build our account of major transitions in social order.

MINDS IN THE LOOP

Scientists summarize, but it is not only as dispassionate observers that we attempt to simplify, and thereby predict and understand, our worlds. To navigate the physical world, for example, we (along with other primates) rely on "folk physics" (Povinelli, 2000), a reasonably predictive account of the coarse-grained physical world of medium-sized dry goods, where fundamental laws such as the conservation of energy are routinely violated. Similarly for how we think about the biological world: when we study informal human reasoning, even among preverbal infants, we find a folk biology (Keil, 2013; Setoh et al., 2013) that includes, among other things, a notion of an *élan vital*, or vital force, permeating living things.

Physical and biological laws remain constant over the course of an individual's life. Not so for social phenomena, and the (approximate) laws that connect them. In the modern era, new rules of behavior can emerge overnight; in the past, cultural change of this form might have been slower, but still far more rapid than the ten-thousand-year time scales of biological evolution.

The fundamental units of social laws are what we might call social facts: coarse-grained summaries of the beliefs and actions of the vast numbers of people. Without such summaries in hand, we are lost: we cannot follow norms unless we learn their essence from the behavior of others; we cannot respect authority if we cannot perceive it. We use these coarse-grained summaries to predict and understand the actions and beliefs of others.

Informal examples abound, but one of the clearest quantitative examples can be found in theories of social power. Whether in a modern high school or the banking world of Renaissance Florence, some individuals are perceived to have more power – of the relevant sort – than others. Some bankers are considered more reliable, even if they have little or no capital to back their debts (Padgett and McLean, 2011); some high school students have more power even if their talents and intrinsic charm might argue otherwise (Vaillancourt et al., 2003).

Power both is created by and summarizes the interactions of a society. A vast body of literature in the social sciences has repeatedly returned to this basic phenomenon: how the manifold interactions within a social group lead to hierarchy of status that bears some – but often not very much – relationship to the original intrinsic properties of the individuals themselves (Mann, 1986). Power thus provides our first explicit example of a socially relevant coarse-graining. To know social power is to know more than just facts about individuals; it is to summarize innumerable facts about the thoughts individuals have about each other, and thoughts about those thoughts, and so forth.

In the modern era, and driven by advances in our studies of non-human behavior, we have come to quantify these hierarchies by *power score*: a single number that summarizes a group consensus on the basis of individual interactions. As they are used in these contemporary studies, power scores compress an $n \times n$ matrix of dyadic interactions to an n-element list. There are many individual-level patterns consistent with any particular ranking, but these scores often predict crucial features of an individual's future (Brush et al., 2013) and evolve over time in predictable ways. Extensions of the basic idea – that relative status can be quantified by reference to pairwise interactions – have proven their worth far beyond the academic arena. Among other things, it forms the core of the original algorithms used by Google to summarize collective opinions about the rank-order value of webpages (Page, 2001). These algorithms are fundamentally recursive: to have power is to be seen to have power by those who are themselves powerful.[3]

An observer equipped with panoptic and high-resolution data, and an algorithm such as eigenvector centrality, can measure social power. Individuals in the society itself, tasked with the day-to-day problem of decision-making, and operating under biological constraints of both memory and perception, face a much harder task. The models they make of their social worlds must not only strive for accuracy. Models must lead to representations that are intelligible to, and computable by, the agents themselves (Krakauer et al., 2010).

In the final analysis, it is the individual who uses these representations to decide what to do. Of course, in doing so, she and her fellows alter the very coarse-grained representations that they rely on. Understanding the process of belief formation in the presence of an overabundance of information is a key challenge in understanding how the loop between individual behavior and group-level facts is closed.

One of the ways in which individuals collectively understand their social worlds is through the use of novel signaling channels that allow for a collective summarization of a more rapid and complex series of individual-level events. These new signal channels can smooth out irrelevant noise and make the underlying social patterns visible to the group as a whole. This account, and its supporting

---

[3] Brush et al. (2013) has distinguished between "breadth" and "depth" measures of social consensus. Breadth measures measure the power of individual X simply by reference just to the beliefs others have about X. Depth measures, by contrast, also make reference to higher-order facts such as the beliefs others hold about those who hold opinions about X. At least some work has confirmed the greater predictive power of depth measures (Hobson and DeDeo, 2015), providing additional evidence that social facts are not simply compressions of individual-level beliefs, but complex, nondecomposable compressions where every $n(n-1)$ dyadic interaction influences each power score.

empirical evidence, was developed by Flack and colleagues (Flack, 2012; Flack and de Waal, 2007; Flack and Krakauer, 2006; Flack et al., 2013), with the example of the social construction of power in pig-tailed macaques. Rather than fight, an individual of this species can send a unidirectional subordination signal, "silent bared teeth" (SBT), which both inhibits conflict, should it be imminent, and provides information about time averages over past outcomes. The coarse-graining here is over time, summarizing the outcomes of multiple conflicts with a single binary variable. The work of Brush et al. (2013) ties these same signals to the distributed consensus in the system as a whole, making the coarse-graining over the social network as well.[4]

A study of a different, though still socially complex, species, the monk parakeet (Hobson et al., 2014), provides another view on how individuals come to know, and act on, coarse-grained facts. Recent collaborative research (Hobson and DeDeo, 2015) shows evidence for emergent loop closure in this species as group behavior develops over time. When parakeets first encounter each other during group formation, aggressive behavior appears strategically unstructured. Over time, however, and as individuals become aware of rank order, they appear to direct individual aggressions strategically and based on relative rank.

High-resolution data on this *knowledge–behavior* loop provides a dynamical picture of how individuals come to know the implicit hierarchies of their world and alter their behavior in response.

---

[4] The role that SBT plays in primate societies seems to meet the main criteria for what John Searle (2008a) refers to as a status function. SBT is not intrinsically an act of subordination; it does not put the user at an immediate physical disadvantage as, say, similar signals in the canine case. Furthermore, its function is made possible by the collective acceptance of this signal. It allows sender and recipient to avoid conflict in part, presumably, because it is understood as such not only by the pair themselves, but – given the public nature of power and the role of third-party interactions – by the group as a whole. This account of SBT in primate society pushes Searle's (somewhat fanciful) account of the origin of status function a few hundred thousand years further back. The conjectured contextual meaning of the SBT – that it functions, in part, to indicate facts about a pair-wise relationship to third parties – distinguishes it from simpler cases such as that of the alarm call or warning signal.

In contrast to the pig-tailed macaques of the example above, monk parakeets appear, so far, to lack a separate signaling system. The density of interactions, however, may allow for participants in this second example to use small, cognitively accessible network motifs to predict the relevant aspects of these coarse-grained power scores.

Macaques short-circuit violent conflict by signaling social consensus on power; parakeets use the same variables to strategically direct aggression against rank peers. The work of Padgett and McLean (2011) provides an instructive version in the human case drawn from the early years of merchant banking in Renaissance Florence. In the absence of open records, Florentine bankers attempted to reconstruct not only the potential solvency of their colleagues but, crucially, the ideas about that solvency held by others. To know whether someone was a good risk was to know, in part, whether others thought they were. In response to this challenge, bankers, in their letters to each other, summarized facts about their own prestige and solvency, and the prestige and solvency of others, through an elaborate system of rhetoric and telling details that, on the surface, appeared highly tangential to the financial matters at hand (McLean, 2007).

When we use machines, in the modern era, to predict features of our society, we often turn, as Google does, to algorithms that rely on successive coarse-grainings of high-resolution data. The recent success of deep learning (Bengio, 2009) is in part due to its ability to adapt, at the same time, its method of coarse-graining and its theory of the logic of those coarse-grained variables. Once we realize that the machine-aided predictors of a system are also participants, it is natural to ask how their use of that knowledge, accurate or not, back-reacts on the society itself. Financial markets provide examples of both positive reinforcement, as in the case of the 2010 Flash Crash (Easley et al., 2011), and negative reinforcement, as traders destroy the very patterns that provide their source of profit (Timmermann and Granger, 2004). We understand very little about how the introduction of these prediction algorithms, on a large scale, will lead to novel feedbacks that affect our political and social worlds; it remains an understudied and entirely open topic.

Whether driven by inference from context or signal, processed by evolved brains or optimized machines, the feedback loop that results from action on the basis of social facts is likely to be a widespread feature of biological complexity. It may extend well beyond the cognitive and even down to molecular scales (Flack et al., 2013; Walker and Davies, 2013). In the case of interacting individuals, the closure of this loop is a precondition for the causal efficacy of high-level descriptions. It represents our first major transition in political order. Empirical work strongly suggests that this transition happens in the prehuman era. Monkeys, and even parakeets, are quite literally political animals.

BROKEN WINDOWS AND THE NORMATIVE PATHWAY

Defusing a conflict by signal alone, using relative power to adaptively guide aggression, lending to a high-prestige bank: in each of these examples, individuals infer social facts and use them for their own advantage. Some species, however, with humans the most notable example, reason not only from wants and needs, but also according to oughts.

In observing a power structure we may learn new strategies to thrive, but we may also perceive it as just or unjust, legitimate or illegitimate, and these latter perceptions hinge not only on what is and what will be but on what *should* be. The modal structure of these beliefs is one not of possible worlds but of deontic logic, how "things are done" by "people like us" or, in the modern era (as we describe below), how things compare to an ideal standard (Chellas, 1980). Norms are, in their most developed form, facts about shared ideals, about what the group believes – or, more formally, a coarse-grained representation and lossy compression of the idiosyncratic beliefs and desires held by individuals. We need not all believe exactly the same thing in order to share a norm; norms constitute a new set of group-level facts.

As with the case of power, facts about norms cannot be reduced to the interactions between two individuals. How a norm of politeness works in a particular commercial transaction, for example, depends very little on what the participants desire. If it is a norm to thank the shopkeeper, a shopkeeper who asks his customer to forego a "thank you" may find his request denied or obeyed at best reluctantly; if his counternormative requests persist, he may find himself shunned by the community as a whole.[5] To be polite is not to respect someone as they desire to be respected but to play out certain patterns of behavior that can reasonably be interpreted as respect in a social context.[6]

As suggested by the example of just and unjust power, the emergence of a norm can provide a novel pathway for individuals to respond to preexisting group-level facts. The normative perception of a hierarchy as unjust should be distinguished from the thought that it might be upended for the agent's benefit. We are able to recognize a situation as unjust, and to respond to this injustice emotionally, even when we have no ability to alter it, and even when we might, for other reasons, consider it an injustice useful on balance.

In humans some normative responses including the ability to invent a game and play by its rules, seem to be acquired very early in life (Tomasello, 2009b). More elaborate norms are learned by observing the community. They are, therefore, predictions: a norm that ceases to have an effect on behavior is unlikely to be so described a few years later. A norm may have an effect without being obeyed – "more honored in the breach than the observance" – but this is exceptional. We can say, with great confidence, that when two men in American society meet to conduct a lengthy business transaction, they will begin by shaking hands.

---

[5] The customer herself may find the request intrinsically unpleasant; norms, once learned, act directly on our feelings. Norm violations can cause both pain and pleasure, over and above the consequences of the action itself.

[6] The hypothetical agent that does this interpretation is, in some philosophical theories (Johnston, 2014; Lacan, 1998), referred to as the "big Other."

Yet we use norms for more than prediction. It is unlikely for the weaker player in an unevenly matched game of tennis to win; it is unlikely for the loser to refuse a handshake at the end. Because of the strength of the handshaking norm, the responses to these two unlikely events will be distinct. We may reevaluate our ratings of the two players based on the final score. Yet even if no formal rule requiring a handshake exists, our responses to the norm violation will involve shunning the individual and group-level shaming; examples of how this (rare) violation is discussed in the press confirm the intuition (Ubha, 2014).

Norms are critical for the maintenance of social stability, and a long tradition in game theory seeks to describe how altruistic norms may emerge from purely self-interested motives (see, e.g., the critical review of Bowles, 2009), or evolutionary group selection (Akçay et al., 2009). In this sense, norms are simply a more elaborate, potentially gene-driven, version of the prudential strategizing described in the previous section. In contrast to individual strategies, however, norms must be shared, and require not just knowledge, but mutual use. Norms play the role of a choreographer that allows multiple individuals to solve joint action problems by coordinating around a specific equilibrium (Gintis, 2009); if we do not share the right norm, for example, having access to a punishment mechanism in a public goods game, it will lead to antisocial, rather than pro-social, results (Herrmann et al., 2008).

In contrast to lab-based experiments, much of the complexity of ethnographic research comes from the parsing out of the layered and often counterintuitive roles that norms play in human society. In part due to this complexity, the underlying cognitive mechanisms required for norms to exist and to influence behavior are hotly debated. As reviewed by de Waal (2014), reconciliation behaviors ("making up" after conflict), responses to unequal rewards, and impartial policing may provide examples of nonhuman normative reasoning. Both reconciliation and responses to inequality are found across multiple taxa. Meanwhile, "knockout" studies have verified the causal role of policing (Flack et al., 2005, 2006) – if it is understood as a norm, it is a norm that matters. While reconciliation and inequality responses may be understood as negotiated one-on-one norms, policing provides an example of a strictly community-based norm, where individuals attempt to preserve group consensus.

A separate school of thought, reviewed in Tomasello (2009a, 2014), ties normative behavior to the ability to act on the basis of a belief about what "we, together, are doing," the capacity for joint intentionality. Joint intentionality is often considered a precondition for human society (Searle, 2010, 2008b); evidence for joint intentionality in non-human animals may come from the example of chimpanzees that engage in group hunting (as opposed to opportunistic, simultaneous chasing) (Bullinger et al., 2014; Call and Tomasello, 1998). To require joint intentionality for norm following, however, may set the standard too high, drawing a firm boundary on the basis of cognitive skills where we might expect shades of gray (Andrews, 2009).

Rather than drill down to the level of these basic mechanisms, we take a particular example from recent empirical work to look for the distinct traces that normative reasoning leaves on the logic of society as a whole. We do so using a series of investigations into the dynamics of conflict in the editing of Wikipedia.

Now more than 14 years old, the community surrounding the online encyclopedia has attracted an enormous amount of scholarly attention, both as a laboratory of human interaction and as a phenomenon in its own right (Bar-Ilan and Aharony, 2014). Ethnographers have studied the culture of Wikipedia editors (Jemielniak, 2014; Reagle, 2010), finding diverse motivations and self-conceptions among the hundreds of thousands of volunteers who massively outnumber the roughly one hundred paid employees of the parent foundation.

Wikipedia is hardly immune to conflict, much of which focuses on article content: what to include in an article and how to represent it. Users who edit pages – particularly, controversial pages associated with political figures such as George W. Bush or Josef Stalin, or conflicts such as Israel–Palestine (Yasseri et al., 2014) – often find they disagree about which facts to include and the prominence those facts should be given. Facts shade naturally into interpretation, and even when all users involved agree on which sources to cite, disagreements do not cease.

Arguments often reduce to competitive editing: one user adds text; a second one modifies it to change the implication, connotation, or weight; the original author, or a new third party, intervenes to shift the tone again. When this process degenerates, and cooperation breaks down completely, editors may resort to what is called a *revert*: completely undoing the work of a previous editor. Reverts are an excellent way to study conflict on large scales because they can be easily identified by machine, rather than by hand-analysis or complex natural-language processing, and we have learned a great deal about collaboration by tracking conflict in this fashion (Yasseri et al., 2012).

Reverts also have the advantage of being a clear norm violation. Multiple policy pages discuss how one ought not to revert: reverts are described as "a complete rejection of the work of another editor" and "the most common causes of an edit war"; rather than revert each other, editor disagreements "should be resolved through discussion;" and editors are "encouraged to work towards establishing consensus, not to have one's own way." Those whose edits are reverted are urged to turn the other cheek: "If you make an edit which is good-faith reverted, do not simply reinstate your edit."[7]

Naturally, reality is far more complicated. Reverts are common, in some periods and for some pages rising to nearly half of all edits made. The very fact that the norm is imperfectly obeyed, however, makes it possible to study the dynamics of how users learn and adjust their behavior in response to the actions of others. In empirical study, we find long-range memory intrinsic to periods of interrevert

[7] Drawn from pages current as of April 1, 2015; see http://en.wikipedia.org/w/index.php?title=Wikipedia:Reverting&oldid=642003221 and http://en.wikipedia.org/w/index.php?title=Wikipedia:Edit_warring&oldid=652860808.

conflict: the more edits a page has had without a revert, the less likely it is to see a revert on the next edit. In DeDeo (2013) a two-parameter model was found for this process, the collective-state model, where the probability of a revert, R, varies as a function of the number of edits, k, since the last revert,

$$P_k(R) = \frac{p}{(k+1)^\alpha},$$ (16.1)

where p and α are constants. When α approaches zero, reverts are uncorrelated and conflict arises without regard for context. Over a wide range of pages, however, we find that α clusters around one-half, leading to a simple *square-root law*: the probability of future conflict declines as the square-root of the amount of conflict seen so far. (In this simple model, the clock resets on the appearance of new conflict.) The law appears robust to a wide range of filters, including the inclusion of partial reverts and the restriction to harder conflicts, where we track conflict by a double-revert, i.e., measure the probability of two reverts in a row. The observed time scales of these runs are short, often only hours or even minutes long, requiring us to refer to intrinsic features of the interaction rather than events in the real world, and involve many users, making these interactions intrinsically social, rather than pairwise (DeDeo, 2014).

One can think of Eq. 16.1 as describing a *reverse broken-windows* effect. In the original account of broken windows, popularized by Wilson and Kelling (1982), minor norm violations led to an increasing likelihood of future violations (a single broken window in an abandoned building attracts more). Here, we find the reverse effect: norm-conformant actions lead to an increasing likelihood of future norm conformance.

Based on this result, in DeDeo (2014) a game-theoretic model was constructed to back-infer the underlying beliefs and desires of the users from their behavior alone. In the spirit of earlier work in inductive game theory (DeDeo et al., 2010, 2011), the fundamental goal was to understand the cognitive complexity of the individuals, and how they reacted to the contexts in which they found themselves.

This was done using an extensive-form public goods game called the stage game. The stage game models the step-by-step pattern of interaction on a single page, where users interact with those who came before, while setting the stage for the editor who comes next.

Analysis of the stage game showed that, under the assumption of a self-reinforcing equilibrium, a very simple model can explain the behavioral data if and when users have context-sensitive utility. Put another way, a parsimonious model is possible when what other people have done in the past affects not only what a user *does* but what a user *wants*. In order to explain why users edit the way they do, we cannot simply describe them as learning how to maximize utility under a fixed tolerance for conflict; we must allow that tolerance to change. Rather than describe a population with a mixture of mutualists and defectors, we have a population whose individuals become mutualists as they see others around them shift toward cooperation themselves.

This is what we expect if the underlying behavior is truly driven by a normative injunction, where the adherence to the norm by others increases our own desire to conform. On the one hand, we can describe this result in the folk-psychological language of wants and desires. On the other hand, however, we have long known that successful cooperation in public goods games requires mechanisms such as punishment and reputational damage for those who violate norms (see Bowles and Gintis, 2011, and references therein). Extensions to those classic results include those of Burton-Chellew et al. (2015), who note that changing preferences for cooperation can in some cases be explained by individuals learning how they may, or may not, be punished for behaving badly.

In the language of our model, changing utility functions can represent either shifts in intrinsic desire or the expectation of future punishment through other pathways. Our inability to split this atom, when the punishment pathway is hidden from view, is a limitation of utility theory itself, which quantifies desires along a single axis. In DeDeo (2014) it was found that norm-conformity accumulates faster when individuals interact (α driven toward one), suggesting that reputation drives learning. A "cheap-talk" result – norm-conformity is not affected by use of associated discussion pages – further complicates the analysis.

Whether or not this increasing cooperativity is to be referenced to good citizens (changing desires) or good laws (effective incentives) (Bowles, 2015), we are firmly in the world of norms: patterns of behavior, understood as group-level standards, and enforced by both community action and individual desire, forced or free. Individuals adjust their behavior in response to what they observe in others; in the example here, simple coarse-grained heuristics on overall levels of cooperation can provide knowledge of the implicit standard. The feedback effects of their responses to this knowledge provide an example of a fundamentally normative form of loop closure, and our second major transition in political order.

GOING TOGETHER TO GET ALONG: NORM BUNDLES

Should Israeli settlements be described as "key obstacles to a peaceful resolution" or "a major issue of contention"? On July 2, 2007, three Wikipedia editors debated these six words on the "talk" page of the article on the Israel–Palestine conflict. Over the next 11 days, the discussion grew to include over 20 editors and ran to more than 16,000 words. On July 13, the last arguments were made, and two of the three original participants had come to agreement on the final wording.

As might be expected, much of the debate centered around the details of the conflict itself, becoming at times only tangentially related to the wording in question. About 30 hours in to the argument, however, the user Jayjg wrote, succinctly, "WP:NPOV says that opinions cannot be stated as fact, and must be attributed to those who hold them. WP:V says that opinions must be sourced. That should solve the problem; follow policy." WP:NPOV is a community abbreviation for a norm that urges editors to "adopt a neutral point of view toward article subjects"; WP:V an abbreviation for the norm that all statements in the encyclopedia be verifiable, particularly when challenged by others.

These abbreviations are more than shorthand; in Jayjg's comment, they linked to pages in a separate space of the encyclopedia where the norms are discussed in detail. The pages describing basic principles of neutrality and verifiability are only a small fraction of the nearly 2,000 norm-related pages that users have created over the lifetime of the encyclopedia (Heaberlin and DeDeo, 2016). Themselves under continual discussion and revision, they have grown to encompass nearly every aspect of the mechanics of article writing, interpersonal interaction, and a small "administrative" class given special privileges within the system as a whole.

Those in conflict on Wikipedia may encounter the norm to assume good faith (AGF). Users might remind each other of this norm when they believe conflicts are driven by unfair assumptions about the other party. The associated page describing AGF links, among other things, to a (collectively written) essay entitled "Don't Call a Spade a Spade" (don't label other users as norm violators; abbreviated NOSPADE); NOSPADE couples, by its own out-link, the AGF norm to both the CIVIL norm ("be respectful and considerate") and the NPOV norm, urged by Jayjg in his original comment, where NOSPADE violations are likely to occur.

The connections between these norms are not logically necessary; one can imagine a different pattern, where the NPOV norm is supported by a strong (here fictional) PROSPADE norm, with users encouraged to identify and critique each other's underlying motivations. In the Wikipedian bundle, however, AGF, NOSPADE, CIVIL, and NPOV are understood as reinforcing structures that provide coherence to a user's expectations. Given the difficulties of text-based communication, the Wikipedia community choice is likely to be adaptive.

Not every normative injunction can be uniquely related to core practices; the LONDONDERRY norm, for examples, describes an internal consensus from 2004 on a controversial naming decision. Examples of potentially adaptive clusters abound: Wikipedia's encouragements for users to undertake creative action without interference include networked norms such as OWN ("no one is the owner of any page"), BUILDER ("don't hope the house will build itself") and even DHTM ("don't help too much") and MYOB ("mind your own business").

This is an example of a more general principle: once created, norms rarely stay as isolated oughts. We want to make sense of our world and constraints on cognitive load naturally lead to the formation of *norm bundles*. These networks of interacting and self-supporting norms reinforce each other by providing logical or emotive support. One norm is now understood as a natural consequence, or a subcase, of another. We regularize – i.e., simplify and systematize – in a variety of linguistic and nonlinguistic domains (Ferdinand et al., 2013; Lieberman et al., 2007). Norm bundling may be driven, as well, by this same instinct to avoid the costs of memory through the systematization of exceptional cases.

In the case of Wikipedia, we can build a network from how norms interact, reinforce, and modify each other. We see the emergence of clusters, where basic principles form high-degree cores within distinct communities and serve as a common point of reference for more peripheral subgraphs; see Figure 16.3 for a representation of the full network, as well as the largest subbundle, which includes both NPOV and verifiability.

Wikipedia may be unusual in its ratio of norm to action. It is difficult not to be impressed by the thousands of pages users have created presenting, discussing, and interpreting their community's standards. It may even be worrying: many wiki-like systems appear to fall into a "policy trap," where content creation is replaced by policy discussion dominated by a smaller, in-group elite: a modern, electronic version of the Iron Law of Oligarchy (Shaw and Hill, 2014).

Wikipedia is not, however, unusual in the complex ways in which its norms cross-link, how it draws on a set of core principles to carry the periphery along, or how a bundle may be more than the sum of its parts. An example at the national level is provided by the U.S. Supreme Court, which in *Griswold v. Connecticut* (381 U.S. 479, 1965) described a right to privacy. This right, nowhere explicitly stated in the Constitution, is what we could call an implication of

FIG. 16.3 Norm bundles on Wikipedia. Nodes refer to policies, guidelines, and essays; links indicate cross-references. Dense clusters of cross-referenced norms range from how to decide whether a person, place, or event is notable enough for inclusion to how and when to split articles into subtopics to appropriate and inappropriate ways to handle the stress of online conflict. The largest subcommunity is represented as a darker cluster of nodes in the top-left of the network, as found by Louvain clustering (Blondel et al., 2008); this bundle describes norms of article writing, including the need for neutrality (NPOV) and verifiability. Other bundles describe norms of interpersonal interaction such as civility and the assumption of good faith, norms associated with administrative systems, and norms on the use of intellectual property. These top four bundles include just over 75% of all pages; see Heaberlin and DeDeo (2016).

norm bundle, an example of how, in the words of Justice William O. Douglas, "specific guarantees in the Bill of Rights have penumbras, formed by emanations from those guarantees that help give them life and substance." Psychologically, our moral injunctions do not appear to us as statements that we can analyze in isolation, nor even as as directed chains of derivations; they are, instead, dense networks of social practices, mixing rational arguments of greater or lesser plausibility with central emotional, narrative, and even mythic appeals (Bellah, 2011; Cavell, 1979; Merlin, 1991).

Norms both interact with and drive changing material contexts. Yet because of the reinforcing nature of norm bundles, shifts in behavior are rarely due to the emergence or strengthening of a single norm in isolation. Rather, when studying long-term norm-driven change, we expect signals of multiple, conceptually distinct – but bundled – norms working together. We can see this in our analysis of the Old Bailey that began this chapter. In Figure 16.4A, we show how speech during trials for violent and nonviolent crimes became increasingly distinct, tracking the bureaucracy's increasing concern to manage, specifically, the violence of its population (Klingenstein et al., 2014). This plot tracks the strength of signals, at the 1-gram level, that distinguish transcripts describing crimes the court considered violent from those it did not. In the early years of our data, little to no distinction exists; classifications, at least at the 1-gram level, appear arbitrary. But from 1780 through to the end of our data in 1913, a long-term secular trend becomes clearly visible, showing how this signal first emerged, and then began to strengthen over time.

These cultural shifts in the attention to violence parallel long-term declines in the homicide rate (Eisner, 2003). The majority of the cases in our data, and the majority of the signal in Figure 16.4, concerns crime less serious than murder. The signal we track is tied to an increasing sensitivity to the "dark matter" of violence – the assaults, kidnappings, and violent thefts that do not leave a dead body for demographers to trace.

In Figure 16.4B, we look closer, into the signal structure itself, to see how the words that signaled these distinctions changed over time. To increase the ratio of signal to noise, we group words into synonym sets, so that the set "impulse" includes words such as kick, hit, blow, and strike; the set "remedy" includes words such as hospital and doctor; the set "greatness," words such as very, great, many, much, and so; the set "sharpness," words such as knife, razor, and blade.

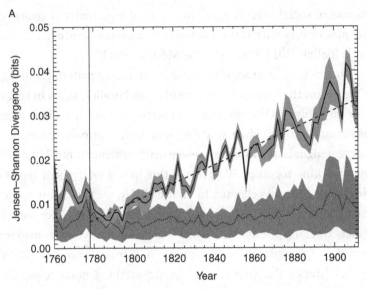

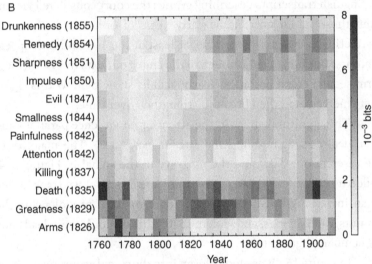

FIG. 16.4 Correlated norm shifts in discussions of violence, 1760–1913. (A) After 1778, trials for violent and nonviolent offenses become increasingly distinct over time, as measured by the Jensen–Shannon distance between spoken text in the two categories (dashed line). (B) The top dozen classes that serve to signal trials for violence. Some, such as references to death, are strong signals of a concern for violence throughout our data. Others, such as those referring to medical evidence ("remedy") and drunkenness, appear much later. Adapted from Klingenstein et al. (2014).

Studying the changing patterns of these signals gives us clues to the nature of the norm bundles that underlie Britain's transition from the eighteenth to the twentieth century. Already by 1770, discussions of death were strong signals that the court had indicted the defendants for a violent crime, as were words associated with firearms. However, words such as knife and cut, or hit and strike, took longer to emerge as signals; the case of Junque, Smith, and Leethorp that opens this chapter provides an example of how, early on, assault and the use of a knife, openly discussed before both judge and jury, were able to appear in a trial ostensibly for the nonviolent offense of grand larceny.

As the court paid greater attention to more minor forms of violence, parallel shifts occur in other, related domains. The sets "smallness" and "attention," containing words associated with (among other things) observation and measurement, also come to prominence: violence must not only be minimized, it must also be measured. Doctors were called on to provide medical evidence, showing how concerns with lesser forms of aggression led to demands for a scientific account of its effects. In the final decades of study, words associated with drunkenness emerge, both because the state increasingly attends to the opportunistic violence associated with drinking and because it is used as an explicit excuse by the defendants themselves: participants attend to violence's external, material causes.

This is what we expect from normative bundling, and a general theory should provide new insight into other phenomena as well. Some norms are extremely adaptive, but many are simply epiphenomenal, like the ritual handshakes of the tennis match. Handshakes can be faked and are costless forms of cheap talk; norm bundles explain the persistence and pervasiveness of these epiphenomenal norms by reference to the role they play in the larger structure.

Figure 16.4, by selecting only those topics that contribute to the distinction in question, should not be understood as promoting a Whiggish account (Butterfield, 1965) of norms in concert combining to produce the modern world. Norms within a bundle do not always work in the same direction, and we expect frustration and disagreement. Incipient conflict can be seen in a graph-theoretic analysis of the Wikipedian bundles shown in Figure 16.3, where norms encouraging users to "ignore all rules" (IAR) in seeking creative ways to improve the encyclopedia maintain a large topological distance from norms specifying, in microscopic detail, conventions for transliterating Belarusian (BELARUSIANNAMES). The NPOV norm links, among other things, to pages describing how to resolve naming conflicts, but also to a user essay entitled "Civil POV Pushing," describing concerns about users who, through persistence and careful adherence to interpersonal norms such as AGF and CIVIL, tilt pages in ways that violate NPOV.

A more serious example of intrabundle conflict, in the case of the common law, can be found in the theory of felony murder and related practices where courts punish people for the unintended consequences of a crime. A death caused by the negligent, but accidental, destruction of a traffic signal may be treated as a civil matter. Conversely, a teenager who steals a stop sign and thereby causes a fatal accident may be tried for manslaughter.[8] The general principle – that one can be punished for an unintended consequence of a conceptually distinct crime – is a sufficiently ancient part of common law that as early as 1716 it was treated as a self-evident fact (Hawkins, 1824). It persists in the United States today but is widely seen to be in fundamental conflict with co-bundled injunctions against strict liability and in favor of the need for *mens rea* (Binder, 2007).

Over a decade ago, Ehrlich and Levin (2005) posed a series of questions for scientists interested in the emergence of norms. Referring to the "regrettably infertile" notion of the meme, they urged

---

[8] An example of the former is *Dixie Drive It Yourself System v. American Beverage Co.* (Louisiana Supreme Court, 1962), where a negligent driver knocked over signal flags, leading to a fatal accident. An example of the latter is the *State of Florida v. Christopher Cole, Nissa Baillie and Thomas Miller* (1997) where the three defendants received 15-year sentences for a (confessed) stop-sign theft that, 2 or 3 weeks later, led to a fatal accident. Notably, review of the Florida case focused on the question of whether it was that stop sign in particular that had been stolen, on whether too much time had elapsed, and on inappropriate behavior by the prosecutor – not on the fundamental linking of the theft and unintended death.

renewed attention to the development of theories that would allow us to both quantify and explain the process of cultural evolution. A decade later, the 23 questions they posed remain unresolved – despite massive progress in the development of meme-contagion models and game theoretic accounts of multiagent interaction. Many of their questions focused on individual-level cognition, including the origin of novel ideas in a mind, the decision to adopt ideas from others, and the covariance of these cognitive processes with other facts about an individual.

If our account is correct, we can make new progress by combining the multiagent approach common to both contagion and game-theoretic models with the cognitive questions Ehrlich and Levin propose. Studies of individual-level cognition, however, must be used for more than simply fixing the parameters of an agent-based model. Groups, not individuals, construct norm bundles, and individuals must then learn them both from direct inspection and from watching how others behave and extrapolating a mental representation that may differ a great deal from the massively complex structure of Figure 16.3.

We understand little of what is required for norm bundling to begin. If non-human animals have norms, do they have bundles? Is norm bundling a gradual transition, as groups begin by pairing norms, or do large bundles emerge suddenly, at a critical point? Gradual or sudden, the emergence of norm bundling represents a new level of complexity in how individuals perceive, and respond to, their social worlds. It provides our third example of a major transition in political order.

CONCLUSIONS

The most ambitious theories of cultural evolution extend into biological time. When they do so, they often divide history into the epochs marked by dramatic shifts in cognitive complexity (Donald, 2012; Merlin, 1991). Drawing on this tradition, we have focused on transitions in the causal pathways between group-level facts and the individual. When minds are in the loop, coarse-graining is no longer just a method for understanding the material world. It becomes a constitutive part of what it means for a collection of individuals to become a society. If this is correct, the origins of society may share a great deal of their causal structure with the origins of life itself (Walker and Davies, 2013).

Reference to the capacities of the individual mind is a common theme in political theory; the *Leviathan* of Thomas Hobbes opened, in 1651, with a theory of cognitive science. Once we recognize the importance of the feedback loop between the individual and the group, however, it becomes harder to distinguish between changes in an individual's ability and the social scaffolding necessary to support it (Zawidzki, 2013). Are Wikipedian norms supported by coordinated punishments and rewards that manipulate simple, self-interested utility maximizers? Or do they involve a desire to conform, pride in one's reasonableness, or the notion of an ideal standard for an electronic public sphere? It is hard to imagine an ideal that everyone holds but no one rewards or enforces. Yet it is hard, also, to imagine people shunning and shaming, praising and rewarding, without adopting the norms themselves and with an eye solely on the causal outcomes of each individual act; the cognitive burden is too high.

In the past, mathematical theories of social behavior have oversimplified both the human mind and the societies it creates. To counter that tendency we have, in this chapter, attempted to provide vivid portraits of some of the systems under study. Social worlds, like biological ones, are intrinsically messy. They build themselves through bricolage, constantly repurposing small details for new ends (Balkin, 2002; Levi-Strauss, 1966). Details abound, may later come to matter, and should be respected; at the very least, we expect their statistical properties will play a role in future mathematical accounts.

Conversely, to mathematize a problem is to allow its examples to be compared across context and scale. If we understand the ecological rationality (Gigerenzer and Todd, 1999) of signaling systems that naturally coarse-grain noisy mechanisms, we may find common explanations for how signals work across culture and species. If we study the fine-grained dynamics of cooperation in a contemporary system, we may be able to reverse-engineer how cultures in the past bundled norms to govern the commons. If we can build a network theory of the emergence of norm bundles, we may be able to compare vastly different societies to find common patterns in cultural evolution.

Deep histories of social complexity (Fukuyama, 2011; Smail, 2007) are often narratives. This does not mean, however, that quantitative accounts must be restricted to system-specific studies. At such a high level, dynamical laws are expected to be probabilistic, but they may be laws nonetheless, able to accurately describe, explain, and even predict the world at a particular resolution. This chapter suggests that these dynamics are driven in part by top-down causation between the individual and the group. It suggests a new role for interdisciplinary collaboration between the cognitive and social sciences. And the complexity of these systems suggests a critical role for the interpretive scholarship of political theorists, ethnographers, and historians.

Acknowledgments

I am grateful to audiences at the Interacting Minds Center of Aarhus University, Denmark; the Ostrom Workshop in Political Theory and Policy Analysis of Indiana University; the Global Brain Institute of Vrije Universiteit Brussel, Belgium; and the Santa Fe Institute, where early versions of this work were presented. I thank Merlin Donald, Tim Hitchcock, Dan Smail, Colin Allen, Jerry Sabloff, John Miller, and Alexander Barron for readings of this work in draft form. This work was supported in part by National Science Foundation Grant #EF-1137929, by a Santa Fe Institute Omidyar Fellowship, and by the Emergent Institutions project.

REFERENCES

Akçay, E., Van Cleve, J., Feldman, M. W., and Roughgarden, J. 2009. A theory for the evolution of other-regard integrating proximate and ultimate perspectives. *Proceedings of the National Academy of Sciences*, **106**(45), 19061–19066.

Andrews, K. 2009. Understanding norms without a theory of mind. *Inquiry*, **52**(5), 433–448.

Baker, M. K. 1990. *Inventing the French Revolution: essays on French political culture in the eighteenth century*. Cambridge University Press.

Balkin, J. M. 2002. *Cultural software: a theory of ideology*. Yale University Press.

Bar-Ilan, J., and Aharony, N. 2014. Twelve years of Wikipedia research. Pages 243–244 of *Proceedings of the 2014 ACM Conference on Web Science*.

Bellah, R. N. 2011. *Religion in human evolution: from the Paleolithic to the Axial Age*. Harvard University Press.

Bengio, Y. 2009. Learning deep architectures for AI. *Foundations and Trends in Machine Learning*, **2**(1), 1–127.

Binder, G. 2007. The culpability of felony murder. *Notre Dame Law Review*, **83**, 965.

Blanton, R., and Fargher, L. 2007. *Collective action in the formation of pre-modern states*. Springer Science & Business Media.

Blondel, V. D., Guillaume, J., Lambiotte, R., and Lefebvre, E. 2008. Fast unfolding of communities in large networks. *Journal of Statistical Mechanics: Theory and Experiment*, **2008**(10), P10008.

Bowles, S. 2008. Policies designed for self-interested citizens may undermine the moral sentiment: evidence from economic experiments. *Science*, **320**(5883), 1605–1609.

Bowles, S. 2009. *Microeconomics: behavior, institutions, and evolution*. Princeton University Press.

Bowles, S. 2015. *Machiavelli's mistake: why good laws are no substitute for good citizens*. Yale University Press.

Bowles, S., and Gintis, H. 2011. *A cooperative species: human reciprocity and its evolution*. Princeton University Press.

Brush, E. R., Krakauer, D. C., and Flack, J. C. 2013. A family of algorithms for computing consensus about node state from network data. *PLOS Computational Biology*, **9**(7), e1003109.

Bullinger, A. F., Melis, A. P., and Tomasello, M. 2014. Chimpanzees (*Pan troglodytes*) instrumentally help but do not communicate in a mutualistic cooperative task. *Journal of Comparative Psychology*, **128**(3), 251.

Burton-Chellew, M. N., Nax, H. H., and West, S. A. 2015. Payoff-based learning explains the decline in cooperation in public goods games. *Proceedings of the Royal Society B: Biological Sciences*, **282**(1801), 20142678.

Butterfield, H. 1965. *The Whig interpretation of history*. W. W. Norton.

Call, J., and Tomasello, M. 1998. Distinguishing intentional from accidental actions in orangutans (*Pongo pygmaeus*), chimpanzees (*Pan troglodytes*) and human children (*Homo sapiens*). *Journal of Comparative Psychology*, **112**(2), 192.

Carballo, D. M., Roscoe, P., and Feinman, G. M. 2014. Cooperation and collective action in the cultural evolution of complex societies. *Journal of Archaeological Method and Theory*, **21**(1), 98–133.

Cavell, S. 1979. *The claim of reason*. Oxford University Press.

Chellas, B. F. 1980. *Modal logic: an introduction*. Cambridge University Press.

Correia, R. B., Chan, K. N., and Rocha, L. M. 2015. Polarization in the US Congress. Paper presented at the 8th Annual Conference of the Comparative Agendas Project (CAP). Lisbon, Portugal, June 23–24, 2015.

De Waal, F. B. M. 2014. Natural normativity: the "is" and "ought" of animal behavior. *Behaviour*, **151**(2–3), 185–204.

DeDeo, S. 2011. Effective theories for circuits and automata. *Chaos: An Interdisciplinary Journal of Nonlinear Science*, **21**(3), 037106.

DeDeo, S. 2013. Collective phenomena and non-finite state computation in a human social system. *PLOS ONE*, **8**(10), e75818.

DeDeo, S. 2014. Group minds and the case of Wikipedia. *Human Computation*, 1:1:5–29. *arXiv preprint arXiv:1407.2210*.

DeDeo, S., Krakauer, D. C., and Flack, J. C. 2010. Inductive game theory and the dynamics of animal conflict. *PLOS Computational Biology*, **6**(5), e1000782.

DeDeo, S., Krakauer, D., and Flack, J. 2011. Evidence of strategic periodicities in collective conflict dynamics. *Journal of the Royal Society Interface*, **8**(62), 1260–1273.

Donald, M. 2012. An evolutionary approach to culture. In Bellah, R. N., and Joas, H. (eds), *The Axial Age and its consequences*. Harvard University Press.

Easley, D., Lopez de Prado, M., and O'Hara, M. 2011. The microstructure of the "Flash Crash": flow toxicity, liquidity crashes and the probability of informed trading. *Journal of Portfolio Management*, **37**(2), 118–128.

Edlund, E., and Jacobi, N. M. 2010. Renormalization of cellular automata and self-similarity. *Journal of Statistical Physics*, **139**(6), 972–984.

Ehrlich, P. R., and Levin, S. A. 2005. The evolution of norms. *PLOS Biol*, **3**(6), e194.

Eisner, M. 2003. Long-term historical trends in violent crime. *Crime and Justice*, 83–142.

Elias, N. 1982. *The civilizing process*. Blackwell.

Ferdinand, V., Thompson, B., Kirby, S., and Smith, K. 2013. Regularization behavior in a non-linguistic domain. In Knauff, M., Sebanz, N., Pauen, M., and Wachsmuth, I. (eds), *Proceedings of the 35th Annual Cognitive Science Society*. Bielefeld University.

Fisher, M. E. 2004. Renormalization group theory: its basis and formulation in statistical physics. In Cao, T. Y. (ed), *Conceptual foundations of quantum field theory*. Cambridge University Press.

Flack, J. C. 2012. Multiple time-scales and the developmental dynamics of social systems. *Philosophical Transactions of the Royal Society B: Biological Sciences*, **367**(1597), 1802–1810.

Flack, J. C., and de Waal, F. 2007. Context modulates signal meaning in primate communication. *Proceedings of the National Academy of Sciences*, **104**(5), 1581–1586.

Flack, J. C., and Krakauer, D. C. 2006. Encoding power in communication networks. *American Naturalist*, **168**(3), E87–E102.

Flack, J. C., Krakauer, D. C., and de Waal, F. B. M. 2005. Robustness mechanisms in primate societies: a perturbation study. *Proceedings of the Royal Society B: Biological Sciences*, **272**(1568), 1091–1099.

Flack, J. C., Girvan, M., De Waal, F. B. M., and Krakauer, D. C. 2006. Policing stabilizes construction of social niches in primates. *Nature*, **439**(7075), 426–429.

Flack, J. C., Erwin, D., Elliot, T., and Krakauer, D. C. 2013. Timescales, symmetry, and uncertainty reduction in the origins of hierarchy in biological systems. Pages 45–74 of Sterelny, K., Joyce, R., Calcott, B., and Fraser, B. (eds), *Evolution, cooperation, and complexity*. MIT Press.

Fortunato, S. 2010. Community detection in graphs. *Physics Reports*, **486**(3–5), 75–174.

Fukuyama, F. 2011. *The origins of political order: from prehuman times to the French revolution*. Farrar, Straus and Giroux.

Gigerenzer, G., and Todd, P. M. 1999. *Fast and frugal heuristics: the adaptive toolbox*. Oxford University Press.

Gintis, H. 2009. *The bounds of reason: game theory and the unification of the behavioral sciences*. Princeton University Press.

Gneezy, U., and Rustichini, A. 2000. A fine is a price. *J. Legal Stud.*, **29**, 1.

Hawkins, W. 1824. *A treatise of the pleas of the crown; or, A system of the principal matters relating to that subject, digested under proper heads*. Printed for S. Sweet. Eighth edition, first published 1717. Edited by John Curwood. Book One, Section 11.

Heaberlin, B., and DeDeo, S. 2016. The evolution of Wikipedia's norm network. *Future Internet*, **8**, 14.

Herrmann, B., Thoni, C., and Gachter, S. 2008. Antisocial punishment across societies. *Science*, **319**(5868), 1362–1367.

Hitchcock, T., Shoemaker, R., Emsley, C., Howard, S., McLaughlin, J., et al. 2012. *The Old Bailey Proceedings Online, 1674–1913*. www.oldbaileyonline.org. Version 7.0, 24 March 2012; Junque & Hall trial Reference Number t17790404-40; www.oldbaileyonline.org/browse.jsp?id=t17790404-40&div=t17790404-40.

Hobson, E., and DeDeo, S. 2015. Social feedback and the emergence of rank in animal society. *PLOS Computational Biology*, **11**(9), e1004411.

Hobson, E. A., Avery, M. L., and Wright, T. F. 2014. The socioecology of monk parakeets: insights into parrot social complexity. *The Auk*, **131**(4), 756–775.

Hughes, J. M., Foti, N. J., Krakauer, D. C., and Rockmore, D. N. 2012. Quantitative patterns of stylistic influence in the evolution of literature. *Proceedings of the National Academy of Sciences*, **109**(20), 7682–7686.

Jemielniak, D. 2014. *Common knowledge?: An ethnography of Wikipedia.* Stanford University Press.

Johnston, A. 2014. Jacques Lacan. In Zalta, E. N. (ed), *The Stanford encyclopedia of philosophy,* summer 2014 edn., sec. 2.3.

Kadanoff, L. P. 2000. *Statistical physics: statics, dynamics and renormalization.* World Scientific.

Keil, F. C. 2013. The roots of folk biology. *Proceedings of the National Academy of Sciences,* **110**(40), 15857–15858.

Klingenstein, S., Hitchcock, T., and DeDeo, S. 2014. The civilizing process in London's Old Bailey. *Proceedings of the National Academy of Sciences,* **111**(26), 9419–9424.

Krakauer, D. C., Flack, J. C., DeDeo, S., Farmer, D., and Rockmore, D. 2010. Intelligent data analysis of intelligent systems. Pages 8–17 of *Advances in intelligent data analysis IX.* Springer.

Lacan, J. 1998. *The Seminar. Book II. The ego in Freud's theory and in the tecnique of psychoananlysis 1954–55.* Translated by S. Tomaseli. Cambridge University Press.

Levi-Strauss, C. 1966. *The savage mind.* University of Chicago Press.

Lieberman, E., Michel, J., Jackson, J., Tang, T., and Nowak, M. A. 2007. Quantifying the evolutionary dynamics of language. *Nature,* **449**(7163), 713–716.

Mann, M 1986. *The sources of social power,* Vol. 1.: *A history of power from the beginnings to AD 1760.* Cambridge University Press.

Marzen, S. E., and DeDeo, S. 2015. The evolution of lossy compression. *arXiv preprint arXiv:1506.06138.*

Maynard Smith, J., and Szathmáry, E. 1997. *The Major Transitions in Evolution.* Oxford University Press.

McLean, P. D. 2007. *The art of the network: strategic interaction and patronage in Renaissance Florence.* Duke University Press.

Merlin, D. 1991. *Origins of the modern mind: three stages in the evolution of culture and cognition.* Harvard University Press.

Ortman, S. G., Cabaniss, A. H. F., Sturm, J. O., and Bettencourt, L. M. A. 2014. The pre-history of urban scaling. *PLoS ONE,* **9**(2), e87902.

Padgett, J. F., and McLean, P. D. 2011. Economic credit in Renaissance Florence. *Journal of Modern History,* **83**(1), 1–47.

Page, L. 2001 (September). *Method for node ranking in a linked database.* US Patent 6,285,999. Filing Date January 9, 1998; Google's "PageRank" algorithm based on measurement of first eigenvector of a transition matrix.

Povinelli, D. J. 2000. *Folk physics for apes: the chimpanzee's theory of how the world works.* Oxford University Press.

Reagle, J. M. 2010. *Good faith collaboration: the culture of Wikipedia.* History and Foundations of Information Science. MIT Press.

Rule, A., Cointet, J. and Bearman, P. S. 2015. Lexical shifts, substantive changes, and continuity in State of the Union discourse, 1790–2014. *Proceedings of the National Academy of Sciences*, **112**(35), 10837–10844.

Searle, J. 2010. *Making the social world: the structure of human civilization*. Oxford University Press.

Searle, J. R. 2008a. *Freedom and neurobiology: reflections on free will, language, and political power*. Columbia University Press.

Searle, J. R. 2008b. Language and social ontology. *Theory and Society*, **37**(5), 443–459.

Setoh, P., Wu, D., Baillargeon, R., and Gelman, R. 2013. Young infants have biological expectations about animals. *Proceedings of the National Academy of Sciences*, **110**(40), 15937–15942.

Shaw, A., and Hill, B. M. 2014. Laboratories of oligarchy? How the iron law extends to peer production. *Journal of Communication*, **64**(2), 215–238.

Smail, D. L. 2007. *On deep history and the brain*. University of California Press.

Timmermann, A., and Granger, C. W. J. 2004. Efficient market hypothesis and forecasting. *International Journal of Forecasting*, **20**(1), 15–27.

Tomasello, M. 2009a. *The cultural origins of human cognition*. Harvard University Press.

Tomasello, M. 2009b. *Why we cooperate*. Boston Review Book. MIT Press. Including responses by Carol Dweck, Joan Silk, Brian Skyrms and Elizabeth Spelke.

Tomasello, M. 2014. *A natural history of human thinking*. Harvard University Press.

Ubha, R. 2014 (June). *What's in a handshake? In tennis, a lot*. http://edition.cnn.com/2014/06/30/sport/tennis/tennis-handshakes-murray/.

Vaillancourt, T., Hymel, S., and McDougall, P. 2003. Bullying is power: implications for school-based intervention strategies. *Journal of Applied School Psychology*, **19**(2), 157–176.

Walker, S. I., and Davies, P. C. W. 2013. The algorithmic origins of life. *Journal of the Royal Society Interface*, **10**(79), 20120869.

Wilson, J. Q., and Kelling, G. L. 1982. Broken windows. *Atlantic Monthly*, **249**(3), 29–38.

Wolpert, D. H., Grochow, J. A., Libby, E., and DeDeo, S. 2014. Optimal high-level descriptions of dynamical systems. *ArXiv e-prints*, Sept. arXiv:1409.7403. Sept. SFI Working Paper 15-06-017.

Yasseri, T., Sumi, R., Rung, A., Kornai, A., and Kertesz, J. 2012. Dynamics of conflicts in Wikipedia. *PLOS ONE*, **7**(6), e38869.

Yasseri, T., Spoerri, A., Graham, M., and Kertesz, J. 2014. The most controversial topics in Wikipedia: a multilingual and geographical analysis. In: Fichman, P., and Hara, N. (eds), *Global Wikipedia: international and cross-cultural issues in online collaboration*. Scarecrow Press.

Zawidzki, T. W. 2013. *Mindshaping: a new framework for understanding human social cognition*. MIT Press.

17 Bits from Brains: Analyzing Distributed Computation in Neural Systems

Michael Wibral, Joseph Lizier, and Viola Priesemann

Artificial computing systems are a pervasive phenomenon in today's life. While traditionally such systems were employed to support humans in tasks that required mere number-crunching, there is an increasing demand for systems that exhibit autonomous, intelligent behavior in complex environments. These complex environments often confront artificial systems with ill-posed problems that have to be solved under constraints of incomplete knowledge and limited resources. Tasks of this kind are typically solved with ease by biological computing systems, as these cannot afford the luxury to dismiss any problem that happens to cross their path as "ill-posed." Consequently, biological systems have evolved algorithms to approximately solve such problems – algorithms that are adapted to their limited resources and that just yield "good enough" solutions quickly. Algorithms from biological systems may, therefore, serve as an inspiration for artificial information processing systems to solve similar problems under tight constraints of computational power, data availability, and time.

One naive way to use this inspiration is to copy and incorporate as much detail as possible from the biological into the artificial system, in the hope to also copy the emergent information processing. However, already small errors in copying the parameters of a system may compromise success. Therefore, it may be useful to derive

This chapter has been adapted from Wibral, M., Lizier, J. T., and Priesemann, V. (2015), Bits from brains for biologically inspired computing, Frontiers in Robotics and AI, 2(5), under the Creative Commons Attribution License (CC BY).

inspiration also in a more abstract way, that is *directly linked to the information processing* carried out by a biological system. But how can can we gain insight into this information processing without caring for its biological implementation?

The formal language to quantitatively describe and dissect information processing – in any system – is provided by information theory. For our particular question we can exploit the fact that information theory does not care about the nature of variables that enter the computation or information processing. Thus, it is in principle possible to treat all relevant aspects of biological computation, and of biologically inspired computing systems, in one natural framework.

In Wibral et al. (2015) we systematically presented how to analyze biological computing systems, especially neural systems, using methods from information theory and discussed how these information-theoretic results can inspire the design of artificial computing systems. Specifically, we focused on three types of approaches to characterizing the information processing undertaken in such systems and on what this tells us about the algorithms they implement. First, we showed how to analyze the information encoded in a system (responses) about its environment (stimuli). Second, we described recent advances in quantifying how much information each response variable carries about the stimuli either uniquely or redundantly, or synergistically together with others. Third, we reviewed the framework of local information dynamics, which partitions information processing into component processes of information storage, transfer, and modification, and in particular measures these processes locally in space and time.

In this chapter we present a summarized form of the discussion in Wibral et al. (2015), beginning with an outline of how we can consider neural information processing at a computational, algorithmic, and implementation level, and the three aproaches we consider for how information theory may be used to relate these (section entitled "Information Theory and Neuroscience"). The reader is referred back to Wibral et al. (2015) for a detailed review of information-theoretic preliminaries, as well as the first two approaches to characterizing information processing: quantifying how information is encoded in a system about its environment, and recent advances in quantifying uniqueness, redundancy, and synergy among response variables about stimuli. We focus here on presenting the information dynamics approach (section entitled "Analyzing Distributed Computation in Neural Systems"), which is particularly useful in gaining insights into the information processing of system components that are far removed from direct stimulation by the outer environment. We also present a brief review of studies (section entitled "Application Examples") where this information-theoretic point of view has served the goal of characterizing information processing in neural and other biological information-processing systems.

INFORMATION THEORY AND NEUROSCIENCE
Preliminaries

To analyze neural systems and biologically inspired computing systems (BICS) alike, and to show how the analysis of one can inspire the design of the other, we have to establish a common framework to approach these with information theory. Neural systems and BICS have the common property that they are composed of various smaller parts that interact. These parts will be called *agents* in general, but we will also refer to them as *neurons* or *brain areas* where appropriate. The collection of all agents will be referred to as the *system*.

We define that an agent X in a system produces an observed time series $\{x_1, \ldots, x_t, \ldots, x_N\}$, which is sampled at time intervals Δ. For simplicity we choose $\Delta = 1$, and index our measurements by $t \in \{1 \ldots N\} \subseteq \mathbb{N}$. The time series is understood as a realization of a *random process* $\mathbf{X}$. The random process is a collection of random variables (RVs) X_t, sorted by an integer index (t). Each RV X_t, at a specific time t, is described by the set of all its J possible outcomes $\mathcal{A}_{X_t} = \{a_1, \ldots a_j \ldots a_J\}$ and their associated probabilities $p_{X_t}(x_t = a_j)$.

Since the probabilities of an outcome $p_{X_t}(x_t = a_j)$ may change with t in nonstationary random processes, we indicate the RV that the probabilities belong to by subscript: $p_{X_t}(\cdot)$. In sum, the physical agent $\mathcal{X}$ is conceptualized as a random process $\mathbf{X}$, composed of a collection of RVs X_t, which produce realizations x_t, according to the probability distributions $p_{X_t}(x_t)$. When referring to more than one agent, the notation is generalized to $\mathcal{X}, \mathcal{Y}, \mathcal{Z}, \ldots$.

The random processes that we analyze in the agents of a computing system usually have memory. This means that the RVs that form the process are no longer independent but depend on variables in the past. In this setting, a proper description of the process requires us to look at the present and past RVs jointly. In general, if there is any dependence between the X_t, we have to form the smallest collection of variables $\mathbf{X_t} = (X_t, X_{t_1}, X_{t_2}, \ldots, X_{t_i}, \ldots)$ with $t_i < t$ that jointly make X_{t+1} conditionally independent of all X_{t_k} with $t_k < \min(t_i)$, i.e.:

$$p(x_{t+1}, x_{t_k} | \mathbf{x_t}) = p(x_{t+1} | \mathbf{x_t}) p(x_{t_k} | \mathbf{x_t})$$

$$\text{i.e. } p(x_{t+1} | x_{t_k}, \mathbf{x_t}) = p(x_{t+1} | \mathbf{x_t}) \tag{17.1}$$

$$\forall t_k < \min(t_i), \ \forall x_{t+1} \in \mathcal{A}_{X_{t+1}}, \ \forall x_{t_k} \in \mathcal{A}_{X_{t_k}}, \ \forall \mathbf{x_t} \in \mathcal{A}_{\mathbf{x_t}}$$

A realization $\mathbf{x_t}$ of such a sufficient collection $\mathbf{X_t}$ of past variables is called a *state* of the random process $\mathbf{X}$ at time t. A sufficient collection of past variables, also called a delay embedding vector, can always be reconstructed from scalar observations for low dimensional deterministic systems (Takens, 1981), while the behavior of scalar observables of most other systems can be approximated well by a finite collection of such past variables for all practical purposes (Ragwitz and Kantz, 2002). Without proper state space reconstruction, information-theoretic analyses will almost inevitably miscount information in the random process (Lizier et al., 2008b; Smirnov, 2013; Vicente et al., 2011). In the remainder of the text we therefore assume proper state space reconstruction.

A more detailed introduction of the notation can be found in Wibral et al. (2015), including how the relevant probability distributions may be estimated.

Basic Information-Theoretic Terms

The Shannon information content is a measure of the reduction in uncertainty, or information gained, from learning the value x of an RV X:

$$h(x) = \log \frac{1}{p(x)} \tag{17.2}$$

Typically, we take $\log_2$ giving units in *bits*. The *average* information content of an RV X is called the *entropy H*:

$$H(X) = \sum_{x \in \mathcal{A}_x} p(x) \log \frac{1}{p(x)} \tag{17.3}$$

The information content of a specific realization x of X, given we already know the outcome y of another variable Y, which is not necessarily independent of X, is called *conditional information content*:

$$h(x|y) = \log \frac{1}{p(x|y)} \tag{17.4}$$

Averaging this for all possible outcomes of X yields the *conditional entropy*:

$$H(X|Y) = \sum_{y \in \mathcal{A}_Y} p(y) \sum_{x \in \mathcal{A}_X} p(x|y) \log \frac{1}{p(x|y)}$$

$$= \sum_{x \in \mathcal{A}_X, y \in \mathcal{A}_Y} p(x, y) \log \frac{1}{p(x|y)} \tag{17.5}$$

The conditional entropy $H(X|Y)$ can be described from various perspectives: $H(X|Y)$ is the average amount of information that we get from making an observation of X after having already made an observation of Y. In terms of uncertainties $H(X|Y)$ is the average remaining uncertainty in X once Y was observed.

The mutual information of two variables X, Y, $I(X : Y)$ is the total average information in one variable $(H(X))$ minus the average information in this variable that cannot be obtained from the other variable $(H(X|Y))$:

$$I(X : Y) = H(X) - H(X|Y) = H(Y) - H(Y|X) \tag{17.6}$$

Similarly to conditional entropy we can also define a *conditional mutual information* between two variables X, Y, given the value of a third variable Z is known:

$$I(X : Y|Z) = H(X|Z) - H(X|Y, Z) \tag{17.7}$$

It is perfectly valid also to inspect local values for mutual information (like the information content h, above). This "localizability" was in fact a requirement that both Shannon and Fano postulated for proper information-theoretic measures (Fano, 1961; Shannon, 1948), and there is a growing trend in neuroscience (Lizier et al., 2011b) and in the theory of distributed computation (Lizier, 2013, 2014b) to return to local values. The *local* mutual information $i(x : y)$ is defined as:

$$i(x : y) = \log \frac{p(x, y)}{p(x)p(y)} = \log \frac{p(x|y)}{p(x)} \tag{17.8}$$

while the *local conditional* mutual information is defined as:

$$i(x : y|z) = \log \frac{p(x|y, z)}{p(x|z)} \tag{17.9}$$

When we take the expected values of these local measures, we obtain mutual and conditional mutual information. These measures are called local, because they allow one to quantify mutual and conditional mutual information between *single realizations*. Note, however, that the probabilities $p(\cdot)$ involved in Eqs. (17.8) and (17.9) are *global* in the sense that they are representative of all possible outcomes. In other words, a valid probability distribution has to be estimated irrespective of whether we are interested in average or local information measures. We also note that local mutual information and local conditional mutual information may be negative, unlike their averaged forms (Fano, 1961; Lizier, 2014b). This occurs for the local mutual information where the measurement of one variable is *misinformative* about the other variable, i.e. where the realization y *lowers* the probability $p(x|y)$ below the initial probability $p(x)$. This means that the observer expected x less after observing y than before, but x occurred nevertheless. Therefore, y was misinformative about x.

Why Information Theory in Neuroscience?

It is useful to organize our understanding of neural (and biologically inspired) computing systems into three major levels, originally proposed by David Marr (1982), and to then see at which level information theory provides insights:

- At the level of the task the neural system or the BICS is trying to solve (*task* level[1]) we ask what information processing problem a neural system (or a part of it) tries to solve. Such problems could, for example, be the detection of edges or objects in a visual scene, or maintaining information about an object after the object is no longer in the visual scene. It is important to note that questions at the task level typically revolve around entities that have a direct meaning to us, e.g., objects or specific object properties used as stimulus categories, or operationally defined states, or concepts such as attention or working memory. An example of an analysis carried out purely at this level is the investigation of whether a person behaves as an optimal Bayesian observer (see references in Knill and Pouget, 2004).

- At the *algorithmic* level we ask what entities or quantities of the task level are represented by the neural system and how the system operates on these representations using algorithms. For example, a neural system may represent either absolute luminance or changes of luminance of the visual input. An algorithm operating on either of these representations may, for example, then try to identify an object in the input that is causing the luminance pattern by a brute force comparison to all luminance patterns ever seen (and stored by the neural system). Alternatively, it may try to further transform the luminance representation via filtering, etc., before inferring the object via a few targeted comparisons.

- At the (biophysical) *implementation* level, we ask how the representations and algorithms are implemented in neural systems. Descriptions at this level are given in terms of the relationship between various biophysical properties of the neural system or its components, e.g., membrane currents or voltages, the morphology of neurons, spike rates, or chemical gradients. A typical study at this level might aim, for example, at reproducing observed physical behavior of neural circuits,

---

[1] Called the "computational level" by Marr originally. This terminology, however, collides with other meanings of computation used in this text.

such as gamma-frequency (>40 Hz) oscillations in local field potentials by modeling the biophysical details of these circuits from the ground up (Markram, 2006).

This separation of levels of understanding served to resolve important debates in neuroscience, but there is also growing awareness of a specific shortcoming of this classic view: results obtained by careful study at any of these levels do not constrain the possibilities at any other level (see the afterword by Poggio in Marr, 1982). For example, the task of winning a game of tic-tac-toe (task level) can be reached by a brute force strategy (algorithmic level) that may be realized in a mechanical computer (implementation level) (Dewdney, 1989). Alternatively, the same task can be solved by flexible rule use (algorithmic level) realized in biological brains (implementation level) of young children (Crowley and Siegler, 1993).

Wibral et al. (2015) described how missing relationships between Marr's levels can be filled in by information theory. We showed how to link the task level and the implementation level by computing various forms of mutual information between variables at these two levels. We also described how these mutual information terms can be further decomposed into the contributions of each agent in a multiagent system, as well as information carried jointly. In this chapter, in the section entitled "Analyzing Distributed Computation in Neural Systems" we present the use of local information measures to link neural activity at the implementation level to components of information processing at the algorithmic level, such as information storage and transfer. This will be done *per agent and time step* and thereby yields a sort of information-theoretic "footprint" of the algorithm in space and time. To be clear, such an analysis will only yield this "footprint" – not identify the algorithm itself. Nevertheless, this footprint is a useful constraint when identifying algorithms in neural systems, because various possible algorithms to solve a problem will clearly differ with respect to this footprint. We close by a short review in the section entitled "Application Examples" of some example applications of information-theoretic analyses of neural data, and describe how they relate to Marr's levels.

ANALYZING DISTRIBUTED COMPUTATION IN NEURAL SYSTEMS

Analyzing Neural Coding and Goal Functions in a Domain- Independent Way

The analysis of neural coding strategies presented for the first two approaches considered in Wibral et al. (2015) (quantifying encoding in a system about its environment, and decomposing this encoding) relies on our a priori knowledge of the set of task-level (e.g., stimulus) features that is encoded in neural responses at the implementation level. If we have this knowledge, information theory will help us to link the two levels. This is somewhat similar to the situation in cryptography where we consider a code "cracked' if we obtain a human-readable plaintext message, i.e., we move from the implementation level (encrypted message) to the task level (meaning). However, what happens if the plaintext was in a language that one never heard of?[2] In this case, we would potentially crack the code without ever realizing it, as the plaintext still has no meaning for us.

The situation in neuroscience bears resemblance to this example in at least two respects. First, most neurons do not have direct access to any properties of the outside world; rather, they receive nothing but input spike trains. All they ever learn and process must come from the structure of these input spike trains. Second, if we as researchers probe the system beyond early sensory or motor areas, we have little knowledge of what is actually encoded by the neurons deeper inside the system. As a result, proper stimulus sets get hard to choose. In this case, the gap between the task and the implementation levels may actually become too wide for meaningful analyses, as noticed recently by Carandini (2012).

---

[2] See, e.g., the Navajo code during World War II that was never broken (Fox, 2014).

Instead of relying on descriptions of the outside world (and thereby involve the task level), we may take the point of view that information processing in a neuron is nothing but the transformation of input spike trains to output spike trains. We may then try to use information theory to link the implementation and algorithmic level, by retrieving a 'footprint" of the information processing carried out by a neural circuit. This approach builds on only a very general agreement that neural systems perform at least *some kind of* information processing. This information processing can be partitioned into the component processes that determine or predict the next RV of a process Y at time t, Y_t: (1) information storage, (2) information transfer, and (3) information modification. A partition of this kind had already been formulated by Turing (see Langton, 1990) and was recently formalized by Lizier et al. (2014) (see also Lizier, 2013):

- **Information storage** quantifies the information contained in the past state variable $\mathbf{Y_{t-1}}$ of a process that is used by the process at the next RV at t, Y_t (Lizier et al., 2012b). This relatively abstract definition means that an observer will see at least a part of the past information in the process" past again in its future, but potentially transformed. Hence, information storage can be naturally quantified by mutual information between the past and the future[3] of a process.
- **Information transfer** quantifies the information contained in the state variables $\mathbf{X_{t-u}}$ (found u time-steps into the past) of one source process X that can be used to predict information in the future variable Y_t of a target process Y, in the context of the past state variables $\mathbf{Y_{t-1}}$ of the target process (Paluš, 2001; Schreiber, 2000; Vicente et al., 2011).
- **Information modification** quantifies the combination of information from various source processes into a new form that is not trivially predictable from any subset of these source processes (for details of this definition, see also Lizier et al., 2010, 2013).

Based on Turing's general partition of information processing (Langton, 1990), Lizier and colleagues proposed an information-theoretic framework to quantify distributed computations in terms of

---

[3] We consider ourselves having information up to time $t - 1$, predicting the future values at t.

all three component processes *locally*, i.e., for each part of the system (e.g., neurons or brain areas) and each time-step (Lizier et al., 2008b, 2010, 2012b). This framework is called *local information dynamics* and has been successfully applied to unravel computation in swarms (Wang et al., 2011), in Boolean networks (Lizier et al., 2011a), and in neural models (Boedecker et al., 2012) and data (Wibral et al., 2014a) (see also the section entitled "Application Examples" for details on these example applications).

Crucially, information dynamics is the *perspective* of an observer who measures the processes X, Y, Z, etc., and tries to *partition* the information in Y_t into the apparent contributions from stored, transferred, and modified information, without necessarily knowing the true underlying system structure. For example, such an observer would label any recurring information in Y as information storage, even where such information *causally* left the system and reentered Y at a later time (e.g., a stigmergic process).

Other partitions are possible (James et al., 2011), for example, partition information in the present of a process in terms of its relationships to the semi-infinite past and semi-infinite future. In contrast, we focus on the information dynamics perspective laid out above since it quantifies terms that can be specifically identified as information storage, transfer, and modification, which aligns with many qualitative descriptions of dynamics in complex systems (Mitchell, 1998). In particular, the information-dynamics perspective is novel in focusing on quantifying these operations on a local scale in space and time.

In the following we present both average and local measures of information transfer and storage, while referring the reader to discussion of information modification in Wibral et al. (2015).

Information Transfer
The analysis of information transfer was formalized initially by Schreiber (2000) and Paluš (2001) and has seen a rapid surge of interest

in neuroscience[4] and general physiology.[5] Information transfer as measured by the transfer entropy introduced below has recently also been given a thermodynamic interpretation by Prokopenko and Lizier (2014), continuing general efforts to link information theory and thermodynamics (Landauer, 1961; Szilárd, 1929), highlighting the importance of the concept.

Definition

Information transfer from a process X (the *source*) to another process Y (the *target*) is measured by the transfer entropy (TE) functional[6] (Schreiber, 2000):

$$
\begin{aligned}
TE(X_{t-u} \to Y_t) &= I(\mathbf{X}_{t-u} : Y_t | \mathbf{Y}_{t-1}) \\
&= \sum_{\substack{y_t \in \mathcal{A}_{Y_t}, \\ \mathbf{y}_{t-1} \in \mathcal{A}_{\mathbf{Y}_{t-1}}, \\ \mathbf{x}_{t-u} \in \mathcal{A}_{\mathbf{X}_{t-u}}}} p(y_t, \mathbf{y}_{t-1}, \mathbf{x}_{t-u}) \log \frac{p(y_t | \mathbf{y}_{t-1}, \mathbf{x}_{t-u})}{p(y_t | \mathbf{y}_{t-1})}
\end{aligned}
$$

where $I(\cdot : \cdot | \cdot)$ is the conditional mutual information, Y_t is the RV of process Y at time t, and $\mathbf{X}_{t-u}, \mathbf{Y}_{t-1}$ are the past state RVs of processes X and Y, respectively. The delay variable u in $\mathbf{X}_{t-u}$ indicates that the past state of the source is to be taken u time-steps into the past to account for a potential physical interaction delay between the processes. This parameter need not be chosen ad hoc, as it was recently proven

---

[4] Amblard and Michel, 2011; Barnett et al., 2009; Battaglia, 2014; Battaglia et al., 2012; Bedo et al., 2014; Besserve et al., 2010; Buehlmann and Deco, 2010; Butail et al., 2014; Chavez et al., 2003; Chicharro, 2014; Garofalo et al., 2009; Gourevitch and Eggermont, 2007; Hadjipapas et al., 2005; Ito et al., 2011; Kawasaki et al., 2014; Leistritz et al., 2006; Li and Ouyang, 2010; Lindner et al., 2011; Liu and Pelowski, 2014; Lizier et al., 2011b; Ludtke et al., 2010; Marinazzo et al., 2014a,b,c; McAuliffe, 2014; Montalto et al., 2014; Neymotin et al., 2011; Orlandi et al., 2014; Paluš, 2001; Porta et al., 2014; Razak and Jensen, 2014; Rowan et al., 2014; Sabesan et al., 2009; Shimono and Beggs, 2015; Staniek and Lehnertz, 2009; Stetter et al., 2012; Thivierge, 2014; Untergehrer et al., 2014; Vakorin et al., 2009, 2010, 2011; van Mierlo et al., 2014; Varon et al., 2015; Vicente et al., 2011; Wibral et al., 2011; Yamaguti and Tsuda, 2015; Zubler et al., 2015.

[5] Faes and Nollo, 2006; Faes and Porta, 2014; Faes et al., 2011, 2012, 2014a,b.

[6] A functional maps from the relevant probability distribution (i.e., functions) to the real numbers. In contrast, an estimator maps from empirical data, i.e., a set of real numbers, to the real numbers.

for bivariate systems that the above estimator is maximized if the parameter u is equal to the true delay δ of the information transfer from X to Y (Wibral et al., 2013). This relationship allows one to estimate the *true* interaction delay δ from data by simply scanning the assumed delay u:

$$\delta = \underset{u}{\operatorname{argmax}}\ [TE\,(\mathbf{X}_{t-u} \to Y_t, u)] \tag{17.10}$$

The TE functional can be linked to Wiener–Granger type causality (Barnett et al., 2009; Granger, 1969; Wiener, 1956). More precisely, for systems with *jointly* Gaussian variables, TE is equivalent[7] to *linear* Granger causality (see Barnett et al., 2009, and references therein). However, whether the assumption of jointly Gaussian variables is appropriate in a neural setting must be checked carefully for each case (note that Gaussianity of each marginal distribution is not sufficient). In fact, electroencephalogram (EEG) source signals were found to be non-Gaussian (Wibral et al., 2008) as indicated by a match of independent components and physical sources.

Transfer Entropy Estimation

When the probability distributions required are known (e.g., in an analytically tractable neural model), TE can be computed directly. However, in most cases the probability distributions have to be derived from data. When probabilities are estimated naively from the data via counting, and when these estimates are then used to compute information-theoretic quantities such as the TE, we speak of a "plug in" estimator. Indeed such plug-in estimators have been used in the past, but they come with serious bias problems (Panzeri et al., 2007). Therefore, newer approaches to TE estimation rely on a more direct estimation of the entropies that TE can be decomposed into (Gomez-Herrero et al., 2015; Kraskov et al., 2004; Vicente et al., 2011; Wibral et al., 2014b). These estimators still suffer from bias problems but to a lesser degree (Kraskov et al., 2004). We describe the

---

[7] To a constant factor of 2.

development of such estimators in detail in Wibral et al. (2015), in particular the extension of the Kraskov et al. (2004) estimator to TE.

Finally, we note that we must reconstruct the states of the processes (see section entitled "Information Theory and Neuroscience") before estimating TE. One approach to state space reconstruction is time-delay embedding (Takens, 1981). It uses past variables $X_{t-n\tau}$, $n = 1, 2, \ldots$ that are spaced in time by an interval τ. The number of these variables and their optimal spacing can be determined using established criteria (Faes et al., 2012; Lindner et al., 2011; Ragwitz and Kantz, 2002; Small and Tse, 2004). The realizations of the state variables can be represented as vectors of the form:

$$\mathbf{x}_t^d = (X_t, X_{t-\tau}, X_{t-2\tau}, \ldots, X_{t-(d-1)\tau}) \tag{17.11}$$

where d is the dimension of the state vector. Using this vector notation, transfer entropy can be written as:

$$TE_{SPO}(\mathbf{X}_{t-u} \to Y_t) = I(\mathbf{X}_{t-u}^{d_x} : Y_t | \mathbf{Y}_{t-1}^{d_y}) \tag{17.12}$$

where the subscript SPO (for self-prediction optimal) is a reminder that the past states of the target, $\mathbf{y}_{t-1}^{d_y}$, have to be constructed such that conditioning on them is optimal in the sense of taking the active information storage in the target correctly into account (Wibral et al., 2013).

Interpretation of Transfer Entropy as a Measure at the Algorithmic Level (*Or: Why Notions of Information Transfer and Causality Are Distinct*)

TE describes computation at the algorithmic level, not at the level of a physical dynamical system. As such it is not optimal for inference about *causal* interactions – although it has been used for this purpose in the past. The fundamental reason for this is that information transfer relies on causal interactions, but nonzero TE can occur without direct causal links, and causal interactions do not necessarily lead to nonzero information transfer (Ay and Polani, 2008; Chicharro and Ledberg, 2012; Lizier and Prokopenko, 2010). Instead, causal interactions may serve active information storage alone (see the next section) or force two systems into identical synchronization, where information transfer becomes effectively zero. This might be summarized by stating that TE is limited to effects of a causal interaction from a source to a target process that are unpredictable given the past of the target process alone. In this sense, TE may be seen as quantifying causal interactions currently *in use for the communication aspect* of distributed computation. Therefore, one may say that TE measures *predictive* or *algorithmic information transfer*.

A simple thought-experiment may serve to illustrate this point. When one plays an unknown record, a chain of causal interactions serve the transfer of information about the music from the record to your brain. Causal interactions happen between the record's grooves and the needle, the magnetic transducer system behind the needle, and so on, up to the conversion of pressure modulations to neural signals in the cochlea that finally activate your cortex. In this situation, there undeniably is information transfer, as the information read out from the source, the record, at any given moment is not yet known in the target process, i.e. the neural activity in the cochlea. However, this information transfer ceases if the record has a crack, making the needle skip and repeat a certain part of the music. Obviously, no new information is transferred, which under certain mild conditions is equivalent to no information transfer at all. Interestingly, an analysis of TE between sound and cochlear activity will yield the same result: the repetitive sound leads to repetitive neural activity (at least after a while). This neural activity is thus predictable by its own past, under the condition of vanishing neural "noise," leaving no room for a prediction improvement by the sound source signal. Hence, we obtain a TE of zero, which is the correct result from a conceptual point of view. Remarkably, at the same time the chain of causal interactions remains practically unchanged. Therefore, a causal model able to fit the data from the original situation will have no problem to fit the data of the situation with the cracked record, as well. Again, this is conceptually the correct result, but this time from a causal point of view.

The difference between an analysis of information transfer in a computational sense and causality analysis based on interventions has been demonstrated convincingly in a recent study by Lizier and Prokopenko (2010). The same authors also demonstrated why an analysis of information transfer can yield better insight than the analysis of causal interactions if the *computation* in the system is to be understood. The difference between causality and information transfer is also reflected in the fact that a single causal structure can support a diverse pattern of information transfer (*functional multiplicity*), and the same pattern of information transfer can be realized with different causal structures (*structural degeneracy*), as shown by Battaglia (2014).

Local Information Transfer

As TE is formally just a conditional mutual information, we can obtain the corresponding local conditional mutual information (Eq. 17.9) from Eq. (17.12). This quantity is called the local transfer entropy (*te*) (Lizier et al., 2008b). For realizations x_t, y_t of two processes X, Y at time t it reads:

$$te\left(\mathbf{X}_{t-u} = \mathbf{x}_{t-u} \rightarrow Y_t = y_t\right) = \log \frac{p\left(y_t | \mathbf{y}_{t-1}^{d_y}, \mathbf{x}_{t-u}^{d_x}\right)}{p\left(y_t | \mathbf{y}_{t-1}^{d_y}\right)} \qquad (17.13)$$

As said earlier in the section on basic information theory, the use of local information measures does not eliminate the need for an appropriate estimation of the probability distributions involved. Hence, for a nonstationary process these distributions will still have to be estimated via an ensemble approach for each time point for the RVs involved – e.g., via physical replications of the system or via enforcing cyclostationarity by design of the experiment.

The analysis of local transfer entropy has been applied with great success in the study of cellular automata to confirm the conjecture that certain coherent spatio-temporal structures traveling through the network are indeed the main carriers of information transfer (Lizier et al., 2008b) (see further discussion in Wibral et al., 2015). Similarly, local transfer entropy has identified coherent propagating wave structures in flocks as information cascades (Wang et al., 2012) and indicated impending synchronization among coupled oscillators (Ceguerra et al., 2011).

Common Problems and Solutions

Typical problems in TE estimation encompass (1) finite sample bias, (2) the presence of non-stationarities in the data, and (3) the need for multivariate analyses. In recent years all of these problems have been addressed at least in isolation, as summarized below:

1. Finite sample bias can be overcome by statistical testing using surrogate data, where the observed realizations $y_t, \mathbf{y}_{t-1}^{d_y}, \mathbf{x}_{t-u}^{d_x}$ of the RVs $Y_t, \mathbf{Y}_{t-1}^{d_y}, \mathbf{X}_{t-u}^{d_x}$ are reassigned to other RVs of the process, such that the temporal order underlying the information transfer is destroyed (for an example, see the procedures suggested in Lindner et al., 2011). This reassignment should conserve as many data features of the single process realizations as possible.

2. As already explained in the section on basic information theory above, nonstationary random processes in principle require that the necessary estimates of the probabilities for TE (e.g., see Eq. 17.13) are based on physical replications of the systems in question. Where this is impossible, the experimenter should design the experiment in such a way that the processes are repeated in time. If such cyclostationary data are available, then TE should be estimated using ensemble methods as described in Gomez-Herrero et al. (2015) and implemented in the TRENTOOL toolbox (Lindner et al., 2011; Wollstadt et al., 2014).

3. So far, we have restricted our presentation of TE estimation to the case of just two interacting random processes X, Y, i.e., a bivariate analysis. In a setting that is more realistic for neuroscience, one deals with large networks of interacting processes X, Y, Z, ... In this case various complications arise if the analysis is performed in a bivariate manner. For example, a process Z could transfer information with two different delays $\delta_{Z \to X}, \delta_{Z \to Y}$ to two other processes X, Y. In this case, a pairwise analysis of TE between X, Y will yield an apparent information transfer from the process that receives information from Z with the shorter delay to the one that receives it with the longer delay (common driver effect).

A similar problem arises if information is transferred first from a process X to Y, and then from Y to Z. In this case, a bivariate analysis will also indicate information transfer from X to Z (cascade effect). Moreover, two sources may transfer information purely synergistically, i.e. the TE from each source alone to the target is zero, and only considering them jointly reveals the information transfer.[8]

From a mathematical perspective this problem seems to be easily solved by introducing the *complete transfer entropy*, which is defined in terms of a *conditional transfer entropy* (Lizier et al., 2008b, 2010):

$$TE\left(\mathbf{X_{t-u}} \rightarrow Y_t | \mathbf{Z^-}\right) = \sum_{\substack{y_t \in \mathcal{A}_{Y_t}, \\ y_{t-1} \in \mathcal{A}_{Y_{t-1}}, \\ \mathbf{x_{t-u}} \in \mathcal{A}_{X_{t-u}}, \\ \mathbf{z^-} \in \mathcal{A}_{Z^-}}} p\left(y_t, \mathbf{y_{t-1}}, \mathbf{x_{t-u}}, \mathbf{z^-}\right)$$

$$\log \frac{p\left(y_t | \mathbf{y_{t-1}}, \mathbf{x_{t-u}}, \mathbf{z^-}\right)}{p\left(y_t | \mathbf{y_{t-1}}, \mathbf{z^-}\right)} \qquad (17.14)$$

where the state RV $\mathbf{Z^-}$ is a collection of the past states of *one or more* processes in the network other than X, Y. We label Eq. (17.14) a complete transfer entropy $TE^{(c)}\left(\mathbf{X_{t-u}} \rightarrow Y_t\right)$ when we take $\mathbf{Z^-} = \mathbf{V^-}$, the set of *all* processes in the network other than X, Y.

It is important to note that TE and conditional/complete TE are *complementary* (see mathematical description of this at the section entitled "Analyzing Distributed Computation in Neural Systems") – each can reveal aspects of the underlying dynamics that the other does not and both are required for a full description. While conditional TE removes redundancies and includes synergies, knowing that redundancy is present may be important, and local pairwise TE additionally reveals interesting cases when a source is misinformative about the dynamics (Lizier et al., 2008a, 2008b).

---

[8] Again, cryptography may serve as an example here. If an encrypted message is received, there will be no discernible information transfer from the encrypted message to plaintext without the key. In the same way, there is no information transfer from the key alone to the plaintext. It is only when the encrypted message and key are combined that the relation between the combination of encrypted message and key, on the one side, and the plaintext, on the other side, is revealed.

Furthermore, even for small networks of random processes the joint state space of the variables $Y_t, \mathbf{Y_{t-1}}, \mathbf{X_{t-u}}, \mathbf{V}^-$ may become intractably large from an estimation perspective. Moreover, the problem of finding all information transfers in the network (a form of *effective network inference*), either from single source variables into the target or sviaynergistic transfer from collections of source variables to the target, is a combinatorial problem and can therefore typically not be solved in a reasonable time.

Therefore, Faes et al. (2012), Lizier and Rubinov (2012), and Stramaglia et al. (2012) suggested to analyze the information transfer in a network iteratively, selecting information sources for a target in each iteration based on either the magnitude of apparent information transfer (Faes et al., 2012) or its significance (Lizier and Rubinov, 2012; Stramaglia et al., 2012). In the next iteration, already selected information sources are added to the conditioning set ($\mathbf{Z}^-$ in Eq. 17.14), and the next search for information sources is started. The approach of Stramaglia and colleagues is particular here in that the conditional mutual information terms are computed at each level as a series expansion, following a suggestion by Bettencourt et al. (2008). This allows for an efficient computation as the series may truncate early, and the search can proceed to the next level. Importantly, these approaches also consider synergistic information transfer from more than one source variable to the target. For example, a variable transferring information purely synergistically with $\mathbf{Z}^-$ may be included in the next iteration, given that the other variables it jointly transfers information with are already in the conditioning set $\mathbf{Z}^-$. However, there is currently no explicit indication in the approaches of Faes et al. (2012) and Lizier and Rubinov (2012) as to whether multivariate information transfer from a set of sources to the target is in fact synergistic; in addition, redundant links will not be included. In contrast, both redundant and synergistic multiplets of variables transferring information into a target may be identified in the approach of Stramaglia et al. (2012) by looking at the sign of the contribution of the multiplet.

Unfortunately, there is also the possibility of cancellation if both types of multivariate information (redundant, synergistic) are present.

Information Storage

Before we present explicit measures of active information storage, a few comments may serve to avoid misunderstanding. Since we analyze neural *activity* here, measures of active information storage are concerned with information stored in this activity – rather than in synaptic properties, for example.[9] This is the perspective of what an observer of that activity (not necessarily with any knowledge of the underlying system structure) would attribute as information storage at the algorithmic level, even if the causal mechanisms at the level of a physical dynamical system underpinning such apparent storage were distributed externally to the given variable (Lizier et al., 2012b). As laid out above, storage is conceptualized here as a mutual information between past and future states of neural activity. From this it is clear that there will not be much information storage if the information contained in the future states of neural activity is low in general. If, on the other hand, these future states are rich in information but bear no relation to past states, i.e., are unpredictable, again information storage will be low. Hence, large information storage occurs for activity that is rich in information but, at the same time, predictable.

Thus, information storage gives us a way to define the predictability of a process that is independent of the prediction error: information storage quantifies how much future information of a process can be predicted from its past, whereas the prediction error measures how much information cannot be predicted. If both are quantified via information measures, i.e., in bits, the error and the predicted information add up to the total amount of information in a random variable of the process. Importantly, these two measures may lead to quite different views about the predictability of a process. This

---

[9] See the distinction made between passive storage in synaptic properties and active storage in dynamics by Zipser et al. (1993).

is because the total information can vary considerably over the process, and as such the predictable and the unpredictable information may vary almost independently. Thus, it is an important distinction whether a natural system or a BICS is designed to minimize prediction errors or to maximize the amount of predicted information in the long run. The latter goal may lead to the system searching for inputs or situations with high total information, whereas the first goal is well served by searching for inputs or situations with minimal information.[10]

Before turning to the explicit definition of measures of information storage, it is worth considering which temporal extent of "past" and "future" states we are interested in. To the largest extent, *predictive information* (Bialek et al., 2001) or *excess entropy* (Crutchfield and Feldman, 2003; Crutchfield and Packard, 1982; Grassberger, 1986) is the mutual information between the *semi-infinite* past and *semi-infinite* future of a process before and after time point t. In contrast, if we are interested in the information currently used for the *next step* of the process, the mutual information between the *semi-infinite* past and the next step of the process, the *active information storage* (Lizier et al., 2012b) is of greater interest. Both measures are defined in the next paragraphs.

Predictive Information/Excess Entropy

Excess entropy is formally defined as:

$$E_{X_t} = \lim_{k \to \infty} I(\mathbf{X}_t^{k-} : \mathbf{X}_t^{k+}) \tag{17.15}$$

where $\mathbf{X}_t^{k-} = \{\mathbf{X}_t, \mathbf{X}_{t-1}, \ldots, \mathbf{X}_{t-k+1}\}$, and $\mathbf{X}_t^{k+} = \{\mathbf{X}_{t+1}, \ldots, \mathbf{X}_{t+k}\}$ indicate collections of the past and future k variables of the process $\mathbf{X}$.[11] These collections of RVs ($\mathbf{X}_t^{k-}$, $\mathbf{X}_t^{k+}$), in the limit $k \to \infty$, span the semi-infinite past and future, respectively. In general, the

---

[10] In predictive coding theories this is also known as the "dark room argument" against error minimization (Friston et al., 2012); see also the use of predictive information to drive self-organized behavior of robots (Martius et al., 2013).

[11] In principle these could harness embedding delays, as defined in Eq. (17.11).

mutual information in Eq. (17.15) has to be evaluated over multiple realizations of the process. For stationary processes, however, E_{X_t} is not time dependent, and Eq. (17.15) can be rewritten as an average over time points t and computed from a single realization of the process – at least in principle (we have to consider that the process must run for an infinite time to allow the limit $\lim_{k \to \infty}$ for all t):

$$E_X = \langle \lim_{k \to \infty} i(\mathbf{x}_t^{k-} : \mathbf{x}_t^{k+}) \rangle_t \qquad (17.16)$$

Here, $i(\cdot : \cdot)$ is the local mutual information from Eq. (17.9), and $\mathbf{x}_t^{k-}$, $\mathbf{x}_t^{k+}$ are realizations of $\mathbf{X}_t^{k-}$, $\mathbf{X}_t^{k+}$. The limit of $k \to \infty$ can be replaced by a finite k_{max} if a k_{max} exists such that conditioning on $\mathbf{X}_t^{k_{max}-}$ renders $\mathbf{X}_t^{k_{max}+}$ conditionally independent of any X_l with $l \leq t - k_{max}$.

Even if the process in question is nonstationary, we may look at values that are local in time as long as the probability distributions are derived appropriately (Lizier et al., 2012b; Shalizi, 2001):[12]

$$e_{X_t} = \lim_{k \to \infty} i(\mathbf{x}_t^{k-} : \mathbf{x}_t^{k+}) \qquad (17.17)$$

Active Information Storage

From a perspective of the dynamics of information processing, we might be interested in information that is used by a process not at some time far in the future, but at the next point in time, i.e. information that is said to be "currently in use" for the computation of the next step (the realization of the next RV) in the process (Lizier et al., 2012b). To quantify this information, a different mutual information is computed, namely, the *active information storage* (AIS) (Lizier et al., 2007, 2012b):

$$A_{X_t} = \lim_{k \to \infty} I(\mathbf{X}_{t-1}^{k-} : X_t) \qquad (17.18)$$

AIS is similar to a measure called "regularity" introduced by Porta et al. (2000) and was also labeled as ρ_μ ("redundant portion" of information in X_t) by James et al. (2011).

---

[12] See further comments in Wibral et al., 2015.

Again, if the process in question is stationary, then $A_{X_t} = $ const. $= A_X$ and the expected value can be obtained from an average over time – instead of an ensemble of realizations of the process – as:

$$A_X = \langle \lim_{k \to \infty} i(\mathbf{x}_{t-1}^{k-} : x_t) \rangle_t \tag{17.19}$$

which can be read as an average over local active information storage (LAIS) values a_{X_t} (Lizier et al., 2012b):

$$A_X = \langle a_{X_t} \rangle_t$$
$$a_{X_t} = \lim_{k \to \infty} i(\mathbf{x}_{t-1}^{k-} : x_t)$$

Even for nonstationary processes we may investigate local active storage values, given the corresponding probability distributions are properly obtained from an ensemble of realizations of X_t, $\mathbf{X}_{t-1}^{k-}$.

Again, the limit of $k \to \infty$ can be replaced by a finite $k_{\max}$ if a $k_{\max}$ exists such that conditioning on $\mathbf{X}_{t-1}^{k_{\max}}$ renders X_t conditionally independent of any X_l with $l \leq t - k_{\max}$ (see Eq. (17.1)).

Interpretation of Information Storage as a Measure at the Algorithmic Level

As laid out above information storage is a measure of the amount of information in a process that is predictable from its past. As such it quantifies, e.g., how well activity in one brain area A can be predicted by another area, by learning its statistics. Hence, questions about information storage arise naturally when asking about the generation of predictions in the brain, e.g., in predictive coding theories (Friston et al., 2006; Rao and Ballard, 1999).

Combining the Analysis of Local Active Information Storage and Local Transfer Entropy

To demonstrate how information storage and transfer are complementary operations, we consider how the overall information $H(X_t)$ in the future of the target process (or its local form, $h(x_t)$) can be explained by looking at *all* sources of information and the history of the target *jointly*, at least up to the remaining stochastic part (the intrinsic innovation of the random process) in the target, as shown by Lizier et al. (2010). It is crucial to note that, in general, we cannot decompose this information into *pairwise* mutual information terms only.

To see the differences between a partition considering variables jointly or only in pairwise terms, consider a series of subsets formed from the set of all variables $\mathbf{Z}_{t-,i}$ (defined above; ordered by i here) that can transfer information into the target, except variables from the target's own history. The bold typeface in $\mathbf{Z}_{t-,i}$ is a reminder that we work with a state space representation where necessary. Following the derivation by Lizier et al. (2010), we create a series of subsets $\mathbf{V}_{X_t}^g \setminus \mathbf{X}_{t-1}$ such that $\mathbf{V}_{X_t}^g \setminus \mathbf{X}_{t-1} = \{\mathbf{Z}_{t-,1}, \ldots, \mathbf{Z}_{t-,g-1}\}$; i.e., the gth subset contains only the first $g-1$ sources. We can decompose the collective transfer entropy from all our source variables, $TE(\mathbf{V}_{X_t} \setminus \mathbf{X}_{t-1} \to X_t)$, as a series of conditional mutual information terms, incrementally increasing the set that we condition on:

$$TE(\mathbf{V}_{X_t} \setminus \mathbf{X}_{t-1} \to X_t) = \sum_{g=1}^{G} I(X_t : \mathbf{Z}_{t-,g} | \mathbf{X}_{t-1}, \mathbf{V}_{X_t}^g \setminus \mathbf{X}_{t-1}) \quad (17.20)$$

These conditional mutual information terms are all transfer entropies – starting for $g = 1$ with a pairwise transfer entropy $TE(\mathbf{Z}_{t-,1} \to X_t)$, then with conditional transfer entropies for $g = 2 \ldots G - 1$, and finishing with a complete TE for $g = G$, $TE(\mathbf{Z}_{t-,G} \to X_t | \mathbf{V}_{X_t}^G \setminus \mathbf{X}_{t-1})$. The total entropy of the target $H(X_t)$ can then be written as:

$$H(X_t) = A_{X_{t-1}} + \sum_{g=1}^{G} I(X_t : \mathbf{Z}_{t-,g} | \mathbf{X}_{t-1}, \mathbf{V}_{X_t}^g \setminus \mathbf{X}_{t-1}) + W_{X_t} \quad (17.21)$$

where W_{X_t} is the innovation in (or remaining stochastic part of) X_t. If we rewrite the partition in Eq. (17.21) in its local form:

$$h(x_t) = a_{X_{t-1}} + \sum_{g=1}^{G} i(x_t : \mathbf{z}_{t-,g} | \mathbf{x}_{t-1}, \mathbf{v}_{X_t}^g \setminus \mathbf{x}_{t-1}) + w_{X_t} \quad (17.22)$$

then these equations show, first, that information storage and (various orders of) transfer terms are complementary components of the computation of the future of the target process X_t. Next, we see that

$H(X_t)$ (and correspondingly $h(x_t)$) is in general not a sum of pairwise terms of individual contributions from the sources to the target. The difference is that the conditioning of the TE terms incorporates synergies and eliminates redundancies among the sources – a phenomenon explained by the partial information decomposition framework (see, e.g., Barrett, 2014; Bertschinger et al., 2014; Griffith and Koch, 2014; Griffith et al., 2014; Harder et al., 2013; Lizier et al., 2013; Stramaglia et al., 2012, 2014; Timme et al., 2014; Williams and Beer, 2010; and the review in Wibral et al., 2015). This observation led Lizier and colleagues to propose a measure of information modification based on the *synergistic* part of the information transfer from the source variables $Z_{t^-,g}$, and the target's history X_{t-1} to the target X_t (Lizier et al., 2013). This definition of information modification is more rigorously defined than previous attempts, but awaits the establishment of a measure of synergy accepted by the field (refer to the review in Wibral et al., 2015).

Finally, we note that the two measures of local active information storage and local TE may be fruitfully combined by *pairing* storage and transfer values at each point in time and for each agent. The resulting space has been termed the "local information dynamics state space" and has been used to investigate the computational capabilities of cellular automata, by pairing $a(y_{j,t})$ and $te\left(x_{i,t-1} \rightarrow y_{j,t}\right)$ for each pair of source and target x_i, y_j at each time point (Lizier et al., 2012a). We suggested (Wibral et al., 2015) that this concept may be used to disentangle various neural processing strategies, and refer the reader to that paper for details.

APPLICATION EXAMPLES

In this section we review several applications of the information dynamics approach to empirical data sets, discussing how the measures allow us to derive constraints on possible algorithms served by the observed dynamics and to narrow in on the algorithm(s) being implemented in the neural system. Additional examples, including identifying emergent information-processing structures in cellular automata, information cascades in flocks, and guiding self-organisation using TE, are reviewed in Wibral et al. (2015).

Active Information Storage in Neural Data

Here, we present two very recent applications of (L)AIS to neural data and their estimation strategies for the probability density functions (PDFs). In both, estimation of (L)AIS was performed using the Java Information Dynamics Toolkit (JIDT) (Lizier, 2012, 2014a), and state space reconstruction was performed in TRENTOOL (Lindner et al., 2011) (for details, see Gomez et al., 2014; Wibral et al., 2014a). The first study investigated AIS in magnetoencephalographic (MEG) source signals from patients with autism spectrum disorder (ASD) and reported a reduction of AIS in the hippocampus in patients compared with healthy controls (Gomez et al., 2014) (Figure 17.1). In this study, the strategy for obtaining an estimate of the PDF was to use only baseline data (between stimulus presentations) to guarantee stationarity of the data. Results from this study align well with predictive coding theories (Friston et al., 2006; Rao and Ballard, 1999) of ASD (see also Gomez et al., 2014, and references therein).

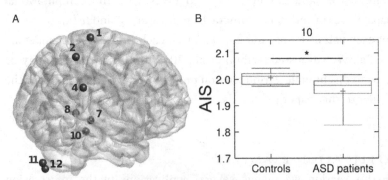

FIG. 17.1 AIS in ASD patients compared with controls. (A) Investigated magnetoencephalographic source locations (spheres). (B) Box and whisker plot for LAIS in source 10 (hippocampus), the one source where AIS for ASD patients was found to be significantly lower than controls. Modified from Gomez et al. (2014); under Creative Commons Attribution License (CC BY).

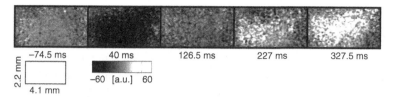

The significance of this study in the current context lies in the fact that it explicitly sought to measure the *information processing consequences* at the algorithmic level of changes in neural dynamics in ASD at the implementation level.

The second study (Wibral et al., 2014a) analyzed LAIS in voltage-sensitive dye (VSD) imaging data from cat visual cortex. The study found low LAIS in the baseline before the onset of a visual stimulus, negative LAIS directly after stimulus onset, and sustained increases in LAIS for the whole stimulation period, despite changing raw signal amplitude (Figure 17.2). These observed information profiles constrain the set of possible underlying algorithms being implemented in the cat's visual cortex. In this study all available data were pooled, from both baseline and stimulation periods, and also across all recording sites (VSD image pixels). Pooling across time is unusual, but reasonable insofar as neurons themselves also have to deal with nonstationarities as they arise, and a measure of *neurally accessible* LAIS should reflect this. Pooling across all sites in this study was motivated by the argument that all neural pools seen by VSD pixels are capable of the same dynamic transitions, as they were all in the same brain area. Thus, pixels were treated as physical replications for the estimation of the PDF. In sum, the evaluation strategy of this study is applicable to nonstationary data, but delivers results that strongly depend on the data included. Its future application therefore needs to be informed by precise estimates of the time scales at which neurons may sample their input statistics.

Active Information Storage in a Robotic System

Recurrent neural networks (RNNs) consist of a reservoir of nodes or artificial neurons connected in some recurrent network structure (Jaeger and Haas, 2004; Maass et al., 2002). Typically, this structure is constructed at random, with only the output neurons' connections trained to perform a given task. This approach is becoming increasingly popular for nonlinear time-series modeling and robotic applications (Boedecker et al., 2012; Dasgupta et al., 2013). The use of intrinsic plasticity–based techniques (Schrauwen et al., 2008) is known to assist performance of such RNNs in general, although this method is still outperformed on memory capacity tasks, for example, by the implementation of certain changes to the network structure (Boedecker et al., 2009). To address this issue, Dasgupta et al. (2013) added an online rule to adapt the "leak rate" of each neuron based on the AIS of its internal state. The leak rate is reduced where the AIS is below a certain threshold and increased where it is above. The technique was shown to improve performance on delayed memory tasks, both for benchmark tests and in embodied wheeled and hexapod robots. Dasgupta et al. (2013) described the effect of their technique as speeding up or slowing down the dynamics of the reservoir based on the time scale(s) of the input signal. In terms of Marr's levels, we can also view this as an intervention at the algorithmic level, directly adjusting the level of information storage in the system in order to affect the higher-level computational goal of enhanced performance on memory capacity tasks. It is particularly interesting to note the connection in information storage features across these different levels here.

Balance of Information Processing Capabilities Near Criticality

It has been conjectured that the brain may operate in a self-organized critical state (Beggs and Plenz, 2003), and recent evidence demonstrates that the human brain is at least very close to criticality, albeit slightly subcritical (Priesemann et al., 2013, 2014). This prompts the question of what advantages would be delivered by operating in such a critical state. From a dynamical systems perspective, one may suggest that the balance of stability (from ordered dynamics) with perturbation spreading (from chaotic dynamics) in this regime (Langton, 1990) gives rise to the scale-free correlations and emergent structures that we associate with computation in natural systems. From an information dynamics perspective, one may suggest that the critical regime represents a balance between capabilities of information storage and information transfer in the system, with too much of either one decaying the ability for emergent structures to carry out the complementary function (Langton, 1990; Lizier et al., 2008a, 2011a).

Several studies have upheld this interpretation of maximized but balanced information-processing properties near the critical regime. In a study of random Boolean networks it was shown that TE and AIS are in an optimal balance near the critical point (Lizier et al., 2008a, 2011a). This is echoed by findings for recurrent neural networks (Boedecker et al., 2012) and for maximization of TE in the Ising model (Barnett et al., 2013), and maximization of entropy in neural models and recordings (Haldeman and Beggs, 2005; Shew and Plenz, 2013). From Marr's perspective, we see here that at the algorithmic level the optimal balance of these information-processing operations is tightly coupled with the emergent and scale-free structures associated with the critical regime at the implementation level. This reflects the ties between Marr's levels as described in the section entitled "Information Theory and Neuroscience." These theoretical findings on computational properties at the critical point are of great relevance to neuroscience, due to the aforementioned importance of criticality in this field.

CONCLUSION AND OUTLOOK

Neural systems perform acts of information processing in the form of distributed (biological) computation, and many of the more complex computations and emergent information processing capabilities remain mysterious to date. Information theory can help to advance our understanding in two ways.

On the one hand, *neural information processing can be quantitatively partitioned into its component processes* of information storage, transfer, and modification using information-theoretic tools (section entitled "Analyzing Distributed Computation in Neural Systems"). These observations allow us to derive constraints on possible algorithms served by the observed neural dynamics. That is to say, these measures of how information is processed allow us to narrow down the algorithm(s) being implemented in the neural system. Importantly, this can be done without necessarily understanding the underlying causal structure precisely.

On the other hand, *the representations that these algorithms operate on can be guessed* by analyzing the mutual information between human-understandable descriptions of relevant concepts and quantities in our experiments and indices of neural activity (described in detail in Wibral et al., 2015). This helps to identify which parts of the real world neural systems attend to. However, care must be taken when asking such questions about neural codes or representations, as the separation of how neurons code uniquely, redundantly, and synergistically has not been solved completely to date (see Wibral et al., 2015).

Taken together, the knowledge about representations and possible algorithms describes the operational principles of neural systems at Marr's algorithmic level. Such information-theoretic insights may hint at solutions for solving ill-defined real-world problems that biologically inspired computing systems have to face with their constrained resources.

REFERENCES

Amblard, P. O., and Michel, O. J. 2011. On directed information theory and Granger causality graphs. *J. Comput. Neurosci.*, **30**(1), 7–16.

Ay, N., and Polani, D. 2008. Information flows in causal networks. *Adv. Complex Syst.*, **11**, 17.

Barnett, L., Barrett, A. B., and Seth, A. K. 2009. Granger causality and transfer entropy are equivalent for Gaussian variables. *Phys. Rev. Lett.*, **103**(23), 238701.

Barnett, L., Lizier, J. T., Harre, M., Seth, A. K., and Bossomaier, T. 2013. Information flow in a kinetic ising model peaks in the disordered phase. *Physical Review Letters*, **111**(17), 177203+.

Barrett, A. B. 2015. *An exploration of synergistic and redundant information sharing in static and dynamical Gaussian systems*. Physical Review E, 91, 052802.

Battaglia, D. 2014. Function follows dynamics: state-dependency of directed functional influences. Pages 111–135 of Wibral, M., Vicente, R., and Lizier, J. T. (eds), *Directed information measures in neuroscience*. Understanding Complex Systems. Springer.

Battaglia, D., Witt, A., Wolf, F., and Geisel, T. 2012. Dynamic effective connectivity of inter-areal brain circuits. *PLOS Comput. Biol.*, **8**(3), e1002438.

Bedo, N., Ribary, U., and Ward, L. M. 2014. Fast dynamics of cortical functional and effective connectivity during word reading. *PLOS ONE*, **9**(2), e88940.

Beggs, J. M., and Plenz, D. 2003. Neuronal avalanches in neocortical circuits. *J. Neurosci.*, **23**(35), 11167–11177.

Bertschinger, N., Rauh, J., Olbrich, E., Jost, J., and Ay, N. 2014. Quantifying unique information. *Entropy*, **16**(4), 2161–2183.

Besserve, M., Scholkopf, B., Logothetis, N. K., and Panzeri, S. 2010. Causal relationships between frequency bands of extracellular signals in visual cortex revealed by an information theoretic analysis. *J. Comput. Neurosci.*, **29**(3), 547–566.

Bettencourt, L. M. A., Gintautas, V., and Ham, M. I. 2008. Identification of functional information subgraphs in complex networks. *Physical Review Letters*, **100**(23), 238701.

Bialek, W., Nemenman, I., and Tishby, N. 2001. Predictability, complexity, and learning. *Neural. Comput.*, **13**(11), 2409–2463.

Boedecker, J., Obst, O., Mayer, N. M., and Minoru, A. 2009. Initialization and self-organized optimization of recurrent neural network connectivity. *HFSP Journal*, **3**(5), 340–349.

Boedecker, J., Obst, O., Lizier, J. T., Mayer, M. N., and Asada, M. 2012. Information processing in echo state networks at the edge of chaos. *Theory in Biosciences*, **131**(3), 205–213.

Buehlmann, A., and Deco, G. 2010. Optimal information transfer in the cortex through synchronization. *PLOS Comput. Biol.*, **6**(9), e1000934.

Butail, S., Ladu, F., Spinello, D., and Porfiri, M. 2014. Information flow in animal–robot interactions. *Entropy*, **16**(3), 1315–1330.

Carandini, M. 2012. From circuits to behavior: a bridge too far? *Nat. Neurosci.*, **15**(4), 507–509.

Ceguerra, R. V., Lizier, J. T., and Zomaya, A. Y. 2011. Information storage and transfer in the synchronization process in locally-connected networks. Pages 54–61 of *2011 IEEE Symposium on Artificial Life (ALIFE)*. IEEE.

Chavez, M., Martinerie, J., and Le Van Quyen, M. 2003. Statistical assessment of nonlinear causality: application to epileptic EEG signals. *J. Neurosci. Methods*, **124**(2), 113–28.

Chicharro, D. 2014. Parametric and non-parametric criteria for causal inference from time-series. Pages 195–219 of Wibral, M., Vicente, R., and Lizier, J. T. (eds), *Directed information measures in neuroscience*. Springer.

Chicharro, D., and Ledberg, A. 2012. When two become one: the limits of causality analysis of brain dynamics. *PLOS ONE*, **7**(3), e32466.

Crowley, K., and Siegler, R. S. 1993. Flexible strategy use in young children's tic-tac-toe. *Cognitive Science*, **17**(4), 531–561.

Crutchfield, J. P., and Feldman, D. P. 2003. Regularities unseen, randomness observed: levels of entropy convergence. *Chaos*, **13**(1), 25–54.

Crutchfield, J. P., and Packard, N. H. 1982. Symbolic dynamics of one-dimensional maps: entropies, finite precision, and noise. *International Journal of Theoretical Physics*, **21**(6–7), 433–466.

Dasgupta, S., Worgotter, F., and Manoonpong, P. 2013. Information dynamics based self-adaptive reservoir for delay temporal memory tasks. *Evolving Systems*, **4**(4), 235–249.

Dewdney, A. K. 1989. A Tinkertoy computer that plays tic-tac-toe. *Scientific American*, **261**(4), 120–123.

Faes, L., and Nollo, G. 2006. Bivariate nonlinear prediction to quantify the strength of complex dynamical interactions in short-term cardiovascular variability. *Med. Biol. Eng. Comput.*, **44**(5), 383–392.

Faes, L., and Porta, A. 2014. Conditional entropy-based evaluation of information dynamics in physiological systems. Pages 61–86 of Wibral, M., Vicente, R., and Lizier, J. T. (eds), *Directed information measures in neuroscience*. Springer.

Faes, L., Nollo, G. & Porta, A. 2011. Information-based detection of nonlinear Granger causality in multivariate processes via a nonuniform embedding technique. *Physical Review E*, **83**(5):051112.

Faes, L., Nollo, G., and Porta, A. 2012. Non-uniform multivariate embedding to assess the information transfer in cardiovascular and cardiorespiratory variability series. *Comput. Biol. Med.*, **42**(3), 290–297.

Faes, L., Nollo, G., Jurysta, F., and Marinazzo, D. 2014a. Information dynamics of brain–heart physiological networks during sleep. *New Journal of Physics*, **16**(10), 105005.

Faes, L., Marinazzo, D., Montalto, A., and Nollo, G. 2014b. Lag-specific transfer entropy as a tool to assess cardiovascular and cardiorespiratory information transfer. *IEEE Trans. Biomed. Eng.*, **61**(10), 2556–2568.

Fano, R. 1961. *Transmission of information*. MIT Press.

Fox, M. 2014. Chester Nez, 93, dies; Navajo words washed from mouth helped win war. *New York Times*, June 5.

Friston, K., Kilner, J., and Harrison, L. 2006. A free energy principle for the brain. *J. Physiol. Paris*, **100**(1–3), 70–87.

Friston, K. J., Thornton, C., and Clark, A. 2012. Free-energy minimization and the dark-room problem. *Frontiers in Psychology*, **3**, 130.

Garofalo, M., Nieus, T., Massobrio, P., and Martinoia, S. 2009. Evaluation of the performance of information theory-based methods and cross-correlation to estimate the functional connectivity in cortical networks. *PLOS ONE*, **4**(8), e6482.

Gomez, C., Lizier, J. T., Schaum, M., Wollstadt, P., Grutzner, C., Uhlhaas, P., Freitag, C. M., Schlitt, S., Bolte, S., Hornero, R., et al. 2014. Reduced predictable information in brain signals in autism spectrum disorder. *Frontiers in Neuroinformatics*, **8**, 9.

Gomez-Herrero, G., Wu, W., Rutanen, K., Soriano, M., Pipa, G., and Vicente, R. 2015. Assessing coupling dynamics from an ensemble of time series. *Entropy*, **17**(4), 1958–1970.

Gourevitch, B., and Eggermont, J. J. 2007. Evaluating information transfer between auditory cortical neurons. *J. Neurophysiol.*, **97**(3), 2533–2543.

Granger, C. W. J. 1969. Investigating causal relations by econometric models and cross-spectral methods. *Econometrica*, **37**, 424–438.

Grassberger, P. 1986. Toward a quantitative theory of self-generated complexity. *International Journal of Theoretical Physics*, **25**(9), 907–938.

Griffith, V., and Koch, C. 2014. Quantifying synergistic mutual information. Pages 159–190 of Prokopenko, M. (ed), *Guided self-organization: inception*. Emergence, Complexity and Computation, vol. 9. Springer.

Griffith, V., Chong, E. K. P., James, R. G., Ellison, C. J., and Crutchfield, J. P. 2014. Intersection information based on common randomness. *Entropy*, **16**(4), 1985–2000.

Hadjipapas, A., Hillebrand, A., Holliday, I. E., Singh, K. D., and Barnes, G. R. 2005. Assessing interactions of linear and nonlinear neuronal sources using MEG beamformers: a proof of concept. *Clin. Neurophysiol.*, **116**(6), 1300–1313.

Haldeman, C., and Beggs, J. M. 2005. Critical branching captures activity in living neural networks and maximizes the number of metastable states. *Physical Review Letters*, **94**(5), 058101.

Harder, M., Salge, C., and Polani, D. 2013. Bivariate measure of redundant information. *Phys. Rev. E*, **87**(1), 012130.

Ito, S., Hansen, M. E., Heiland, R., Lumsdaine, A., Litke, A. M., and Beggs, J. M. 2011. Extending transfer entropy improves identification of effective connectivity in a spiking cortical network model. *PLOS ONE*, **6**(11), e27431.

Jaeger, H., and Haas, H. 2004. Harnessing nonlinearity: predicting chaotic systems and saving energy in wireless communication. *Science*, **304**(5667), 78–80.

James, R. G., Ellison, C. J., and Crutchfield, J. P. 2011. Anatomy of a bit: information in a time series observation. *Chaos: An Interdisciplinary Journal of Nonlinear Science*, **21**(3), 037109+.

Kawasaki, M., Uno, Y., Mori, J., Kobata, K., and Kitajo, K. 2014. Transcranial magnetic stimulation-induced global propagation of transient phase resetting associated with directional information flow. *Front. Hum. Neurosci.*, **8**, 173.

Knill, D. C., and Pouget, A. 2004. The Bayesian brain: the role of uncertainty in neural coding and computation. *Trends Neurosci.*, **27**(12), 712–719.

Kraskov, A., Stoegbauer, H., and Grassberger, P. 2004. Estimating mutual information. *Phys. Rev. E. Stat. Nonlin. Soft Matter Phys.*, **69**(6 Part 2), 066138.

Landauer, R. 1961. Irreversibility and heat generation in the computing process. *IBM Journal of Research and Development*, **5**, 183–191.

Langton, C. G. 1990. Computation at the edge of chaos: phase transitions and emergent computation. *Physica D: Nonlinear Phenomena*, **42**(1), 12–37.

Leistritz, L., Hesse, W., Arnold, M., and Witte, M. 2006. Development of interaction measures based on adaptive non-linear time series analysis of biomedical signals. *Biomed. Tech. (Berl)*, **51**(2), 64–69.

Li, X., and Ouyang, G. 2010. Estimating coupling direction between neuronal populations with permutation conditional mutual information. *NeuroImage*, **52**(2), 497–507.

Lindner, M., Vicente, R., Priesemann, V., and Wibral, M. 2011. TRENTOOL: a Matlab open source toolbox to analyse information flow in time series data with transfer entropy. *BMC Neurosci.*, **12**(119)(1), 1–22.

Liu, T., and Pelowski, M. 2014. A new research trend in social neuroscience: towards an interactive-brain neuroscience. *PsyCh Journal*, **3**, 177–188.

Lizier, J. T. 2012. *JIDT: An information-theoretic toolkit for studying the dynamics of complex systems*. http://jlizier.github.io/jidt/.

Lizier, J. T. 2013. *The local information dynamics of distributed computation in complex systems*. Springer Theses. Springer.

Lizier, J. T. 2014a. JIDT: an information-theoretic toolkit for studying the dynamics of complex systems. *Frontiers in Robotics and AI*, **1**, 11.

Lizier, J. T. 2014b. Measuring the dynamics of information processing on a local scale in time and space. Pages 161–193 of Wibral, M., Vicente, R., and Lizier, J. T. (eds), *Directed Information Measures in Neuroscience*. Understanding Complex Systems. Springer.

Lizier, J. T., and Prokopenko, M. 2010. Differentiating information transfer and causal effect. *Eur. Phys. J. B*, **73**, 605–615.

Lizier, J. T., and Rubinov, M. 2012. *Multivariate construction of effective computational networks from observational data*. Preprint 25/2012. Max Planck Institute for Mathematics in the Sciences.

Lizier, J. T., Prokopenko, M., and Zomaya, A. Y. 2007. Detecting non-trivial computation in complex dynamics. Pages 895–904 Almeida e Costa, F., Rocha, L. M., Costa, E., Harvey, I., and Coutinho, A. (eds), *Proceedings of the 9th European Conference on Artificial Life (ECAL 2007)*. Lecture Notes in Computer Science, vol. 4648. Springer.

Lizier, J. T., Prokopenko, M., and Zomaya, A. Y. 2008a. The information dynamics of phase transitions in random Boolean networks. Pages 374–381 of Bullock, S., Noble, J., Watson, R., and Bedau, M. A. (eds), *Proceedings of the Eleventh International Conference on the Simulation and Synthesis of Living Systems (ALife XI), Winchester, UK*. MIT Press.

Lizier, J. T., Prokopenko, M., and Zomaya, A. Y. 2008b. Local information transfer as a spatiotemporal filter for complex systems. *Phys. Rev. E*, **77**(2 Part 2), 026110.

Lizier, J. T., Prokopenko, M., and Zomaya, A. Y. 2010. Information modification and particle collisions in distributed computation. *Chaos*, **20**(3), 037109.

Lizier, J. T., Pritam, S., and Prokopenko, M. 2011a. Information dynamics in small-world Boolean networks. *Artif. Life*, **17**(4), 293–314.

Lizier, J. T., Heinzle, J., Horstmann, A., Haynes, J. D., and Prokopenko, M. 2011b. Multivariate information-theoretic measures reveal directed information structure and task relevant changes in fMRI connectivity. *J. Comput. Neurosci.*, **30**(1), 85–107.

Lizier, J. T., Prokopenko, M., and Zomaya, A. Y. 2012a. Coherent information structure in complex computation. *Theory Biosci.*, **131**(3), 193–203.

Lizier, J. T., Prokopenko, M., and Zomaya, A. Y. 2012b. Local measures of information storage in complex distributed computation. *Information Sciences*, **208**, 39–54.

Lizier, J. T., Flecker, B., and Williams, P. L. 2013. Towards a synergy-based approach to measuring information modification. Pages 43–51 of *2013 IEEE Symposium on Artificial Life (ALIFE)*. IEEE.

Lizier, J. T., Prokopenko, M., and Zomaya, A. Y. 2014. A framework for the local information dynamics of distributed computation in complex systems. Pages 115–158 of Prokopenko, M. (ed), *Guided self-organization: inception*. Emergence, Complexity and Computation, vol. 9. Springer.

Ludtke, N., Logothetis, N. K., and Panzeri, S. 2010. Testing methodologies for the nonlinear analysis of causal relationships in neurovascular coupling. *Magn. Reson. Imaging*, **28**(8), 1113–1119.

Maass, W., Natschläger, T., and Markram, H. 2002. Real-time computing without stable states: a new framework for neural computation based on perturbations. *Neural Computation*, **14**(11), 2531–2560.

Marinazzo, D., Pellicoro, M., Wu, G., Angelini, L., Cortés, Jesús, M., and Stramaglia, S. 2014a. Information transfer and criticality in the Ising model on the human connectome. *PLOS ONE*, **9**(4), e93616.

Marinazzo, D., Wu, G., Pellicoro, M., and Stramaglia, S. 2014b. Information transfer in the brain: insights from a unified approach. Pages 87–110 of Wibral, M., Vicente, R., and Lizier, J. T. (eds), *Directed information measures in neuroscience*. Understanding Complex Systems. Springer.

Marinazzo, D., Gosseries, O., Boly, M., Ledoux, D., Rosanova, M., Massimini, M., Noirhomme, Q., and Laureys, S. 2014c. Directed information transfer in scalp electroencephalographic recordings: insights on disorders of consciousness. *Clin. EEG Neurosci.*, **45**(1), 33–39.

Markram, H. 2006. The blue brain project. *Nat. Rev. Neurosci.*, **7**(2), 153–160.

Marr, D. 1982. *Vision: a computational investigation into the human representation and processing of visual information.* Henry Holt.

Martius, G., Der, R., and Ay, N. 2013. Information driven self-organization of complex robotic behaviors. *PLOS ONE*, **8**(5), e63400+.

McAuliffe, J. 2014. 14. The new math of EEG: symbolic transfer entropy, the effects of dimension. *Clinical Neurophysiology*, **125**, e17.

Mitchell, M. 1998. Computation in cellular automata: a selected review. Pages 95–140 of Gramss, T., Bornholdt, S., Gross, M., Mitchell, M., and Pellizzari, T. (eds), *Non-standard computation*. Weinheim: Wiley-VCH Verlag GmbH & Co. KGaA.

Montalto, A., Faes, L., and Marinazzo, D. 2014. MuTE: A Matlab toolbox to compare established and novel estimators of the multivariate transfer entropy. *PLOS ONE*, **9**(10), e109462.

Neymotin, S. A., Jacobs, K. M., Fenton, A. A., and Lytton, W. W. 2011. Synaptic information transfer in computer models of neocortical columns. *J. Comput. Neurosci.*, **30**(1), 69–84.

Orlandi, J. G., Stetter, O., Soriano, J., Geisel, T., and Battaglia, D. 2014. Transfer entropy reconstruction and labeling of neuronal connections from simulated calcium imaging. *PLOS ONE*, **9**(6), e98842.

Paluš, M. 2001. Synchronization as adjustment of information rates: detection from bivariate time series. *Phys. Rev. E*, **63**, 046211.

Panzeri, S., Senatore, R., Montemurro, M. A., and Petersen, R. S. 2007. Correcting for the sampling bias problem in spike train information measures. *J. Neurophysiol.*, **98**(3), 1064–72.

Porta, A., Guzzetti, S., Montano, N., Pagani, M., Somers, V., Malliani, A., Baselli, G., and Cerutti, S. 2000. Information domain analysis of cardiovascular variability signals: evaluation of regularity, synchronisation and co-ordination. *Medical and Biological Engineering and Computing*, **38**(2), 180–188.

Porta, A., Faes, L., Bari, V., Marchi, A., Bassani, T., Nollo, G., Perseguini, N. M., Milan, J., Minatel, V., Borghi, S. A., et al. 2014. Effect of age on complexity and causality of the cardiovascular control: comparison between model-based and model-free approaches. *PLOS ONE*, **9**(2), e89463.

Priesemann, V., Valderrama, M., Wibral, M., and Le Van Quyen, M. 2013. Neuronal avalanches differ from wakefulness to deep sleep: evidence from intracranial depth recordings in humans. *PLOS Comput. Biol.*, **9**(3), e1002985.

Priesemann, V., Wibral, M., Valderrama, M., Pröpper, R., Le Van Quein, M., Geisel, T., Triesch, J., Nikolić, D., and Munk, M. H. J. 2014. Spike avalanches in vivo suggest a driven, slightly subcritical brain state. *Frontiers in Systems Neuroscience*, **8**, 108.

Prokopenko, M., and Lizier, J. T. 2014. Transfer entropy and transient limits of computation. *Sci. Rep.*, **4**, 5394.

Ragwitz, M., and Kantz, H. 2002. Markov models from data by simple nonlinear time series predictors in delay embedding spaces. *Phys. Rev. E Stat. Nonlin. Soft Matter Phys.*, **65** (5 Part 2), 056201.

Rao, R. P., and Ballard, D. H. 1999. Predictive coding in the visual cortex: a functional interpretation of some extra-classical receptive-field effects. *Nat. Neurosci.*, **2**(1), 79–87.

Razak, F. A., and Jensen, H. J. 2014. Quantifying "causality" in complex systems: understanding transfer entropy. *PLOS ONE*, **9**(6), e99462.

Rowan, M. S., Neymotin, S. A., and Lytton, W. W. 2014. Electrostimulation to reduce synaptic scaling driven progression of Alzheimer's disease. *Frontiers in Computational Neuroscience*, **8**, 39.

Sabesan, S., Good, L. B., Tsakalis, K. S., Spanias, A., Treiman, D. M., and Iasemidis, L. D. 2009. Information flow and application to epileptogenic focus localization from intracranial EEG. *IEEE Trans. Neural Syst. Rehabil. Eng.*, **17**(3), 244–253.

Schrauwen, B., Wardermann, M., Verstraeten, D., Steil, J. J., and Stroobandt, D. 2008. Improving reservoirs using intrinsic plasticity. *Neurocomputing*, **71**(7–9), 1159–1171.

Schreiber, T. 2000. Measuring information transfer. *Phys. Rev. Lett.*, **85**(2), 461–464.

Shalizi, C. R. 2001. *Causal architecture, complexity and self-organization in time series and cellular automata*. Ph.D. thesis, University of Wisconsin-Madison.

Shannon, C. E. 1948. A mathematical theory of communication. *Bell System Technical Journal*, **27**, 379–423 and 623–656.

Shew, W. L., and Plenz, D. 2013. The functional benefits of criticality in the cortex. *The Neuroscientist*, **19**(1), 88–100.

Shimono, M., and Beggs, J. M. 2015. Functional clusters, hubs, and communities in the cortical microconnectome. *Cereb. Cortex*, **25**(10), 3743–3757.

Small, M., and Tse, C. K. 2004. Optimal embedding parameters: a modelling paradigm. *Physica D: Nonlinear Phenomena*, **194**, 283–296.

Smirnov, D. A. 2013. Spurious causalities with transfer entropy. *Physical Review E*, **87**(4), 042917.

Staniek, M., and Lehnertz, K. 2009. Symbolic transfer entropy: inferring directionality in biosignals. *Biomed. Tech. (Berl)*, **54**(6), 323–328.

Stetter, O., Battaglia, D., Soriano, J., and Geisel, T. 2012. Model-free reconstruction of excitatory neuronal connectivity from calcium imaging signals. *PLOS Comput. Biol.*, **8**(8), e1002653.

Stramaglia, S., Wu, G. R., Pellicoro, M., and Marinazzo, D. 2012. Expanding the transfer entropy to identify information circuits in complex systems. *Physical Review E*, **86**(6), 066211.

Stramaglia, S., Cortes, J. M., and Marinazzo, D. 2014. Synergy and redundancy in the Granger causal analysis of dynamical networks. *New Journal of Physics*, **16**(10), 105003.

Szilárd, L. 1929. Über die Entropieverminderung in einem thermodynamischen System bei Eingriffen intelligenter Wesen (On the reduction of entropy in a thermodynamic system by the intervention of intelligent beings). *Zeitschrift für Physik*, **53**, 840–856.

Takens, F. 1981. Detecting strange attractors in turbulence. Pages 366–381 of Rand, D., and Young, L. S. (eds), *Dynamical systems and turbulence, Warwick 1980*. Lecture Notes in Mathematics, vol. 898. Springer.

Thivierge, J. P. 2014. Scale-free and economical features of functional connectivity in neuronal networks. *Phys. Rev. E*, **90**(2), 022721.

Timme, N., Alford, W., Flecker, B., and Beggs, J. M. 2014. Synergy, redundancy, and multivariate information measures: an experimentalist's perspective. *Journal of Computational Neuroscience*, **36**(2), 119–140.

Untergehrer, G., Jordan, D., Kochs, E. F., Ilg, R., and Schneider, G. 2014. Fronto-parietal connectivity is a non-static phenomenon with characteristic changes during unconsciousness. *PLOS ONE*, **9**(1), e87498.

Vakorin, V. A., Krakovska, O. A., and McIntosh, A. R. 2009. Confounding effects of indirect connections on causality estimation. *J. Neurosci. Methods*, **184**(1), 152–160.

Vakorin, V. A., Kovacevic, N., and McIntosh, A. R. 2010. Exploring transient transfer entropy based on a group-wise ICA decomposition of EEG data. *Neuroimage*, **49**(2), 1593–1600.

Vakorin, V. A., Mišić, B., Krakovska, O., and McIntosh, A. R. 2011. Empirical and theoretical aspects of generation and transfer of information in a neuromagnetic source network. *Front. Syst. Neurosci.*, **5**, 96.

Van Mierlo, P., Papadopoulou, M., Carrette, E., Boon, P., Vandenberghe, S., Vonck, K., and Marinazzo, D. 2014. Functional brain connectivity from EEG in epilepsy: seizure prediction and epileptogenic focus localization. *Progress in Neurobiology*, **121**, 19–35.

Varon, C., Montalto, A., Jansen, K., Lagae, L., Marinazzo, D., Faes, L., and Van Huffel, S. 2015. Interictal cardiorespiratory variability in temporal lobe and absence epilepsy in childhood. *Physiological Measurement*, **36**(4), 845.

Vicente, R., Wibral, M., Lindner, M., and Pipa, G. 2011. Transfer entropy – a model-free measure of effective connectivity for the neurosciences. *J. Comput. Neurosci.*, **30**(1), 45–67.

Wang, X. R., Lizier, J. T., and Prokopenko, M. 2011. Fisher information at the edge of chaos in random Boolean networks. *Artif. Life*, **17**(4), 315–329.

Wang, X. R., Miller, J. M., Lizier, J. T., Prokopenko, M., and Rossi, L. F. 2012. Quantifying and tracing information cascades in swarms. *PLOS ONE*, **7**(7), e40084.

Wibral, M., Turi, G., Linden, D. E. J., Kaiser, J., and Bledowski, C. 2008. Decomposition of working memory-related scalp ERPs: crossvalidation of fMRI-constrained source analysis and ICA. *Int. J. Psychophysiol.*, **67**(3), 200–211.

Wibral, M., Rahm, B., Rieder, M., Lindner, M., Vicente, R., and Kaiser, J. 2011. Transfer entropy in magnetoencephalographic data: quantifying information flow in cortical and cerebellar networks. *Prog. Biophys. Mol. Biol.*, **105**(1–2), 80–97.

Wibral, M., Pampu, N., Priesemann, V., Siebenhhner, F., Seiwert, H., Lindner, M., Lizier, J. T., and Vicente, R. 2013. Measuring information-transfer delays. *PLOS ONE*, **8**(2), e55809.

Wibral, M., Lizier, J. T., Vögler, S., Priesemann, V., and Galuske, R. 2014a. Local active information storage as a tool to understand distributed neural information processing. *Frontiers in Neuroinformatics*, **8**(1), 78–88.

Wibral, M., Vicente, R., and Lindner, M. 2014b. Transfer entropy in neuroscience. Pages 3–36 of Wibral, M., Vicente, R., and Lizier, J. T. (eds), *Directed information measures in neuroscience: understanding complex systems*. Springer.

Wibral, M., Lizier, J. T., and Priesemann, V. 2015. Bits from brains for biologically inspired computing. *Frontiers in Robotics and AI*, **2**(5), 1–25.

Wiener, N. 1956. The theory of prediction. In: Beckmann, E. F. (ed), *Modern mathematics for the engineer*. McGraw-Hill

Williams, P. L., and Beer, R. D. 2010. Nonnegative decomposition of multivariate information. *arXiv preprint arXiv:1004.2515*.

Wollstadt, P., Martnez-Zarzuela, M., Vicente, R., Daz-Pernas, F. J., and Wibral, M. 2014. Efficient transfer entropy analysis of non-stationary neural time series. *PLOS ONE*, **9**(7), e102833.

Yamaguti, Y., and Tsuda, I. 2015. Mathematical modeling for evolution of heterogeneous modules in the brain. *Neural Networks*, **62**, 3–10.

Zipser, D., Kehoe, B., Littlewort, G., and Fuster, J. 1993. A spiking network model of short-term active memory. *Journal of Neuroscience*, **13**(8), 3406–3420.

Zubler, F., Gast, H., Abela, E., Rummel, C., Hauf, M., Wiest, R., Pollo, C., and Schindler, K. 2015. Detecting functional hubs of ictogenic networks. *Brain Topography*, **28**(2), 305–317.

18 Machine Learning and the Questions It Raises

G. Andrew D. Briggs and Dawid Potgieter

Machine learning is a lively academic discipline and a key player in the continuous pursuit for new technological developments. The editorial in the first issue of the journal *Machine Learning*, published in March 1986, described the discipline as that field of inquiry concerned with the processes by which intelligent systems improve their performance over time (Langley, 1986). A glossary of terms published in the same journal in 1998 refined this to: Machine Learning is the field of scientific study that concentrates on induction algorithms and on other algorithms that can be said to 'learn' (Kohavi and Provost, 1998).

Thomas J. Watson, the brilliant salesman who from 1914 to 1956 oversaw the remarkable growth and success of IBM, serving as both CEO and chairman, was famously quoted as saying in 1943, 'I think there is a world market for maybe five computers.' With more than one billion computers now in use worldwide (Virki, 2008), this quote is often referenced to illustrate how vastly their usefulness had been underestimated. No area of computer science is making progress more rapidly than machine learning, with computers being capable of tasks that were a few decades ago only mentioned in science-fiction stories. Watson brought to IBM from his previous employment his trademark motto 'Think'. It would at the time have been reasonable for Watson to suppose that only humans could really 'think'. While computers could surpass humans in adding, subtracting, multiplying, and dividing, they were hardly thought of as being good at human tasks, such as playing chess, which required thinking. This begs the question, 'What is thinking?' In February 1996 World Chess Champion Garry Kasparov took on the IBM computer Deep Blue in Philadelphia. Even with the IBM engineers allowed to reprogram the computer between games, the world champion won, but only just, losing one game, drawing two, and winning three. His victory was short-lived. The following year he played a rematch. With the score even after the first five of six games, Kasparov allowed Deep Blue to commit a knight sacrifice, which wrecked his Caro–Kann defence and forced him to resign in fewer than twenty moves.

It might be thought that playing chess is the kind of human thinking that is well adapted to a digital computer, in which a computer can employ techniques for which the human brain is ill-suited. A task for which the human eye–brain combination is superbly suited is facial recognition. It is not hard to understand why such expertise should be useful as humans organised themselves into hunter-gatherer groups, or even earlier. It is also not hard to think why such a task should be extraordinarily difficult for a digital computer. There are countless variations of lighting, viewing angle, mouth expressions, and other parameters. The advances have been made possible by a combination of raw computing power, vast data sets, and clever algorithms. At Heathrow Airport iris scans have been superseded by comparison of what the camera sees with the picture stored in an e-passport. This is faster and more reliable than an experienced immigration officer, and as secure as the fingerprint method still used by US immigration authorities. Even voice recognition, which for years was the source of endless frustration for victims of automated telephone systems, can now be more accurate and more versatile than a human listener. The rest of this chapter reflects on two areas of timely enquiry. By drawing from selected findings in neuroscience, the first part explores to what extent we might expect computer processors to mimic the information-processing mechanisms of the brain, an endeavour often referred to as neuromorphic technology. Recent findings in this area may provide deeper insights into information transfer in the brain and how such processes relate to machine learning. The second part reflects on deeper questions, practical, ethical, and philosophical, which will inevitably need to be addressed as the learning capacity of machines continues to surpass that of humans in more and more ways.

The human brain uses learning mechanisms very similar to the brains of other mammals. This chapter includes findings from rodents and nonhuman primates from which more experimental data have been collected.

The human brain is arguably the most complex object in the known universe, the only 'pound of flesh' taking credit for trying to understand itself. The mechanisms by which brains process information have for decades inspired computer scientists. When describing brains in terms of computation, neurons are often likened to wires. This metaphor may be useful in its simplest form: just as electricity flows from one wire to the next, an electrical signal can propagate along one neuron and be passed to another. Such a metaphor can also, within limited context, be extended to describe neural circuits as electric circuits and the brain as a very clever computer. But the metaphor eventually fails. Connectivity between two neurons is not always constant, as it is with wires, but can vary depending on several factors, including the frequency of neurotransmission. This feature, called synaptic plasticity (see Table 18.1 for definitions), allows neurons to store and process information at the same time, thus integrating both memory and processing power, which are handled separately in computers.

Engineers have optimistically tackled the problem of mimicking synaptic plasticity, and memristor-based devices have been able to achieve this to some extent (Li et al., 2014). The prospect of building synapse-like computational devices therefore seems hopeful, but is this enough to build a computer that works like a brain? And if not, then what other mechanisms must be mimicked to achieve such a goal? The following observations may illustrate why the answer to the former question is 'no', and offer some reflections on how we might explore the answer to the latter.

Neurotransmission involves more than 'on' or 'off' signals. The kind of synapse that a memristor-based device might mimic makes use of ionotropic receptors, which, when bound by a neurotransmitter,

TABLE 18.1 Neuroscience Terms and Definitions

Term	Definition
Dopamine	A compound present in the body as a neurotransmitter; dopamine neurotransmission is implicated in learning, decision-making, and movement
En passant varicosities	Swellings along the length of an axon that do not always form a synapse, but they can sometimes release a neurotransmitter
Globus pallidus	Part of the forebrain, located beneath the cerebral cortex towards the middle of the brain
Ionotropic receptors	Receptors that act as ion channels across the neuronal cell membrane, thereby changing the neuron's electrical conductivity
Metabotropic receptors	Receptors that act through signal transduction mechanisms inside the cell; such mechanisms may involve changes to metabolic processes or gene expression
Neural activity	Electrical activity of a neuron or cluster of neurons
Neuron	A specialised cell that carries electrical impulses; neurons make up fewer than half of the cells in a human brain
Neurotransmission	The release of a chemical from a neuron in order to send a message to other cells
Striatum	Part of the forebrain, located in humans between the globus pallidus and the cerebral cortex
Synapse	A structured junction between two neurons consisting of a small gap across which neurotransmitters can diffuse
Synaptic plasticity	A change in the likelihood or efficiency of neurotransmission due to the frequency of its prior occurrence or other factors
Somatodentritic	A subcellular region of a neuron that normally receives signals from external neurotransmitters without sending any signals; the soma contains the cell nucleus where DNA is stored and transcribed

either excite or inhibit the electrical activity of a recipient neuron. However, biological neurons also make use of metabotropic receptors, which may have little or no immediate effect on electrical activity, but do affect the cell over longer time scales. When metabotropic receptors are bound by a neurotransmitter, the downstream effects are mediated through a wide range of mechanisms available to the cell,

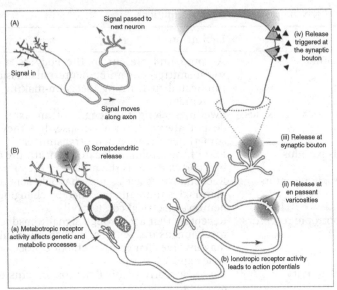

FIG. 18.1 (A) Cartoon to illustrate the simplified model of neuron function. A signal is received at the dendrites or cell body; the signal propagates along an axon and is passed to another neuron at the synaptic junction. (B) An illustration of more complicated signalling mechanisms that are found in biological neurons. Activation of metabotropic receptors (a) leads to a variety of effects, including changes to gene regulation, protein expression, and metabolic processes. Activation of ionotropic receptors (b) leads to excitation or inhibition of action potentials. Neurotransmitter release has been observed from somatodendritic regions (i), and en passant along unmyelinated axons (ii). Different kinds of neurotransmitter release have been observed at synaptic boutons, including release that spills over significantly from the synaptic junction (iii), as well as release triggered at the synapse (iv) through activation of presynaptic membrane receptors.

including signalling cascades, metabolic activity, and changes to gene regulation (Figure 18.1). Neurotransmission through metabotropic receptors plays a crucial role in information processing. Dopamine, for example, a neurotransmitter implicated in learning, decision-making, and movement, acts at D1- and D2-type metabotropic receptors (Beaulieu et al., 2015).

Information spreads beyond the synapse. The traditional metaphor likens neurons to wires and relies on neurotransmission occurring at distinct isolated regions called synapses, which act on one recipient neuron without spreading significantly to others. However, neurotransmission can, in some cases, spill over the synapse to influence multiple neurons in the vicinity. Dopamine released in the striatum, for example, has been shown to diffuse into the extracellular space beyond the confines of the synapse (Cragg and Rice, 2004). Neurotransmission is also not limited to taking place at synaptic structures but has been observed from somatodendritic parts of the neuron (Rice et al., 1997), and it has been suggested to take place at en passant varicosities along the length of an axon (Hattori et al., 1991) (see Figure 18.1). Moreover, the axonal branches of a single dopamine neuron, the kind that might release dopamine from en passant varicosities, can spread out to innervate more than 5% of the volume of striatum (Matsuda et al., 2009). This is because the axon branching is so extensive that if all branches of one such a neuron – from a rat brain – were connected end to end, the resulting structure would be nearly 80 cm in length (Matsuda et al., 2009). Taken together, these observations create a different perspective from one of the brain being like a collection of electrical wires connected end to end. Neurotransmitter release is more varied than that and can, in some neurons, take place beyond the confinement of synapses.

Information is coded at different levels – beyond the scale of the synapse. The traditional metaphor of the brain as a computer implies that information is embodied in an electrical signal that is passed from one neuron to another. However, information can affect different scales in the brain. For example, rhythmic neural activity, synchronised between multiple neurons in the globus pallidus, plays a crucial role in initiating goal-directed movement (Little and Brown, 2014). This kind of activity is composed of a collection of single action potentials, but it is not any action potential in particular, or even the sum of action potentials, but rather the synchronous and rhythmic firing of such action potentials that plays a crucial role. Moreover, different frequencies of rhythmic activity have been shown to play distinct roles in goal-directed actions (Brinkman et al., 2014). Such synchronous activity can therefore be thought of as an emergent phenomenon and demonstrates that information processing within the brain can take place at different scales, which broadens the repertoire of information-processing mechanisms available to the brain.

Neural circuits have large-scale redundancies. Computers and brains both employ error correction mechanisms, but the brain has a remarkable capacity to compensate functionally for the loss of neurons. Parkinson disease is a degenerative condition in which the capacity to initiate goal-directed movements is affected. When a patient exhibits enough movement-related symptoms to be diagnosed, roughly half of the most vulnerable population of neurons have already died (Fearnley and Lees, 1991). By contrast, only a specially designed computer, such as the Hewlett-Packard Teramac built in 1997, would be immune to failure of even a small fraction of its components (Heath et al., 1998). The fault tolerance of Teramac, however, relied on a 24-hour-long procedure during which a separate workstation identified all defective resources and wrote their locations to a configuration table as being 'in use' (Birnbaum and Williams, 2000). In essence, Teramac required the help of another computer to make sure it could start in spite of its faulty components. Brains are of course much more versatile in this sense. Unlike wires and transistors, which tend to need replacing when broken, the connections between neurons can change and grow dynamically without the need for invasive intervention. The biological mechanisms that underlie brain function can compensate for (a certain level of) failures whenever they occur. In the case of some forms of brain damage, such as the kind of cell death suffered during a stroke, the neurons surrounding a damaged area can 'rewire' themselves by sprouting new axon branches and making new connections to other cells (Dancause et al., 2005). In the case of degenerative diseases such as Parkinson, the surviving neurons can increase the amount of dopamine output until they too become dysfunctional. Other neurons that normally release serotonin can also begin to release dopamine, albeit sometimes inappropriately (Tanaka et al., 1999).

Some neurons release more than one type of neurotransmitter. It has been suggested that some neurons release both excitatory and inhibitory neurotransmitters (Gutiérrez, 2005), which may increase the repertoire of possible signals being transmitted between neurons (Seal and Edwards, 2006). Although such a mechanism may not serve both to excite and to inhibit the same recipient cell, as doing so would be counterproductive, it is feasible to imagine the neurotransmission from a single neuron exciting some recipient neurons and inhibiting others. Such a function could be achieved either by the recipient neurons being restricted in the receptors they express or by the different neurotransmitters being located in different and adequately separated axonal branches. The capacity for neurons to release more than one neurotransmitter also extends the potential for different possible computations. Input to a single neuron does not lead to a single output, or even many outputs of the same kind, but to multiple, and possibly a range of, different kinds of outputs.

Not all action potentials lead to neurotransmitter release. A single action potential contributes only to the probability of neurotransmitter release. For some neurons, there is a high probability that a single action potential will result in neurotransmitter release, but for others the probability can be very low, allowing additional factors to influence neurotransmission. For example, neurotransmission from dopamine axons in the striatum is influenced both by the frequency of action potentials and by the concentration of another neurotransmitter around the axon terminal, namely acetylcholine (Zhang and Sulzer, 2004) (see Figure 18.1). The amount of dopamine released by these neurons is strongly proportional to the frequency of action potentials when acetylcholine is absent. The higher the rate of action potentials, the more dopamine will be released. However, when a significant concentration of acetylcholine is present, then the magnitude of dopamine release is changed such that action potentials at both low and high frequencies lead to a moderate amount of dopamine release. Moreover, the synchronised activation of acetylcholine neurons, when surrounding dopamine axon terminals, is sufficient to cause dopamine release without any action potentials being present at all in the dopamine neurons (Threlfell et al., 2012). These findings suggest that some neurons behave almost like a logic gate, or rather – since the neurons in question also branch out across a relatively large volume of the striatum (Matsuda et al., 2009) – like a collection of different logic gates. Whether such a collection acts in unison or whether each neuron acts independently is not completely understood. Taken together, the observations presented here testify that a simple model that describes the brain in terms of a bunch of interconnected electrical wires is outdated, which may in turn suggest new directions for machine learning.

Our understanding of the brain is becoming increasingly sophisticated. Mammalian brains, and primate brains in particular, utilise an immense repertoire of mechanisms for processing information. Engineers who aim to build computers that utilise the same mechanisms as brains do will have to continue developing more complex technologies, but taking on such a task for the purpose of building better computers may be both overambitious and unnecessary. Trying to understand every biological mechanism in the brain for the purpose of building a computer might be a bit like researching the type font and ink composition of a book simply to quote a paragraph from it. How then should scientists approach machine learning in relation to the human brain? The evidence presented here suggests three approaches. First, the different scales at which brains process information remain incompletely understood and warrant continued investigation. How much of the information processing takes place through molecular interactions within a cell? How much takes place through electrical signals passed between individual neurons, and how much takes place through the synchronised rhythmic activity of large groups of neurons? Alongside research into scales of information processing come questions about the hierarchy of algorithmic processes. Computers make use of a hierarchy of algorithmic processes, but the brain's capacity for algorithmic processes remains incompletely understood. Can the function of a healthy human brain can be reduced to a set of distinct algorithmic processes, and how? Information not only is processed by electrochemical activity between neurons but might also involve complex molecular interactions inside single neurons. It may therefore be difficult to determine where the boundaries lie between hierarchies of algorithms.

Second, we need a deeper understanding of the full repertoire of computational mechanisms available to the brain. More accurate models of brain function will provide a foundation to construct new models of neural processes. For example, a neuron that releases only glutamate, and does so only at synaptic structures, might be modelled as a wire with plasticity at the synapse – but a neuron that releases both glutamate (excitatory) and GABA (inhibitory) could be modelled as two circuits with the same inputs but opposite output states and different output connections.

Finally, progress in machine learning may be made by looking at the function of biological mechanisms rather than aiming to build synthetic versions of the same hardware. For example, the memristor-based devices that were created as a synthetic alternative to synaptic plasticity (Li et al., 2014) will not necessarily be able to serve the function of a biological synapse. The latter is diverse, as biological mechanisms tend to be. Neurons can achieve plasticity through a range of signalling molecules binding at a variety of receptors, thereby triggering changes that include genetic expression, the relocation of proteins, and the regulation of their activity (Citri and Malenka, 2008). On the other hand, memristor-based devices – and semiconductor materials in general – tend to be much simpler in their construction and more limited in the variety of actions that are available at a molecular level. An individual neuron may not be just a component of a computation but might perform a complete computation. Conversely, computations take place not just within individual neurons but also between neurons, and sometimes the activity of a single neuron is redundant. The constraints and resources available to biological organisms differ from those that are available in the laboratory. Electronic devices can perform similar information processing as brains can by different – and possibly more efficient – mechanisms. The best goal may be not to build a computer with the same capacities as a brain, but rather to develop a range of machines, each capable of outperforming human brains in different ways.

Quantum computers are utterly different from classical computers, and for certain tasks they can be exponentially better. At the core of machine learning lies probability theory, Bayesian theory in particular. To learn involves inference of probabilities over variables of interest, and this is underpinned by two key operations: optimisation and numerical integration. Both scale poorly with the complexity of the models used and indeed with the amount of data. Classical computation cannot resolve this curse of dimensionality. Quantum computation offers an avoidance for such problems; indeed, work in information engineering at the University of Oxford has shown that classical inference ground energy states in a quantum system are equivalent. Furthermore, quantum algorithms, inspired by classical sample-based inference, can also be developed to replace the most used classical inference algorithms (Fox et al., 2008). Unusually for quantum information processing, these algorithms depend for their effectiveness on decoherence to disperse unwanted information into surrounding coolants. If the algorithm provides a useful computational speed-up, then maybe evolution would have found a way to use it; it is an empirical question whether this is actually the case. There is a subneuronal computation theory that could be extended to perform this kind of Markov random field (MRF) inference, making use of biological coolants flowing through microtubules to provide the necessary decoherence. Quantum methods have been proposed to reduce the vastness of the data sets that have proved crucial for machine learning, using superposition states not only of the information itself but also of the address switches used to retrieve data from classical memories (Lloyd et al., 2013). This would yield a logarithmic reduction in the amount of memory required to store the quantum information.

Having briefly reflected on which lines of enquiry might help us to build better machines, the remainder of this chapter focuses more broadly on a few related questions. Information is not just a set of bits, or even bits that convey semantic meaning, but information can have a causal power given the appropriate conditions. In the context of machine learning, the causal power of information is increasing dramatically, which can have significant implications for humanity. The Industrial Revolution saw a huge increase in our productivity through the power of machines. But humans were still very much in control. We now approach technological advances that allow for humans not necessarily having to be in control; machines are gaining autonomy. Although machines can process information well enough to become less reliant on human decisions, machines lack information for concepts like 'good' and 'bad'. They do not have a framework for morals or values as we understand these concepts or, in fact, as philosophers still debate them. The questions discussed in this section explore some of these broader implications of recent advances in machine learning. The day may be not far off – indeed, it may already be here – when machine learning advances to the point of creating new challenges to humans, for example, by undertaking tasks that traditionally we might have ascribed to human judgement. A machine might learn enough to be able to analyse the information in a company's accounts to the standard of an experienced accountant. It might not only be able to answer the key question of an audit: Are these accounts a true and accurate statement of the financial affairs of the organisation? It might be able to do more: What changes might increase the profitability or reduce the tax liability?

The rapid advances in information processing by machine learning raise questions that are better tackled sooner rather than later.

1. **What will be the implications for human work of machine learning?** Every previous advance in information processing, indeed in technology generally, has resulted in changes to human employment (Uglow, 2003). The mechanisation of agriculture led to urbanisation. The industrialisation of pottery making led to factory working. The results were a mixture of poverty and deprivation for the unemployed in cities, and steady wages and prosperity for many employed in factories. Nineteenth-century England saw both at the same time. The computerisation of banking and financial services in the twentieth century removed the need for large numbers of people in clerical work. At its best it liberated them to use their brainpower for more interesting, knowledge-based work. Will machine learning do that? Certain surgical operations can be performed better by a robot than even the best-trained pair of human hands (Badani et al., 2007), thus enabling more accurate surgery to be performed and liberating the surgeon to concentrate on higher-level tasks of diagnosis and treatment planning. Already people tend to be more honest about their potentially embarrassing activities that led to sickness when answering questions for a computer rather than a clinician. What if the day comes when machines can outperform surgeons at diagnosis, treatment planning, and all surgical procedures? What higher-level activity will then be left to the consultant? More positively, how can technological advancements such as these be leveraged to bring about the most good to society? Much of the work mentioned in this book, and elsewhere (Mirmomeni et al., 2014), explores the idea that information – and, in particular, having the correct information and the capacity to process it well – can be conceptualised as making a key contribution to the survival and evolution of an organism. Biological evolution of humans is a very slow process, and one with which we should not interfere. However, the capacity of machines to process information is growing rapidly, and vastly exceeds the rate of evolution of humans. Can society benefit from engaging with such technological advances as progress that benefits our species as a whole? Can such an approach inspire more people to create helpful

technologies that will change the way that people work even more than the Industrial Revolution did in the eighteenth century?

2. Who will set the hierarchy of goals for machine learning? At present the programmer sets the goals: compare that face with the record in the e-passport; transcribe this speech into text. In 2014, TheySay, a spin-off company from Oxford University, algorithmically analysed the sentiments of Yes and No supporters leading up to the referendum vote about whether Scotland should leave the United Kingdom (Morgan, 2015). Harvesting text from social media, news, blogs, and chat rooms, the machine used more than 68,000 rules to determine the grammatical content of the text. It was then able to extract and assign meaning and provide insights about sentiment (positive or negative), emotions (fear, anger, happiness, and sadness), and sureness (certainty and confidence), thus building up a picture of outrage, nervousness, despondency, and joy. These goals were set by humans for the purposes that they wanted to achieve. What happens when powerful machines like these are programmed for maleficent purposes? The legal and ethical framework that is required to safely use new technology is sometimes developed much later than the technology itself. For example, one of the world's largest mass multiplayer online computer games, Eve Online, has suffered losses of tens of thousands of US dollars because the technology was developed without a system to ensure ethical behaviour and justice. Some players stole online game credits from other players, which were then traded in for cash. The largest recorded theft was worth over US$51,000 (Drain, 2012). How can we learn from this sort of dilemma? What might happen if we develop machines with the information-processing power to set their own goals, but without the necessary framework to do so responsibly? Should goal-setting for machines be preemptively regulated? If so, how should it be done?

3. What is the meaning of 'responsibility' in machine learning? The December 2014 issue of *Nature Nanotechnology* carried a thesis piece entitled 'Could We 3D Print an Artificial Mind?' (Maynard,

2014). The final paragraph concluded, 'Which leads us to a question that is, if anything, more difficult to address than the aforementioned technical hurdles: if our technological capabilities are beginning to shift from the fanciful to the plausible in constructing an artificial mind that has some degree of awareness, how do we begin to think about responsibility in the face of such audacity?' One of us wrote to the author to ask him whether this final paragraph was simply a rhetorical flourish with which to end the piece, or whether these were issues about which he had been thinking deeply. He replied that his intent was to finish with a link to his broader work, which revolves around emerging technologies and responsible innovation. The responsibility he was referring to is that incumbent on the various societal actors who may be involved in the development and use of technologies that could lead to artificial minds or similar. He confessed that he had not thought about responsibility from the perspective of the artificial mind in the piece. He acknowledged that this is an intriguing avenue to go down, which touches on some of the current philosophical work around artificial general intelligence.

4. **Could machine learning constitute a threat to our existence?** The Oxford philosopher Professor Nick Bostrom thinks so: 'If some day we build machine brains that surpass human brains in general intelligence, then this new superintelligence could become very powerful. And, as the fate of the gorillas now depends more on us humans than on the gorillas themselves, so the fate of our species would depend on the actions of the machine superintelligence' (Bostrom, 2014). In January 2014 at an FQXi conference in Vieques on 'The Physics of Information', the director, Professor Max Tegmark, initiated a straw poll among the participants to ask what they considered to be at present the greatest risk to the survival of the human race (http://fqxi.org/conference/2014). The obvious suspects, such as global climate change, came rather far down the list. Second from top was the threat from synthetic biology: the risk that some maverick, or perhaps some future child who had been given a gene-splicing

kit for Christmas, might produce a virus to which humans could not produce sufficient immunity. The highest threat was perceived to come from artificial general intelligence. Dilemmas in this area were foreseen as early as 1942 in Isaac Asimov's short story, 'Runaround', in which machines have to obey three laws loosely paraphrased as: (1) never letting a human come to harm; (2) obeying orders, unless rule 2 conflicts with 1; and (3) preserving itself, unless rule 3 conflicts with 1 or 2. In the story, a robot that was very expensive, and therefore programmed with a priority to preserve itself, is ordered to go on a dangerous mission that leads to a conflict between rules 2 and 3. The robot consequently malfunctions, and does so nearly to the peril of humans. That story was no more than fiction. Is the day approaching when such fiction might become reality? What if machines not only could be cleverer than humans but could adopt goals that were malevolent (again to beg the question) towards humans? So-called algorithmic self-improvement might run away and produce systems that greatly exceed human intelligence in ways that might not be human friendly. The long-established philosophical tools of decision theory may be relevant for addressing questions of goal stability under algorithmic self-improvement. Could we run simulations on machine learning to predict whether such goals will evolve? If machine learning can in principle threaten human existence, then how can we find the best way to prevent such a threat? Could machine learning be programmed to act with character virtues such as humility, forgiveness, and kindness?

5. **Where is wisdom to be found for machine learning?** If machines can learn to be as intelligent as humans, can they also learn to be as wise? What would it mean to describe computing as 'wise'? Would it involve being morally careful or perceptive? Would it make sense to ascribe attributes of wisdom to a machine only if one could also ascribe attributes of foolishness? At the meeting of the American Association for the Advancement of Science in San Jose in February 2015, a session was devoted to 'Wise Computing'. Professor Kazuo Iwano, of the Japan Science and Technology Agency, introduced

the session under its original title of 'Wisdom Computing' (Iwano, 2015). He described how research activities in Japan are working to understand and develop wisdom by sublimating distributed and heterogeneous data and information. Humans are capable of accessing more information than ever in real time, but can we claim that we have become wiser than ever individually or collectively? Machine learning is attaining enormous capabilities in accessing and analysing information and controlling objects such as airplanes and automobiles. Iwano presented a chart with the abscissa indicating on a logarithmic scale the duration of information, and the ordinate indicating the extent of its dissemination. Text messages scored poorly on both scales. Shakespeare scored highly on both. On top of all came the Bible. Could machines indeed acquire wisdom by learning from sources of spiritual information such as the ancient scriptures? If machines can learn to be wise, then what would that look like? Could 'machine wisdom' become an integral part of technology as machines gain greater capacity for performing new tasks, and outperforming humans?

The unprecedented technological breakthroughs of the last few decades have endowed humans with a new world of opportunity and, in some cases, presented new threats to our survival. As progress in machine learning accelerates, there will be new opportunities to grasp and concomitant threats to avoid. We hope this chapter will aid the former without losing sight of the latter.

REFERENCES

Badani, K. K., Kaul, S., and Menon, M. 2007. Evolution of robotic radical prostatectomy. *Cancer*, **110**(9), 1951–1958.

Beaulieu, J. M., Espinoza, S., and Gainetdinov, R. R. 2015. Dopamine receptors–IUPHAR review 13. *British Journal of Pharmacology*, **172**(1), 1–23.

Birnbaum, J., and Williams, R. S. 2000. Physics and the information revolution. *Physics Today*, **53**(1), 38–43.

Bostrom, N. 2014. *Superintelligence: paths, dangers, strategies*. Oxford University Press.

Brinkman, L., Stolk, A., Dijkerman, H. C., de Lange, F. P., and Toni, I. 2014. Distinct roles for alpha- and beta-band oscillations during mental simulation of goal-directed actions. *Journal of Neuroscience*, **34**(44), 14783–14792.

Citri, A., and Malenka, R. C. 2008. Synaptic plasticity: multiple forms, functions, and mechanisms. *Neuropsychopharmacology*, **33**(1), 18–41.

Cragg, S. J., and Rice, M. E. 2004. DAncing past the DAT at a DA synapse. *Trends in Neurosciences*, **27**(5), 270–277.

Dancause, N., Barbay, S., Frost, S. B., Plautz, E. J., Chen, D., Zoubina, E. V., Stowe, A. M., and Nudo, R. J. 2005. Extensive cortical rewiring after brain injury. *Journal of Neuroscience*, **25**(44), 10167–10179.

Drain, B. October 28, 2012. EVE Evolved: top ten ganks, scams, heists and events. www.engadget.com/2012/10/28/eve-evolved-top-ten-ganks-scams-heists-and-events/.

Fearnley, J. M., and Lees, A. J. 1991. Ageing and Parkinson's disease: substantia nigra regional selectivity. *Brain*, **114**(5), 2283–2301.

Fox, C., Rezek, L., and Roberts, S. 2008. Local quantum computing for fast probably MAP inference in graphical models. www.robots.ox.ac.uk/~irezek/Outgoing/Papers/qbayes2 .pdf.

Gutiérrez, R. 2005. The dual glutamatergic–GABAergic phenotype of hippocampal granule cells. *Trends in Neurosciences*, **28**(6), 297–303.

Hattori, T., Takada, M., Moriizumi, T., and Van der Kooy, D. 1991. Single dopaminergic nigrostriatal neurons form two chemically distinct synaptic types: possible transmitter segregation within neurons. *Journal of Comparative Neurology*, **309**(3), 391–401.

Heath, J. R., Kuekes, P. J., Snider, G. S., and Williams, R. S. 1998. A defect-tolerant computer architecture: opportunities for nanotechnology. *Science*, **280**(5370), 1716–1721.

Iwano, K. 2015. Wise computing: collaboration between people and machines. Paper presented at the AAAS 2015 Annual Meeting on Innovations, Information, and Imaging, San Jose, CA. https://aaas.confex.com/aaas/2015/webprogram/Session9386.html.

Kohavi, R., and Provost, F. 1998. Glossary of terms. *Machine Learning*, **30**(2–3), 271–274.

Langley, P. 1986. Editorial: on machine learning. *Machine Learning*, **1**(1), 5–10.

Li, Y., Zhong, Y., Zhang, J., Xu, L., Wang, Q., Sun, H., Tong, H., Cheng, X., and Miao, X. 2014. Activity-dependent synaptic plasticity of a chalcogenide electronic synapse for neuromorphic systems. *Scientific Reports*, **4**(4906).

Little, S., and Brown, P. 2014. The functional role of beta oscillations in Parkinson's disease. *Parkinsonism & Related Disorders*, **20**, S44–S48.

Lloyd, S., Mohseni, M., and Rebentrost, P. 2013. Quantum algorithms for supervised and unsupervised machine learning. *arXiv preprint arXiv:1307.0411*.

Matsuda, W., Furuta, T., Nakamura, K. C., Hioki, H., Fujiyama, F., Arai, R., Kaneko, T. 2009. Single nigrostriatal dopaminergic neurons form widely spread and highly dense axonal arborizations in the neostriatum. *Journal of Neuroscience*, **29**(2), 444–453.

Maynard, A. D. 2014. Could we 3D print an artificial mind? *Nature Nanotechnology*, **9**(12), 955–956.

Mirmomeni, M., Punch, W. F., and Adami, C. 2014. Is information a selectable trait? *arXiv preprint arXiv:1408.3651*.

Morgan, D. March 10, 2015. The Zeitgeist in the machine. http://innovation.ox.ac.uk/wp-content/uploads/2015/03/Theysay-The-Zeitgeist-in-the-Machine.pdf.

Rice, M. E., Cragg, S. J., and Greenfield, S. A. 1997. Characteristics of electrically evoked somatodendritic dopamine release in substantia nigra and ventral tegmental area in vitro. *Journal of Neurophysiology*, **77**(2), 853–862.

Seal, R. P., and Edwards, R. H. 2006. Functional implications of neurotransmitter co-release: glutamate and GABA share the load. *Current Opinion in Pharmacology*, **6**(1), 114–119.

Tanaka, H., Kannari, K., Maeda, T., Tomiyama, M., Suda, T., and Matsunaga, M. 1999. Role of serotonergic neurons in L-DOPA-derived extracellular dopamine in the striatum of 6-OHDA-lesioned rats. *Neuroreport*, **10**(3), 631–634.

Threlfell, S., et al. 2012. Striatal dopamine release is triggered by synchronized activity in cholinergic interneurons. *Neuron*, **75**(1), 58–64.

Uglow, J. 2003. *The lunar men: five friends whose curiosity changed the world*. Macmillan.

Virki, T. June 23, 2008. *Computers in use pass 1 billion mark*. www.reuters.com/article/us-computers-statistics-idUSL2324525420080623.

Zhang, H., and Sulzer, D. 2004. Frequency-dependent modulation of dopamine release by nicotine. *Nature Neuroscience*, **7**(6), 581–582.

Index